Advanced Lectures in Mathematics (ALM)

ALM 1: Superstring Theory
ALM 2: Asymptotic Theory in Probability and Statistics with Applications
ALM 3: Computational Conformal Geometry
ALM 4: Variational Principles for Discrete Surfaces
ALM 6: Geometry, Analysis and Topology of Discrete Groups
ALM 7: Handbook of Geometric Analysis, No. 1
ALM 8: Recent Developments in Algebra and Related Areas
ALM 9: Automorphic Forms and the Langlands Program
ALM 10: Trends in Partial Differential Equations
ALM 11: Recent Advances in Geometric Analysis
ALM 12: Cohomology of Groups and Algebraic K-theory
ALM 13: Handbook of Geometric Analysis, No. 2
ALM 14: Handbook of Geometric Analysis, No. 3
ALM 15: An Introduction to Groups and Lattices: Finite Groups and Positive Definite Rational Lattices
ALM 16: Transformation Groups and Moduli Spaces of Curves
ALM 17: Geometry and Analysis, No. 1
ALM 18: Geometry and Analysis, No. 2
ALM 19: Arithmetic Geometry and Automorphic Forms
ALM 20: Surveys in Geometric Analysis and Relativity
ALM 21: Advances in Geometric Analysis
ALM 22: Differential Geometry: Under the Influence of S.-S. Chern
ALM 23: Recent Developments in Geometry and Analysis
ALM 24: Handbook of Moduli, Volume I
ALM 25: Handbook of Moduli, Volume II
ALM 26: Handbook of Moduli, Volume III
ALM 27: Number Theory and Related Areas
ALM 28: Selected Expository Works of Shing-Tung Yau with Commentary, Volume I
ALM 29: Selected Expository Works of Shing-Tung Yau with Commentary, Volume II
ALM 30: Automorphic Forms and *L*-functions

Advanced Lectures in Mathematics
Volume 30

Automorphic Forms and *L*-functions

edited by

Jianya Liu

International Press
www.intlpress.com

HIGHER EDUCATION PRESS

Advanced Lectures in Mathematics, Volume 30
Automorphic Forms and *L*-functions

Volume Editor:
Jianya Liu (School of Mathematics, Shandong University)

ISBN: 978-1-57146-296-1

Printed in the United States of America.

18 17 16 15 14 1 2 3 4 5 6 7 8 9

ADVANCED LECTURES IN MATHEMATICS

Preface

This is a collection of lecture notes from the CIMPA-UNESCO-CHINA Research School 2010: Automorphic Forms and L-functions, held at Shandong University at Weihai, in August 1–14, 2010.

The scientific directors were W. Duke and P. Sarnak (Chair) who provided detailed advice regarding the important aspects of the research school. The invited speakers were J. Cogdell, G. Harcos, E. Lapid, Xiaoqing Li, Wenzhi Luo, P. Michel, A. Reznikov, F. Shahidi, and Yangbo Ye.

Each of the speakers was scheduled to give a series of four lectures. Shahidi, although could not come to Weihai in person, has given full support to the research school by making his notes available, and Cogdell expertly combined his own lecture notes with Shahidi's material to give eight lectures. Not only have the speakers given excellent lectures, but they have also spent time completing their lecture notes – available for us here. Besides the speakers, there were fifteen scholars from all over the world plus ten international graduate students and over sixty Chinese graduate students who participated in various activities offered by the research school. The diligence and enthusiasm, professional skills and hard work of speakers, participants and organizers have contributed to the great success of the research school.

M. Waldschmidt visited Weihai in 2007 and provided clear instructions on how to organize a CIMPA school. M. Jambu, representative of CIMPA, not only participated the whole school, but also conducted lots of administrative duties on behalf of CIMPA before and after the school.

It is a great pleasure to express my sincere thanks to all the friends who have made this research school possible and who have contributed to its success. Thanks are also due to Lizhen Ji, Liping Wang, and Huaying Li for their effort and patience, and to Taiyu Li for helping me prepare the TeX file for publication. This event would not have been possible without the financial support from Shandong University, the NSFC, and CIMPA-UNESCO, and I would like to express my gratitude and thanks to them.

Jianya Liu
School Coordinator
July 2013

Academic Committee

William Duke, UCLA
Peter Sarnak (Chair), Princeton University and IAS

Speakers

Jim Cogdell, The Ohio State University
Gergely Harcos, The Hungarian Academy of Sciences
Erez Lapid, The Hebrew University of Jerusalem
Xiaoqing Li, SUNY at Buffalo
Wenzhi Luo, The Ohio State University
Philippe Michel, Ecole Polytechnique Fédérale de Lausanne
Andre Reznikov, Bar-Ilan University
Freydoon Shahidi, Purdue University
Yangbo Ye, The University of Iowa

School Coordinator

Jianya Liu, Shandong University

The CIMPA-UNESCO-CHINA Research School 2010: Automorphic Forms and L-functions

Contents

Automorphic Forms and L-functions
ALM 30, pp. 1–47

L-functions and Functoriality

James W. Cogdell*

Abstract

The principle of functoriality is one of the central tenets of the Langlands program. There are two main approaches to functoriality, via the trace formula or via the method of L-functions. In this article we discuss the method of L-functions, pioneered by Piatetski-Shapiro. This method primarily deals with functoriality in the case where the target group is GL_n, the fundamental tool here being the converse theorem for GL_n. As a vehicle for functoriality, the input to the converse theorem must be checked, and this is done by controlling the L-functions of the automorphic representations to be transferred. For the cases presented here these L-functions are controlled by the Langlands-Shahidi method, which we outline as well. We discuss most of the major cases of functoriality that have been obtained by combining the converse theorem for GL_n and the Langlands-Shahidi method.

2010 Mathematics Subject Classification: 11F70, 11R39, 22E50, 22E55.

Keywords and Phrases: Automorphic forms, L-functions, Converse theorem, Eisenstein series, Langlands-Shahidi method, Functoriality.

The principle of functoriality is one of the central tenets of the Langlands program. It is a purely automorphic avatar of Langlands' vision of a non-abelian class field theory. One can find an outline of this in Section 5 below. There are two main approaches to functoriality. The one envisioned by Langlands is through the Arthur-Selberg trace formula. With the recent proof of the Fundamental Lemma by Ngô, Waldspurger, and others this method is now available and will be the subject of a forthcoming book of Jim Arthur [1]. The second method is that of L-functions.

The method of L-functions was pioneered by Piatetski-Shapiro. It primarily deals with functoriality in the case where the target group is GL_n. The fundamental tool here is the converse theorem for GL_n, as explained in Section 6 below. As

*Department of Mathematics, Ohio State University, Columbus, OH 43210, USA, cogdell@math.ohio-state.edu. The author was supported in part by NSF Grants DMS-0654017 and DMS-0968505.

Piatetski-Shapiro said "Arthur's approach is more general, but this approach is much easier." (see Shahidi's contribution to [6]). The converse theorem is a way to tell when a representation of $GL_n(\mathbb{A})$ is automorphic based on the analytic properties of its L-functions (see Section 2 below). As a vehicle for functoriality, the input to the converse theorem must be checked, and this is done by controlling the L-functions of the automorphic representations to be functorially transferred. There are two ways to control these L-functions: via integral representations and via Eisenstein series, or the Langlands-Shahidi method. For the examples discussed here, these are controlled by the Langlands-Shahidi method, which we discuss in Sections 3–5. The functorial liftings themselves are obtained in Sections 7 and 8.

While Piatetski-Shapiro viewed this method as being simpler, it is in many ways more flexible and still mysterious. For example, as presented in Section 8, by this method you can obtain the third and fourth symmetric power lifting from GL_2 and these are still not attainable by the trace formula. In general, whenever you can control enough twisted L-functions, you can apply the converse theorem and obtain functoriality to GL_n. Of course controlling these L-functions is hard and for most of them we do not have a way to analyze them: they either fall outside the range of the Langlands-Shahidi method or we do not yet know if we can find integral representations for them. So there is still much work to be done.

These notes represent the lectures I gave at the CIMPA-UNESCO-CHINA Research School on *Automorphic Forms and L-functions* in August of 2010. Their purpose was to present an introduction to the L-function approach to functoriality. My original lectures were to be on GL_n and functoriality. These comprise Sections 1 and 2 here, which present the theory of automorphic representations and L-functions for GL_n, including the converse theorem, and then Sections 6 and 7 which were devoted to an exposition of functoriality via L-functions and then the example of functoriality for the classical groups. However, Shahidi was not able to attend the conference at the last moment, so I also gave an informal introduction to the Langlands-Shahidi method, based on Shahidi's notes [10]. These appear as Sections 3–5 and 8 here. These sections can be viewed as a "gentle introduction" to [10] and I have tried to indicate where the results here can be found in Shahidi's contribution. However, having the opportunity to give a single self-contained introduction to our approach to functoriality by the method of L-functions, I took the opportunity to integrate both series of lecture into a single contribution. I hope the reader finds this useful.

A word on references. I have surveyed the material in Sections 1,2,6 and 7 many times. Rather than burden this set of informal notes with pages of references, I refer instead to the sources I used for these talks, which are mainly my previous surveys. One can find more extensive references there. Similarly, for Sections 3–5 and 8 I have included in the bibliography the sources that I have used, which were surveys by Kim and Shahidi, Shahidi's new book, and Shahidi's contribution to this volume. I hope the reader does not mind this informality.

Finally I would like to thank Jianya Liu, the faculty at Shandong University, and all the students that attended for giving me the opportunity for presenting these lectures.

I L-functions for $GL(n)$ and Converse Theorems

1 Modular Forms and Automorphic Representations

1.1 Classical Modular Forms and Their L-functions

We begin our tale by recalling the classical results of Hecke and Weil.

Let $\mathfrak{H} = \{z = x + iy \mid y > 0\}$ denote the complex upper half-plane. Then a (classical) modular form of weight k for $\Gamma = SL_2(\mathbb{Z})$ is a function $f : \mathfrak{H} \to \mathbb{C}$ which is holomorphic on $\mathfrak{H}$ and at the cusps of Γ and for each $\gamma = \begin{pmatrix} a & b \\ c & d \end{pmatrix} \in \Gamma$ we have

$$f(\gamma z) = f\left(\frac{az+b}{cz+d}\right) = (cz+d)^k f(z).$$

We denote the space of these by $M_k(\Gamma)$. These arise in the theory of theta series, elliptic modular forms, etc. Since $f(z+1) = f(z)$, by modularity under $T = \begin{pmatrix} 1 & 1 \\ 0 & 1 \end{pmatrix}$, this will have a Fourier expansion

$$f(z) = \sum_{n=0}^{\infty} a_n e^{2\pi i n z}$$

and is called a *cusp form* if

$$a_0 = \int_0^1 f(x+iy)\, dx = 0.$$

Recall that $\Gamma\backslash\mathfrak{H}$ is the Riemann sphere with the "cusp at ∞" removed, so cusp forms are those modular forms that vanish at the cusps. The subspace of cusp forms of weight k is denoted by $S_k(\Gamma)$.

To each cusp form Hecke associated an L-function given by the Dirichlet series

$$L(s,f) = \sum_{n=1}^{\infty} \frac{a_n}{n^s}$$

which is absolutely convergent for $\mathrm{Re}(s) > \frac{k}{2} + 1$. The L-function is related to f through a Mellin transform:

$$\Lambda(s,f) = (2\pi)^{-s}\Gamma(s)L(s,f) = \int_0^{\infty} f(iy) y^s \, d^\times y.$$

By transferring the analytic properties of f to $L(s,f)$, Hecke showed:

Theorem 1.1. *If $f \in S_k(\Gamma)$ then $\Lambda(s,f)$ is nice, i.e.,*

(1) $\Lambda(s,f)$ *extends to an entire function of* s;

(2) $\Lambda(s,f)$ *is bounded in vertical strips* (*BVS*);

(3) $\Lambda(s,f)$ *satisfied the functional equation* (*FE*)

$$\Lambda(s,f) = (i)^k \Lambda(k-s,f).$$

Note that the functional equation results from the modular transformation law of $f(z)$ under $S = \begin{pmatrix} 0 & -1 \\ 1 & 0 \end{pmatrix}$, namely $f(Sz) = f\left(\frac{-1}{z}\right) = z^k f(z)$.

Remark 1.1. In general $L(s,f)$ need not have an *Euler product.* Hecke introduced an algebra of operators $\mathcal{H} = \langle T_p \mid p \text{ prime}\rangle$, the original Hecke algebra, such that if f is a simultaneous eigen-function for all T_n, i.e., $T_n f = \lambda_n f$, and if we normalize f such that $a_1 = 1$ then $\lambda_n = a_n$ and we have

$$L(s,f) = \prod_p (1 - a_p p^{-s} + p^{2k-1} p^{-2s})^{-1}.$$

Hecke was able to invert this process, via the inverse Mellin transform, and prove the "Hecke Converse Theorem".

Theorem 1.2. *Let*

$$D(s) = \sum_{n=1}^{\infty} \frac{a_n}{n^s}$$

be a Dirichlet series that converges in some right half-plane. Set

$$\Lambda(s) = (2\pi)^{-s}\Gamma(s)D(s).$$

If $\Lambda(s)$ is nice *as above, i.e.,* (1) *entire continuation,* (2) *BVS, and* (3) *satisfies the functional equation $\Lambda(s) = (i)^k \Lambda(k-s)$, then*

$$f(z) = \sum_{n=1}^{\infty} a_n e^{2\pi i n z}$$

is a cusp form of weight k for $\Gamma = SL_2(\mathbb{Z})$.

Note that by the Fourier expansion $f(z+1) = f(z)$, so $f(z)$ is modular under the translation matrix T. Since $SL_2(\mathbb{Z})$ is generated by T and the inversion S, we only need the transformation law under S. This follows from the functional equation for $\Lambda(s)$ via Mellin inversion

$$f(iy) = \sum_{n=1}^{\infty} a_n e^{-2\pi n y} = \frac{1}{2\pi i}\int_{\mathrm{Re}(s)=2} \Lambda(s) y^{-s}\, ds.$$

If f is a cusp form not for all of $SL_2(\mathbb{Z})$ but say for the Hecke congruence group of level N

$$\Gamma = \Gamma_0(N) = \left\{ \begin{pmatrix} a & b \\ c & d \end{pmatrix} \in SL_2(\mathbb{Z}) \middle| c \equiv 0 \pmod{N} \right\}$$

then the inversion S is no longer in Γ. Weil still defined $L(s,f)$ for $f \in S_k(\Gamma)$ and proved they were nice, using now $S_N = \begin{pmatrix} 0 & -1 \\ N & 0 \end{pmatrix}$, which normalizes $\Gamma_0(N)$, in place of S. (Now the functional equation relates f and a form g related to f through S_N.) To invert this process, i.e., to establish the "Weil Converse Theorem", Weil showed that besides knowing that $D(s)$ was nice, one had to also control twisted Dirichlet series; one needed that the

$$D_\chi(s) = \sum_{n=1}^{\infty} \frac{\chi(n)a_n}{n^s}$$

were *nice* (essentially) for all Dirichlet characters χ of conductor relatively prime to N. The important thing to note here is that one Dirichlet series no longer suffices for inversion, one must control a family of twisted Dirichlet series as well.

1.2 Automorphic Representations of $GL(n)$

The modern theory of automorphic representations is a theory of functions on adele groups. To make the connection note that

$$\mathfrak{H} = PGL_2^+(\mathbb{R})/PSO_2(\mathbb{R}).$$

So functions of $\mathfrak{H}$ can be lifted to functions on $GL_2(\mathbb{R})$. So we now have functions on a group.

To make an adelic theory, take k a number field ... but for now we can take $k = \mathbb{Q}$. The the adele ring of $\mathbb{Q}$ is

$$\mathbb{A} = \mathbb{A}_{\mathbb{Q}} = \varinjlim_S (\mathbb{R} \times \prod_{p\in S} \mathbb{Q}_p \times \prod_{p \notin S} \mathbb{Z}_p) = \prod_v{}' \mathbb{Q}_v,$$

where the limit is over all finite subsets S of primes of $\mathbb{Z}$. Then $\mathbb{Q} \hookrightarrow \mathbb{A}$ as a discrete, co-compact subgroup with $\mathbb{Q} \cap (\mathbb{R} \times \prod_p \mathbb{Z}_p) = \mathbb{Z}$, so that $\mathbb{Z}\backslash\mathbb{R} \simeq \mathbb{Q}\backslash\mathbb{A}/\prod_p \mathbb{Z}_p$.

Similarly, for $\mathbb{A} = \mathbb{A}_{\mathbb{Q}}$,

$$GL_n(\mathbb{A}) = \varinjlim_S (GL_n(\mathbb{R}) \times \prod_{p\in S} GL_n(\mathbb{Q}_p) \times \prod_{p\notin S} GL_n(\mathbb{Z}_p)) = \prod_v{}' GL_n(\mathbb{Q}_v)$$

and $GL_n(\mathbb{Q}) \hookrightarrow GL_n(\mathbb{A})$ as a canonical discrete subgroup. While the quotient $GL_n(\mathbb{Q})\backslash GL_n(\mathbb{A})$ is no longer compact, it is finite volume modulo the center $Z_n(\mathbb{A}) \simeq \mathbb{A}^\times$.

The analogue of the space of classical modular forms is the space of automorphic forms

$$\mathcal{A}(GL_n(\mathbb{Q})\backslash GL_n(\mathbb{A}); \omega)$$

consisting of functions $\varphi : GL_n(\mathbb{A}) \to \mathbb{C}$ such that $\varphi(\gamma g) = \varphi(g)$ for all $\gamma \in GL_n(\mathbb{Q})$ plus regularity and growth conditions to match the classical conditions of holomorphy and holomorphy at the cusps. Here ω is a fixed unitary central character, i.e., φ satisfies $\varphi(zg) = \omega(z)\varphi(g)$ for $z \in Z_n(\mathbb{A})$. $GL_n(\mathbb{A})$ acts on the space $\mathcal{A}$ by right translations. The irreducible sub-quotients of this representation are the *automorphic representations* (π, V_π). In this passage we have traded the tool of one complex variable (i.e., holomorphy) for the tool of non-abelian harmonic analysis.

The notion of a cusp form models the classical idea that translation integrals are zero.

Definition 1.1. $\varphi \in \mathcal{A}$ is a *cusp form* if for each pair of nonzero integers n_1, n_2 with $n_1 + n_2 = n$ we have

$$\int_{M_{n_1,n_2}(\mathbb{Q})\backslash M_{n_1,n_2}(\mathbb{A})} \varphi\left(\begin{pmatrix} I_{n_1} & X \\ 0 & I_{n_2} \end{pmatrix} g\right) \, dX = 0.$$

Note that the subgroups

$$N_{n_1,n_2} = \left\{ \begin{pmatrix} I_{n_1} & X \\ 0 & I_{n_2} \end{pmatrix} \middle| X \in M_{n_1,n_2} \right\}$$

are the unipotent radicals of the two-block maximal parabolic subgroups P_{n_1,n_2} of GL_n.

Let $\mathcal{A}_0 \subset \mathcal{A}$ denote the space of cusp forms. Then a theorem of Gelfand and Piatetski-Shapiro says that in general the space of cusp forms is completely reducible with finite multiplicities:

$$\mathcal{A}_0(GL_n(\mathbb{Q})\backslash GL_n(\mathbb{A}); \omega) = \bigoplus m(\pi) V_\pi.$$

(In fact, for GL_n, $m(\pi) = 0$ or 1.) The (π, V_π) that occur with nonzero multiplicities are the *cuspidal automorphic representations.* Moreover, just as $GL_n(\mathbb{A}) = \prod_v' GL_n(\mathbb{Q}_v)$ we have

$$(\pi, V_\pi) \simeq \otimes_v{}'(\pi_v, V_{\pi_v}),$$

where each (π_v, V_{π_v}) is an irreducible, smooth, admissible representation of $GL_n(\mathbb{Q}_v)$ and for almost all v, the representation π_v has a unique $K_v = GL_n(\mathbb{Z}_v)$-fixed vector, i.e., is *unramified.* (N.B. This abstract restricted tensor decomposition of an irreducible representation will be responsible for the Euler product of the L-function. It is the adelic avatar of the Hecke operators.)

Key to Hecke's theory was the notion of a Fourier expansion. Here there is a Fourier expansion for cusp forms for GL_n due to Piatetski-Shapiro and Shalika. It is as follows. Let

$$N = \left\{ n = \begin{pmatrix} 1 & x_{1,2} & \cdots & * \\ & 1 & \ddots & \vdots \\ & & \ddots & x_{n-1,n} \\ & & & 1 \end{pmatrix} \right\}$$

be the maximal unipotent subgroup of GL_n If $\psi : \mathbb{Q}\backslash\mathbb{A} \to \mathbb{C}$ is a nontrivial additive character of $\mathbb{A}$, invariant by translations by $\mathbb{Q}$, then ψ defines a character of $N(\mathbb{A})$, left invariant by $N(\mathbb{Q})$, via

$$\psi(n) = \psi(x_{1,2} + \cdots + x_{n-1,n}).$$

If $\varphi \in \mathcal{A}_0$ set

$$W_\varphi(g) = \int_{N(\mathbb{Q})\backslash N(\mathbb{A})} \varphi(ng)\psi^{-1}(n)\, dn.$$

Then $W_\varphi(ng) = \psi(n)W_\varphi(g)$ for all $n \in N(\mathbb{A})$. W_φ is called the *Whittaker function* of φ and it is what occurs in the following Fourier expansion of φ.

Theorem 1.3. *If* $\varphi \in \mathcal{A}_0$ *then*

$$\varphi(g) = \sum_{\gamma \in N_{n-1}(\mathbb{Q})\backslash GL_{n-1}(\mathbb{Q})} W_\varphi\left(\begin{pmatrix} \gamma & 0 \\ 0 & 1 \end{pmatrix} g\right).$$

This turns out to be the proper notion of a Fourier expansion in this context. (The proof isn't hard once you realize this is the correct notion – it is simply an inductive abelian Fourier expansion beginning along the last column of N.)

1.3 Uniqueness of Whittaker Models

The space of functions

$$\mathcal{W}(\pi, \psi) = \left\{ W_\varphi(g) \middle| \varphi \in V_\pi \right\}$$

gives a model of π in the space of functions W on $GL_n(\mathbb{A})$ satisfying $W(ng) = \psi(n)W(g)$, on which the group $GL_n(\mathbb{A})$ acts by right translation. This realization is called a *Whittaker model* for π.

Similarly, if $\pi \simeq \otimes' \pi_v$ then a Whittaker model for a local representation π_v is a realization of π_v in a space of functions

$$\mathcal{W}(\pi_v, \psi_v) = \left\{ W_{\xi_v}(g) \middle| \xi_v \in V_{\pi_v} \right\},$$

where again we require that $W_{\xi_v}(n_v g_v) = \psi_v(n_v) W_{\xi_v}(g_v)$ for $n_v \in N(\mathbb{Q}_v)$, $g_v \in GL_n(\mathbb{Q}_v)$ and where ψ_v is a local component of ψ. It is a theorem of Gelfand-Kazhdan and Shalika that locally an irreducible smooth admissible representation (π_v, V_{π_v}) can have at most one such model (with a continuity assumption at the archimedean place). This result is *the local uniqueness of the Whittaker model.*

From the local statements one can deduce *the uniqueness of the global Whittaker model.* From this it follows that if $\varphi \in V_\pi \subset \mathcal{A}_0$ and under the isomorphism $\pi \simeq \otimes' \pi_v$ we have $\varphi \simeq \otimes \xi_v$ then we have a factorization

$$W_\varphi(g) = \prod_v W_{\xi_v}(g_v).$$

This is a highly nontrivial factorization!

2 L-functions for GL_n and Converse Theorems

Let's recall Hecke: for $f \in S_k(\Gamma)$, a Hecke eigen-form, we have

$$\int_0^\infty f(iy)y^s\, d^\times y = \Lambda(s,f) = (2\pi)^{-s}\Gamma(s)\prod_p (1 - a_p p^{-s} + p^{2k-1}p^{-2s})^{-1}.$$

The analogous integral would be for $\varphi \in \mathcal{A}_0(GL_2(\mathbb{Q})\backslash GL_2(\mathbb{A});\omega)$ and would be

$$\int_{\mathbb{Q}^\times\backslash\mathbb{A}^\times} \varphi\begin{pmatrix} a & 0 \\ 0 & 1\end{pmatrix} |a|^s\, d^\times a = \int_{GL_1(\mathbb{Q})\backslash GL_1(\mathbb{A})} \varphi\begin{pmatrix} a & 0 \\ 0 & 1\end{pmatrix} |a|^s\, d^\times a.$$

However, we will want to include twists as necessary for a Converse Theorem. Here that would be twists by an adelic Dirichlet character χ, i.e., an automorphic form for GL_1:

$$\int_{GL_1(\mathbb{Q})\backslash GL_1(\mathbb{A})} \varphi\begin{pmatrix} a & 0 \\ 0 & 1\end{pmatrix} \chi(a)|a|^s\, d^\times a.$$

2.1 Global Rankin-Selberg Integrals (d'apres Jacquet, Piatetski-Shapiro, and Shalika)

In the higher rank case, the analogous integrals are the Rankin-Selberg integrals for $GL_n \times GL_m$ with $n > m$. The theory which is closest to the classical theory of Hecke is for $GL_n \times GL_{n-1}$. So let (π, V_π) be a cuspidal automorphic representation of $GL_n(\mathbb{A})$ and let $(\pi', V_{\pi'})$ be a cuspidal automorphic representation of $GL_{n-1}(\mathbb{A})$. For cusp forms $\varphi \in V_\pi$ and $\varphi' \in V_{\pi'}$ we set

$$I(s,\varphi,\varphi') = \int_{GL_{n-1}(\mathbb{Q})\backslash GL_{n-1}(\mathbb{A})} \varphi\begin{pmatrix} h & 0 \\ 0 & 1\end{pmatrix} \varphi'(h)|\det(h)|^{s-\frac{1}{2}}\, dh.$$

Note that in fact we have a *family* of such integrals, one for each pair of φ and φ'.

The basic facts about these integrals are the following.

(i) The individual integrals are *nice*. As in the classical case, here we have that the integrals are absolutely convergent for all s and hence entire; they are bounded in vertical strips (BVS); they satisfy a functional equation (FE) of the form $I(s,\varphi,\varphi') \doteq I(1-s,\widetilde{\varphi},\widetilde{\varphi}')$ where $\widetilde{\varphi}(g) = \varphi({}^tg^{-1})$.

(ii) The individual integrals are *Eulerian*. If we insert the Fourier expansion for φ, then the integral unfolds to

$$I(s,\varphi,\varphi') = \int_{N_{n-1}(\mathbb{A})\backslash GL_{n-1}(\mathbb{A})} W_\varphi\begin{pmatrix} h & 0 \\ 0 & 1\end{pmatrix} W'_{\varphi'}(h)|\det(h)|^{s-\frac{1}{2}}\, dh,$$

where $W_\varphi \in \mathcal{W}(\pi,\psi)$ and $W'_{\varphi'} \in \mathcal{W}(\pi',\psi^{-1})$. If we now assume we have decomposable $\varphi \simeq \otimes\xi_v$ and $\varphi' \simeq \otimes\xi'_v$, then the two global Whitaker functions in the integral now factor into a product of local Whittaker functions and then our integral itself factors as

$$I(s,\varphi,\varphi') = \prod_v \int_{N_{n-1}(\mathbb{Q}_v)\backslash GL_{n-1}(\mathbb{Q}_v)} W_{\xi_v}\begin{pmatrix} h_v & 0 \\ 0 & 1 \end{pmatrix} W'_{\xi'_v}(h_v)|\det(h_v)|^{s-\frac{1}{2}}\, dh_v$$
$$= \prod_v \Psi_v(s, W_v, W'_v)$$

which is convergent now only for $\mathrm{Re}(s) \gg 1$, with $W_v = W_{\xi_v} \in \mathcal{W}(\pi_v\psi_v)$ and $W'_v = W'_{\xi'_v} \in \mathcal{W}(\pi'_v, \psi_v^{-1})$. As we see, it is the decomposition $\pi \simeq \otimes' \pi_v$ that is responsible for the factorization of the Whittaker functions and hence ultimately for the Eulerian factorization of the global integral into a product of local integrals.

2.2 Local L-functions (Euler Factors)

Now let's look at a local non-archimedean place v of a number field k. We have the completion k_v, its ring of integers $\mathfrak{o}_v$ and maximal ideal $\mathfrak{p}_v = (\varpi_v)$. We set $q_v = |\mathfrak{o}_v/\mathfrak{p}_v| = |\varpi_v|_v^{-1}$.

Let $W_v \in \mathcal{W}(\pi_v, \psi_v)$ and $W'_v \in \mathcal{W}(\pi'_v, \psi_v^{-1})$. Then the basic properties of the local Rankin-Selberg integrals are the following.

(i) Each integral $\Psi_v(s, W, W'_v)$ is a rational function of q_v^{-s} and in fact their span $\langle \Psi_v(s, W_v, W'_v)\rangle$ forms a $\mathbb{C}[q_v^s, q_v^{-s}]$-fractional ideal of $\mathbb{C}(q_v^{-s})$. This fractional ideal has a (normalized) generator of the form $P_v(q_v^{-s})^{-1}$ with $P_v(X) \in \mathbb{C}[X]$ with $P_v(0) = 1$ and $\deg(P_v) \leqslant nm$.

Definition 2.1. $L(s, \pi_v \times \pi'_v) = P_v(q_v^{-s})^{-1}$.

(ii) For each individual integral we have

$$e(s, W_v, W'_v) = \frac{\Psi_v(s, W_v, W'_v)}{L(s, \pi_v \times \pi'_v)}$$

is entire and for each $s_0 \in \mathbb{C}$ there exist a choice of W_v and W'_v such that $e(s_0, W_v, W'_v) \neq 0$. In other words, the local L-function $L(s, \pi_v \times \pi'_v)$ exactly captures the poles of the family of local integrals.

(iii) The local functional equation. There exists $\gamma(s, \pi_v \times \pi'_v, \psi_v) \in \mathbb{C}(q_v^{-s})$ such that for all choices of W_v and W'_v we have

$$\Psi_v(1-s, \widetilde{W}_v, \widetilde{W}'_v) = \omega_{\pi'_v}(-1)^{n-1}\gamma(s, \pi_v \times \pi'_v, \psi_v)\Psi_v(s, W_v, W'_v)$$

or

$$\frac{\Psi_v(1-s, \widetilde{W}_v, \widetilde{W}'_v)}{L(1-s, \widetilde{\pi}_v \times \widetilde{\pi}'_v)} = \omega_{\pi'_v}(-1)^{n-1}\varepsilon(s, \pi_v \times \pi'_v, \psi_v)\frac{\Psi_v(s, W_v, W'_v)}{L(s, \pi_v \times \pi'_v)}$$

with the local ε-factor $\varepsilon(s, \pi_v \times \pi'_v, \psi_v)$ then being entire without zeroes. More precisely,

$$\varepsilon(s, \pi_v \times \pi'_v, \psi_v) = \varepsilon\left(\frac{1}{2}, \pi_v \times \pi'_v, \psi_v\right) q_v^{-c(\pi_v \times \pi'_v)\left(s-\frac{1}{2}\right)}$$

where $\varepsilon\left(\frac{1}{2}, \pi_v \times \pi'_v, \psi_v\right)$ is the local root number and $c(\pi_v \times \pi'_v)$ the local conductor (exponent). Here $\widetilde{W}(g) = W\left(\begin{pmatrix} & & 1 \\ & \cdot^{\cdot^{\cdot}} & \\ 1 & & \end{pmatrix} {}^t g^{-1}\right)$. For future reference, let us note that the L-, ε-, and γ-factors are related by

$$\gamma(s, \pi_v \times \pi'_v, \psi_v) = \frac{\varepsilon(s, \pi_v \times \pi'_v, \psi_v) L(1-s, \widetilde{\pi}_v \times \widetilde{\pi}'_v)}{L(s, \pi_v \times \pi'_v)}.$$

(iv) If π_v and π'_v are unramified, as well as ψ_v and $k_v/\mathbb{Q}_p$, which is the case at almost all finite places of k, then for the normalized unramified Whittaker functions W_v° and $W_v'^\circ$ we have that the local integral computes the local L-function on the nose and the local ε-factor is 1, i.e.,

$$L(s, \pi_v \times \pi'_v) = \Psi_v(s, W_v^\circ, W_v'^\circ),$$
$$\varepsilon(s, \pi_v \times \pi'_v, \psi_v) = 1.$$

At an archimedean place $v|\infty$ the local theory is similar yet different. Due to the existence of a local Langlands parametrization for real and complex groups, the form of $L(s, \pi_v \times \pi'_v)$ is given as the local archimedean factor of the associated Artin representation. So it will be a product of factors of the form $\Gamma(a_{v,i}s + b_{v,i})$, where now $\Gamma(s)$ is Euler's gamma function, and accounts for the gamma factors in the classical functional equation. Now one must work to show that this "Langlands given" factor behaves well with respect to the local integrals, i.e., that analogues of (ii) and (iii) above hold. This can be quite complex and what we know is mainly due to Jacquet-Shalika and subsequent work of Jacquet.

2.3 When $n - m > 1$

If we consider (π, V_π) a cuspidal representation of $GL_n(\mathbb{A})$ and $(\pi', V_{\pi'})$ a cuspidal representation of $GL_m(\mathbb{A})$ with $m < n-1$ the theory is quite similar. One begins with a global projection operator $\mathbb{P} : V_\pi \longrightarrow U_\pi$ where U_π is a representation of the mirabolic subgroup $P_{m+1} \subset GL_{m+1}$, which has the form

$$P_{m+1} = \left\{ \begin{pmatrix} g & u \\ 0 & 1 \end{pmatrix} \middle| g \in GL_m \right\},$$

with the property that

$$\mathbb{P}\varphi\begin{pmatrix} h & 0 \\ 0 & 1 \end{pmatrix} = \sum_{\gamma \in N_m(k)\backslash GL_m(k)} W_\varphi\left(\begin{pmatrix} \gamma & 0 \\ 0 & I_{n-m} \end{pmatrix}\begin{pmatrix} h & 0 \\ 0 & I_{n-m} \end{pmatrix}\right).$$

$\mathbb{P}$ is essentially given by a partial Whittaker transform, as if one was only working out part of the Fourier expansion, twisted by an appropriate power of the determinant. With this operator in hand, the family of global integrals now has the form

$$I(s, \varphi, \varphi') = \int_{GL_m(\mathbb{Q})\backslash GL_m(\mathbb{A})} \mathbb{P}\varphi\begin{pmatrix} h & 0 \\ 0 & 1 \end{pmatrix} \varphi'(h) |\det(h)|^{s-\frac{1}{2}}\, dh$$

which now unfold to

$$I(s,\varphi,\varphi') = \int_{N_m(\mathbb{A})\backslash GL_m(\mathbb{A})} W_\varphi \begin{pmatrix} h & 0 \\ 0 & I_{n-m} \end{pmatrix} W'_{\varphi'}(h)|\det(h)|^{s-\frac{n-m}{2}}\, dh.$$

These are again Eulerian

$$\begin{aligned} I(s,\varphi,\varphi') &= \prod_v \int_{N_m(k_v)\backslash GL_m(k_v)} W_{\xi_v} \begin{pmatrix} h_v & 0 \\ 0 & I_{n-m} \end{pmatrix} W'_{\xi'_v}(h_v)|\det(h_v)|^{s-\frac{n-m}{2}}\, dh_v \\ &= \prod_v \Psi_v(s,W_v,W'_v) \end{aligned}$$

and from here one establishes the analogues of the results in Section 2.2. The only significant change is in the local functional equation, where the projection operator $\mathbb{P}$ reappears as an unipotent integration in the integral on the left hand side of the functional equation.

2.4 When $m = n$

When $m = n$ the construction is slightly different. It is the automorphic analogue of the classical construction of Rankin and Selberg. Now (π, V_π) and $(\pi', V_{\pi'})$ are both cuspidal representations of $GL_n(\mathbb{A})$. There is an extra player now, a (mirabolic) Eisenstein series. This Eisenstein series is of a special type and is associated to a Schwartz function $\Phi \in \mathcal{S}(\mathbb{A}^n)$; let us denote it by $E(g;s,\Phi)$. Then the family of global integrals has the form

$$I(s,\varphi,\varphi',\Phi) = \int_{GL_n(k)\backslash GL_n(\mathrm{A})} \varphi(g)\varphi'(g)E(g;s,\Phi)\, dg$$

for $\varphi \in V_\pi$ and $\varphi' \in V_{\pi'}$. These integrals converge absolutely for all s and have a functional equation. (The analytic properties, including the functional equation, come from those of the Eisenstein series.) The family is again Eulerian and unfold to a family of local integrals of the form

$$\Psi_v(s,W_v,W'_v,\Phi_v) = \int_{N_n(k_v)\backslash GL_n(k_v)} W_v(g)W'_v(g)\Phi_v(e_n g)|\det(g)|^s\, dg,$$

where $e_n = (0,\ldots,0,1) \in k^n$. Once again, from here one establishes the analogous results as in Section 2.2.

2.5 Global L-functions

If we combine the nice properties of the global integrals from Sections 2.1, 2.3 and 2.4 with the local analysis of Section 2.2 then we arrive at the global analogue of Hecke's theorem. Again take (π, V_π) a cuspidal automorphic representation of $GL_n(\mathbb{A})$ and $(\pi', V_{\pi'})$ a cuspidal representation of $GL_m(\mathbb{A})$.

We define the global L-factor and ε-factors as Euler products:

$$L(s,\pi\times\pi') = \prod_v L(s,\pi_v\times\pi'_v),$$

$$\varepsilon(s,\pi\times\pi') = \prod_v \varepsilon(s,\pi_v\times\pi'_v,\psi_v),$$

where the product for the L-factor converges for $\mathrm{Re}(s) \gg 0$ and the product for the ε-factor is a finite product. Note that we are now using the automorphic L-function convention where the L-factor is the product over *all* places – this we denoted Λ classically.

Theorem 2.1. *$L(s,\pi\times\pi')$ is nice in the sense that*

(1) *$L(s,\pi\times\pi')$ extends to a meromorphic function of s, entire if $m < n$;*

(2) *$L(s,\pi\times\pi')$ is bounded in vertical strips;*

(3) *we have the global functional equation*

$$L(s,\pi\times\pi') = \varepsilon(s,\pi\times\pi')L(1-s,\widetilde{\pi}\times\widetilde{\pi}').$$

We note that in fact our analysis only gives the BVS in (ii) for the cases of $m = n, n-1$. The more general boundedness in vertical strips comes from the work of Shahidi.

2.6 Converse Theorems for GL_n

Converse Theorems for GL_n invert these integral representations much the same way that Hecke and Weil did. However, here we begin with an Euler product rather than a Dirichlet series, since that is what the direct theory, Theorem 2.1, gives us. Conceptually the theorem takes the following form. Given $\pi = \otimes'\pi_v$, an irreducible admissible representation of $GL_n(\mathbb{A})$ on a space V_π, when can we embed

$$V_\pi \hookrightarrow \mathcal{A}(GL_n(k)\backslash GL_n(\mathbb{A});\omega)?$$

To relate to Hecke and Weil, by our local theory, to each local component π_v of π we can associate a local L- and ε-factor

$$\pi_v \mapsto L(s,\pi_v), \quad \varepsilon(s,\pi_v,\psi_v)$$

and so to π itself the Euler products

$$L(s,\pi) = \prod_v L(s,\pi_v), \quad \varepsilon(s,\pi,\psi) = \prod_v \varepsilon(s,\pi_v,\psi_v).$$

We think of π as encoding an L-function, as an Euler product of degree n. We must always assume two things:

(1) $L(s,\pi)$ converges in some right half plane $\mathrm{Re}(s) \gg 0$;

(2) the central character $\omega = \omega_\pi$ is automorphic, i.e., is an idele class character (which guarantees that the global ε-factor is independent of ψ).

Like Weil, our criterion will involve twisting. So if $(\pi', V_{\pi'})$ is a *cuspidal automorphic* representation of $GL_m(\mathbb{A})$ for $m < n$ we can similarly define

$$L(s, \pi \times \pi') = \prod_v L(s, \pi_v \times \pi'_v),$$

$$\varepsilon(s, \pi \times \pi') = \prod_v \varepsilon(s, \pi_v \times \pi'_v, \psi_v).$$

We will say that the $L(s, \pi \times \pi')$ are *nice* if they behave as they would if π were cuspidal automorphic, i.e.,

(1) $L(s, \pi \times \pi')$ extends to an entire function of s;

(2) $L(s, \pi \times \pi')$ is bounded in vertical strips;

(3) we have the global functional equation

$$L(s, \pi \times \pi') = \varepsilon(s, \pi \times \pi') L(1 - s, \widetilde{\pi} \times \widetilde{\pi}').$$

To state the theorems, let us set

$$\mathcal{A}_0(m) = \coprod_\omega \mathcal{A}_0(GL_m(k)\backslash GL_m(\mathbb{A}); \omega),$$

$$\mathcal{T}_0(m) = \coprod_{1 \leqslant d \leqslant m} \mathcal{A}_0(d).$$

Theorem 2.2. *Fix $\pi = \otimes' \pi_v$ an irreducible admissible representation of $GL_n(\mathbb{A})$ as above. If $L(s, \pi \times \pi')$ is nice for all $\pi' \in \mathcal{T}_0(n-1)$, then π is a cuspidal automorphic representation of $GL_n(\mathbb{A})$, i.e., we can embed π into $\mathcal{A}_0(GL_n(k)\backslash GL_n(\mathbb{A}); \omega_\pi)$.*

Conceptually, the proof of this is very similar to that of Hecke and Weil. We can also reduce the twists with more work. This requires a delicate local construction and has no classical analogue.

Theorem 2.3. *Fix $\pi = \otimes' \pi_v$ an irreducible admissible representation of $GL_n(\mathbb{A})$ as above, with $n \geqslant 3$. If $L(s, \pi \times \pi')$ is nice for all $\pi' \in \mathcal{T}_0(n-2)$, then π is a cuspidal automorphic representation of $GL_n(\mathbb{A})$, i.e., we can embed π into $\mathcal{A}_0(GL_n(k)\backslash GL_n(\mathbb{A}); \omega_\pi)$.*

There is a very useful variant of these results. Let S be a finite set of finite places of k and let η be a *fixed* idele class character of $\mathbb{A}^\times$. Let

$$\mathcal{T}_0^S(m) = \{\pi' \in \mathcal{T}_0(m) \mid \pi'_v \text{ is unramified for all } v \in S\}.$$

Theorem 2.4. *Fix $\pi = \otimes' \pi_v$ an irreducible admissible representation of $GL_n(\mathbb{A})$ as above and fix an idele class character η. If $L(s, \pi \times \pi')$ is nice for all $\pi' \in \mathcal{T}_0^S(n-2) \otimes \eta$, then π is quasi-automorphic, i.e., there exists an automorphic representation π_1 of $GL_n(\mathbb{A})$ such that $\pi_{1,v} \simeq \pi_v$ for all $v \notin S$.*

Note that we can no longer claim that π is cuspidal, and indeed it may not be. We can summarize the ideas of this section as follows.

Moral Theorem. *Any suitably nice L-function of degree n must be automorphic (or modular), i.e., associated to an automorphic representation of* GL_n.

In this Moral Theorem, *suitably nice* means that the L-function must be given by a convergent Euler product and that it and suitable twists must be entire, BVS, and satisfy an appropriate functional equation, all the standard properties that L-functions arising from arithmetic or geometry are conjectured to satisfy.

2.7 On Integral Representations

The theory of integral representation for the Rankin-Selberg L-functions, particularly the $n = m$ case, is the paradigm for the study of automorphic L-functions by integral representations. The general outline of the method as practiced are the following steps.

(i) Write down a family of global integrals which have nice analytic properties, including continuation, BVS and functional equation.

In general the integrals are variants on the $n = m$ integrals here, where the analytic properties follow from those of an Eisenstein series. In this sense, the Hecke type integrals for $m < n$, and through them the Converse Theorem, are special.

(ii) Show that the global integrals are Eulerian, i.e., factor into a product of local integrals.

As is the case here, this unfolding and then factorization often relies on a local and global uniqueness principle, here the uniqueness of the Whittaker models. It is only in rare instances, such as the "new way" of Rallis and Piatetski-Shapiro, that this is not the case.

(iii) Perform the local unramified calculation.

At almost all places, the data that goes into the local integral is unramified and the integral can be computed explicitly. This is a nontrivial calculation, usually involving an explicit formula for the unramified functions in the integral, here the formula of Shintani for the unramified Whittaker function for GL_n. This is then followed by a calculation in invariant theory which expresses the result in terms of a Langlands L-function coming from an appropriate representation of the L-group. This "identifies" the L-function represented by the integrals in terms of Langlands' paradigm.

Often the process stops here. But for many purposes, such as functoriality, this is not enough. There remain three steps.

(iv) Analyze and compute the local L-function at the finite ramified places, including continuation and functional equations.

(v) Analyze the local L-function at the archimedean places, again including the local functional equation.

(vi) Combine steps (i)–(v) to complete the analysis of the global L-function, including continuation, BVS, and global functional equation.

It is only for the Rankin-Selberg L-functions for GL_n that we have completed these last three steps. (iv) was carried out by Jacquet, Piatetski-Shapiro, and Shalika for Rankin-Selberg convolutions. (v) was initiated by Jacquet and Shalika, but the final papers are by Jacquet alone. These involve passing to the Casselman-Wallach completions and introduces another layer of technical difficulty. (vi) is almost completed for GL_n. There remains the question of BVS, which currently can be done for $m = n$ and $m = n - 1$ within the method. (The difficulty comes from the passage to the Casselman-Wallach completion in the archimedean theory.) For BVS in the other cases, we must rely on the Langlands-Shahidi method, which follows.

II L-functions via Eisenstein Series

In many ways, this part serves as a reader's guide to the lectures of Shahidi [10]. I will try to indicate where the results I mention are to be found in his contribution.

3 The Origins: Langlands

3.1 A Classical Example

The Eisenstein series for $\Gamma = SL_2(\mathbb{Z})$ is defined as follows.

Let $B = \left\{ \begin{pmatrix} a & b \\ 0 & d \end{pmatrix} \right\} \subset SL_2(\mathbb{R})$ be the Borel subgroup of $SL_2(\mathbb{R})$. For $z \in \mathfrak{H}$ set

$$\begin{aligned} E(z,s) &= \sum_{(B\cap\Gamma)\backslash\Gamma} \operatorname{Im}(\gamma z)^{\frac{s}{2}+\frac{1}{2}} \qquad \text{for } \operatorname{Re}(s) > 1 \\ &= \sum_{\substack{(c,d)\in\mathbb{Z}^2 \\ (c,d)=1}} \frac{y^{\frac{s+1}{2}}}{|cz+d|^{s+1}} \\ &= \frac{1}{\zeta(s+1)} \sum_{(m,n)\neq(0,0)} \frac{y^{\frac{s+1}{2}}}{|mz+d|^{s+1}}. \end{aligned}$$

The Fourier expansion of $E(z,s)$ has been computed classically. We find, if we set $Z(s) = \pi^{-\frac{s}{2}}\Gamma\left(\frac{s}{2}\right)\zeta(s)$, then

$$E(z,s) = 2\left(y^{\frac{s+1}{2}} + \frac{Z(s)}{Z(s+1)} y^{\frac{1-s}{2}}\right) + 4\sum_{n=1}^{\infty} \frac{|n|^s \sigma_{-s}(n)}{Z(s+1)} \sqrt{y} K_{\frac{s}{2}}(2\pi|n|y) e^{2\pi i n x},$$

where $\sigma_{-s}(n) = \sum_{d|n} d^{-s}$.

Classically, one can use the *known* meromorphic continuation and functional equation of the Riemann zeta function $\zeta(s)$, and its completion $Z(s)$, to obtain the meromorphic continuation and functional equation of the Eisenstein series:

$$E(z,s) = \frac{Z(s)}{Z(s+1)} E(z,-s).$$

The idea of Langlands was to reverse this process: from the analytic properties of an Eisenstein series, deduce analytic properties of the L-functions appearing in its Fourier expansion.

3.2 Induced Representations

In terms of automorphic representations, Eisenstein series correspond to induced representations.

Let k be a number field. (You can take $k = \mathbb{Q}$ if you want.) Let G be a connected reductive linear algebraic group over k, which we can assume to be split for simplicity, so we can view G as a subgroup of a suitably large GL.

Example. As an example to carry along consider the symplectic group $G = Sp_{2n}$:

$$Sp_{2n} = Isom\left(\begin{pmatrix} 0 & J_n \\ -J_n & 0 \end{pmatrix}\right) \subset GL_{2n}, \quad \text{where} \quad J_n = \begin{pmatrix} & & 1 \\ & \cdot^{\cdot^{\cdot}} & \\ 1 & & \end{pmatrix}.$$

G will contain a Borel subgroup B, which we can take to be

$$B = G \cap \left\{ \begin{pmatrix} * & \cdots & * \\ & \ddots & \vdots \\ & & * \end{pmatrix} \right\}.$$

Then $B = TU$ where T is a maximal torus, which will be the diagonal elements in B, and U is the unipotent radical of B,

$$U = G \cap \left\{ \begin{pmatrix} 1 & \cdots & * \\ & \ddots & \vdots \\ & & 1 \end{pmatrix} \right\}.$$

We will need the following structural objects:

$$\begin{aligned} X(T) &= \mathrm{Hom}_k(T, \mathbb{G}_m): \text{ the } k\text{-rational characters of } T; \\ \Phi(T) &= \text{the characters of } T \text{ acting on } \mathrm{Lie}(G)\text{: the } \textit{roots} \text{ of } T \text{ in } G; \\ \Phi^+(T) &= \text{the characters of } T \text{ acting on } \mathrm{Lie}(U)\text{: the } \textit{positive roots}; \\ \Delta &= \text{a basis of } \Phi^+(T)\text{: the } \textit{simple roots}; \\ W &= N(T)/Z(T): \text{ the } \textit{Weyl group}. \end{aligned}$$

Note that $\Delta \subset \Phi^+(T) \subset \Phi(T) \subset X(T)$. Also $N(T)$ denotes the normalizer of T in G and $Z(T)$ the centralizer.

The Weyl group acts as a reflection group on $\Phi(T)$. It will contain a "longest element" $w_\ell \in W$ with the property that if we let $\Phi^-(T)$ denote the characters of T acting on the Lie(U^-), U^- being the opposite or lower triangular unipotent subgroup, then $w_\ell(\Phi^+) = \Phi^-$. In GL_n, the longest element of W is represented by $w_\ell = \begin{pmatrix} & & 1 \\ & \cdot^{\cdot^{\cdot}} & \\ 1 & & \end{pmatrix}$.

Let P be a parabolic subgroup of G containing B. We will only need to consider maximal parabolic subgroups and will also restrict to those that are "self-associate". Just as $B = TU$, P will have a decomposition $P = MN$ with $M \supset T$ the reductive Levi component of P and $N \subset U$ the unipotent radical of P.

Example. In the example of Sp_{2n}, P will be a block upper triangular subgroup of the form

$$P = \left\{ \begin{pmatrix} * & * & * \\ & * & * \\ & & * \end{pmatrix} \right\}$$

with

$$M = \left\{ \begin{pmatrix} A & & \\ & B & \\ & & A^* \end{pmatrix} \in Sp_{2n} \middle| A \in GL_r;\ B \in Sp_{2t} \right\} \simeq GL_r \times Sp_{2t},$$

where $2r + 2t = 2n$, and

$$N = \left\{ \begin{pmatrix} I_r & * & * \\ & I_{2t} & * \\ & & I_r \end{pmatrix} \in Sp_{2n} \right\}.$$

Let

$$\delta_P(mn) = |\det(\mathrm{Ad}(m)|\mathrm{Lie}(N))|$$

be the modulus character of P. Then $\delta_P : M \to \mathbb{R}^\times$ and $\delta_P \in X(M) \otimes \mathbb{R}$. If (σ, V_σ) is a (reasonable) representation of $M(k_v)$ in the local situation (resp. an automorphic representation of $M(\mathbb{A})$ in the global situation) we can construct a representation of $G = G(k_v)$ (resp. $G = G(\mathbb{A})$) by

$$I(s, \sigma) = \mathrm{Ind}_P^G(\sigma \delta_P^{s/c_P}),$$

where $s \in \mathbb{C}$ and $c_P \in \mathbb{Z}$ is a normalizing denominator depending on P. This is the representation of G by right translation on

$$V(s, \sigma) = \{f : G \to \mathbb{C} \,|\, f(mng) = \delta_P(m)^{\frac{s}{c_P} + \frac{1}{2}} \sigma(m) f(g), \text{ plus regularity conditions}\}.$$

If $\sigma \simeq \otimes' \sigma_v$ is an automorphic representation of $M(\mathbb{A})$ then we have

$$I(s, \sigma) \simeq \otimes' I(s, \sigma_v)$$

as a representation of $G(\mathbb{A})$, but $I(s,\sigma)$ is not yet automorphic; it is only left invariant under $P(k)$, not all of $G(k)$.

3.3 Eisenstein Series

We will use the theory of Eisenstein series to embed $I(s,\sigma)$ into the space of automorphic forms. We assume that σ is a unitary cuspidal automorphic form for $M(\mathbb{A})$.

Definition 3.1. If $f_s \in I(s,\sigma)$ we define the Eisenstein series associated to f_s by

$$E(g,s,f_s) = \sum_{\gamma \in P(k)\backslash G(k)} f_s(\gamma g).$$

The series which defines the Eisenstein series converges for $\mathrm{Re}(s) \gg 0$. It intertwines $I(s,\sigma)$ with the space of automorphic forms on $G(\mathbb{A})$, to give the automorphic model of the induced representation. The basic properties of the Eisenstein series, due to Selberg, Harish-Chandra, and Langlands, are as follows.

(i) $E(g,s,f_s)$ converges for $\mathrm{Re}(s) \gg 0$ and has a meromorphic continuation to all of $\mathbb{C}$.

(ii) $E(g,s,f_s)$ satisfies a functional equation of the form

$$E(g,s,f_s) = E(g,-s,M(s,\sigma)f_s),$$

where

$$M(s,\sigma) : V(s,\sigma) \to V(-s,w_0(\sigma))$$

is an intertwining operator defined as follows. First set

$$w_0 = w_\ell w_\ell^M \in W \quad \text{and} \quad w_0(\sigma)(g) = \sigma(w_0^{-1} g w_0).$$

Part of P being "self-associate" is that $w_0(M) = M$. Then

$$M(s,\sigma)f_s(g) = \int_{N(\mathbb{A})} f_s(w_0^{-1} n g)\, dn$$

for $\mathrm{Re}(s) \gg 0$.

(iii) The constant term of the Eisenstein series along P is

$$\int_{N(k)\backslash N(\mathbb{A})} E(ng,s,f_s) = f_s + M(s,\sigma)f_s.$$

(iv) The analytic properties, and particularly the poles, of $E(g,s,f_s)$ and $M(s,\sigma)$ are the same. There are only a finite number of simple poles for $\mathrm{Re}(s) \geqslant 0$; there are no poles on $\mathrm{Re}(s) = 0$.

3.4 The L-functions in the Constant Term

Langlands computed the Euler products that appear in the constant term of the Eisenstein series. "Recall", from what we said about GL_n, that if $\sigma \simeq \otimes' \sigma_v$ is a cuspidal representation of $M(\mathbb{A})$, then for almost all places v, σ_v will be unramified. This will persist for $I(s,\sigma)$ acting on $V(s,\sigma)$, i.e, if σ_v is unramified, so will be $I(s,\sigma_v)$ and in $V(s,\sigma_v)$ there will be a unique (normalized) K_v-fixed vector $f^\circ_{s,v}$, where K_v is a (reasonable) maximal compact subgroup of $G(k_v)$. So if $f_s \in V(s,\sigma)$ is decomposable, then we can write

$$f_s \simeq \otimes f_{s,v} = \left(\bigotimes_{v\in S} f_{s,v}\right) \otimes \left(\bigotimes_{v\notin S}{}' f^\circ_{s,v}\right).$$

When we apply the global intertwining operator it will factor into a product of local intertwining operators, given by similar integral expressions, which is written as

$$M(s,\sigma) f_s = \otimes A(s,\sigma_v,w_0) f_{s,v}.$$

Now for $v \notin S$, $A(s,\sigma_v,w_0) f^\circ_{s,v}$ will again be unramified and hence be a multiple of the unramified vector $\widetilde{f}^\circ_{-s,v} \in V(-s, w_0(\sigma_v))$, i.e.,

$$A(s,\sigma_v,w_0) f^\circ_{s,v} = c_v(s) \widetilde{f}^\circ_{-s,v}$$

with $c_v(s)$ a scalar function of s (which turns out to be a rational function of q_v^s). Using the method of Gindikin-Karpelevich, which reduces to a succession of SL_2 calculations, Langlands computed $c_v(s)$. Without explaining the notation involved, the result of Langlands' calculation was

$$c_v(s) = \prod_{\substack{\beta\in\Phi^+ \\ w_0(\beta)\in\Phi^-}} \frac{1-\chi_v\circ\beta^\vee(\varpi_v) q_v^{s\langle\widetilde{\alpha},\beta\rangle+1}}{1-\chi_v\circ\beta^\vee(\varpi_v) q_v^{s\langle\widetilde{\alpha},\beta\rangle}},$$

which looks like a ratio of products of Euler factors! Here $\widetilde{\alpha} \in \Phi$ is a root associated to P, $\beta^\vee$ is the co-root associated to β (see below), and χ_v is a character associated to the unramified representation σ_v.

How are we to interpret this? Langlands had at his disposal the formalism of Artin L-functions and the Satake parametrization of unramified representations. To make sense of these Euler products, Langlands introduced the the L-group and what we now call Langlands L-functions. (For more details on these, see Part III.)

(i) The L-group. Over an algebraically closed field, the algebraic group G determines and is determined by its *root datum* which is the quadruple $(X(T), \Phi(T), X^\vee(T), \Phi^\vee(T))$ where $X^\vee(T) = \mathrm{Hom}_k(\mathbb{G}_m, T)$ is the group of rational co-characters of T and $\Phi^\vee$ the set of co-roots. If we simply dualize this root datum we obtain $(X^\vee(T), \Phi^\vee(T), X(T), \Phi(T))$ which in turn is the root datum for the *Langlands dual group* $\widehat{G}$, which we can take over $\mathbb{C}$, i.e., set $\widehat{G} = \widehat{G}(\mathbb{C})$. The root datum for G

inherits an action of the local Galois group $\mathrm{Gal}(\overline{k}_v/k_v)$ and we can transfer this to the dual root system and hence $\widehat{G}$. The L-group of G is then ${}^L G = \widehat{G} \rtimes \mathrm{Gal}(\overline{k}_v/k_v)$.

(ii) Satake parameters. Langlands interpreted Satake's parametrization of the unramified representation σ_v as an assignment $\sigma_v \mapsto t_v \in \widehat{T} \subset \widehat{M} \subset {}^L M$.

(iii) Local Langlands L-functions. If $r : {}^L M \to GL_n(\mathbb{C})$ is a finite dimensional complex representation of ${}^L M$ we define a local Euler factor attached to the unramified representation σ_v and the representation r by

$$L(s, \sigma_v, r) = \det(1 - r(t_v) q_v^{-s})^{-1}.$$

Coming back to the constant term of the Eisenstein series, we consider the representation r of ${}^L M$ acting by the adjoint representation (conjugation) on $\mathrm{Lie}({}^L N)$. This decomposes as

$$r = \bigoplus_{j=1}^{m} r_j \qquad \text{with } r_j \text{ irreducible.}$$

The irreducible piece r_j is actually the sum of the the co-root spaces in $\mathrm{Lie}({}^L N)$ spanned by the co-roots $\beta^\vee$ for $\beta \in \Phi^+$ with $\langle \widetilde{\alpha}, \beta \rangle = j$, where $\widetilde{\alpha}$, β and $\beta^\vee$ are as in Langlands' formula for $c_v(s)$. Then Langlands' formula for $c_v(s)$ takes the form

$$c_v(s) = \prod_{j=1}^{m} \frac{L(js, \sigma_v, \widetilde{r}_j)}{L(js+1, \sigma_v, \widetilde{r}_j)}.$$

Thus we arrive at the following result.

Theorem 3.1 ([10, Lemma 4.1]).

$$M(s,\sigma) f_s = \left(\bigotimes_{v \in S} A(s, \sigma_v, w_0) f_{s,v} \right) \otimes \left(\prod_{v \notin S} \prod_{j=1}^{m} \frac{L(js, \sigma_v, \widetilde{r}_j)}{L(js+1, \sigma_v, \widetilde{r}_j)} \bigotimes_{v \notin S} \widetilde{f}^{\,\circ}_{-s,v} \right)$$

for $\mathrm{Re}(s) \gg 0$.

We set

$$L^S(s, \sigma, r) = \prod_{v \notin S} L(s, \sigma_v, r).$$

This is the partial L-function attached to the automorphic representation σ, the representation r, and the set of places S. Langlands proved that this converges absolutely in some right half-plane. So the product

$$\prod_{j=1}^{m} \frac{L^S(js, \sigma, \widetilde{r}_j)}{L^S(js+1, \sigma, \widetilde{r}_j)}$$

is the (partial) Euler product in the constant term of the Eisenstein series. From here Langlands went on to formulate the local and global Langlands Conjectures and Functoriality, which we will come back to in Part III.

Example. In our example of $G = Sp_{2n}$ with maximal parabolic $P = MN$ with $M \simeq GL_r \times Sp_{2t}$, the ${}^LG = SO_{2n+1}(\mathbb{C})$ and ${}^LM = GL_r(\mathbb{C}) \times SO_{2t+1}(\mathbb{C})$. In $\mathrm{Lie}({}^LN) \subset \mathrm{Lie}(SO_{2t+1}(\mathbb{C}))$ the representation r decomposes as follows:

$$\begin{pmatrix} 0 & r_1 & r_2 \\ & 0 & {}^t r_1 \\ & & 0 \end{pmatrix}$$

and we have $m = 2$ and

$$r_1 = St_{GL_r} \otimes St_{SO_{2t+1}},$$
$$r_2 = \Lambda^2(St_{GL_r}),$$

where St_G denotes the standard or defining representation of G. If we decompose $\sigma_v = \sigma_{1,v} \otimes \tau_v$ according to the decomposition of M, then $L(s, \sigma_v, r_1) = L(s, \sigma_{1,v} \times \tau_v)$ gives the Rankin-Selberg convolution L-function of $GL_r \times Sp_{2t}$ and $L(s, \sigma_v, r_2) = L(s, \sigma_{1,v}, \Lambda^2)$ gives the exterior square L-function of GL_r.

Our goal is now to use the meromorphic continuation and functional equation of $E(g, s, f_s)$ to obtain the meromorphic continuation and functional equation of the $L(s, \sigma, \widetilde{r}_j)$. However there are three immediate problems.

(1) The problem of the ratios. Every L-function occurs as part of a ratio

$$\frac{L(js, \sigma, \widetilde{r}_j)}{L(js+1, \sigma, \widetilde{r}_j)}.$$

(2) The problem of the products. Each ratio occurs as part of a product

$$\prod_{j=1}^{m} \frac{L(js, \sigma, \widetilde{r}_j)}{L(js+1, \sigma, \widetilde{r}_j)}.$$

(3) The problem of the missing local factors. How can we define and analyze the local factors $L(s, \sigma_v, \widetilde{r}_j)$ for places $v \notin S$ where σ_v is ramified?

4 The Method: Langlands-Shahidi

4.1 Resolution of the Problem of the Ratios

The resolution of the problem of the ratios is central to the Langlands-Shahidi method and involves Whittaker models.

Fix a nontrivial additive character $\psi = \otimes \psi_v$ of $k \backslash \mathbb{A}$. We can use it to define a non-degenerate character χ of the unipotent radical U of the Borel subgroup B of G through the composition of the following sequence of maps

$$U \longrightarrow U^{ab} = U/[U, U] \xrightarrow{\sim} \bigoplus_{\alpha \in \Delta} k_\alpha \xrightarrow{tr} k \xrightarrow{\psi} \mathbb{C}.$$

The choice of an isomorphism between U^{ab} and $\oplus k_\alpha$ is called a *splitting*. The trace map is simply $(x_\alpha) \mapsto \sum x_\alpha$. Since globally we assume ψ is trivial on k, we will have χ is trivial on $U(k)$ as well.

Definition 4.1. A representation (π, V_π) of G (which we can take to be the local $G(k_v)$ or global $G(\mathbb{A})$) is called χ-generic if there exists a nonzero functional $\lambda_\chi : V_\pi \longrightarrow \mathbb{C}$ such that for all $u \in U$ we have

$$\lambda_\chi(\pi(u)v) = \chi(u)\lambda_\chi(v).$$

Such a functional is called a χ-Whittaker functional. For $v \in V_\pi$ the function $W_v(g) = \lambda_\chi(\pi(g)v)$ is the associated χ-Whittaker function.

Globally, this abstract notation of a generic representation is too weak. What is needed is something seemingly stronger, the notion of *globally χ-generic.*

Definition 4.2. A cuspidal automorphic representation (π, V_π) of $G(\mathbb{A})$ is called globally χ-generic if the explicit χ-Whittaker functional

$$\varphi \mapsto \int_{U(k)\backslash U(\mathbb{A})} \varphi(u)\chi^{-1}(u)\, du$$

is non-vanishing on V_π.

Let us now return to our Eisenstein series. Suppose now that (σ, V_σ) is a globally χ-generic representation of $M(\mathbb{A})$ with respect to the restriction of χ to $U_M = M \cap U$. Denote the globally χ-generic Whittaker functional on σ by λ_χ^M. Then $I(s,\sigma)$ is also χ-generic with explicit Whittaker functional λ_χ on $V(s,\sigma)$ given by

$$\lambda_\chi(f_s) = \int_{N(k)\backslash N(\mathbb{A})} \lambda_\chi^M(f_s(w_0^{-1}n))\chi^{-1}(n)\, dn.$$

If we then form the Eisenstein series $E(g,s,f_s)$ we can compute its χ-Fourier coefficient using the same methods that Langlands used for the constant term.

By definition, the Fourier coefficient is

$$E_\chi(g,s,f_s) = \int_{U(k)\backslash U(\mathbb{A})} E(ug,s,f_s)\chi^{-1}(u)\, du.$$

If f_s is decomposable, $f_s = \otimes f_{s,v}$ we get

$$E_\chi(g,s,f_s) = \otimes \lambda_{\chi_v}(f_{s,v})(g) = \prod_v W_{f_{s,v}}(g)$$

and at the unramified places $v \notin S$, if $f_{s,v}^\circ$ is the unramified vector then $W_{f_{s,v}^\circ} = W_v^\circ$ the associated unramified Whittaker function. The Whittaker functions are not "multiplicative" in the same sense that the constant terms were and the method of Gindikin-Karpelevich no longer applies. However one can evaluate the unramified Whitaker functions by using the Casselman-Shalika formula to obtain the following result.

Theorem 4.1 ([10, Theorem 6.2]).

$$E_\chi(e,s,f_s) = \left(\prod_{v\in S} W_{f_{s,v}}(e)\right) \cdot \prod_{j=1}^{m} \frac{1}{L^S(js+1,\sigma,\widetilde{r}_j)}.$$

Compare this with the classical situation where $Z(s+1)$ occurred in the denominator of the Fourier coefficient. This resolves the problem of the ratios.

As an application, we have the following corollary. Since $E(g, s, f_s)$ has no poles on $\mathrm{Re}(s) = 0$, neither does $E_\chi(g, s, f_s)$. This then leads to non-vanishing on the 1-line!

Corollary 4.1 ([10, Corollary 6.1]). *For* $\mathrm{Re}(s) = 0$

$$\prod_{j=1}^{m} L^S(js+1, \sigma, \widetilde{r}_j) \neq 0.$$

Remark 4.1. If $G \neq GL_n$, not all cuspidal representations are globally generic. This is a restriction on the representations under consideration. However, it is conjectured that every tempered L-packet should contain a globally generic member and that every globally generic representation, or even a representation that is consistently locally generic, should be tempered. So this should be no restriction on the level of the L-functions themselves.

4.2 Resolution of the Problem of the Products

This is resolved by a nice inductive procedure which is embodied in the following proposition.

Proposition 4.1 ([10, Proposition 7.1]). *Given $1 < j \leqslant m$, there exists a (split) group G_j over k and a maximal k-parabolic subgroup $P_j = M_j N_j \subset G_j$ and a globally generic cuspidal automorphic representation σ_j of $M_j(\mathbb{A})$ such that if the adjoint action r' of ${}^L M_j$ on* $\mathrm{Lie}({}^L N_j)$ *decomposes as*

$$r' = \bigoplus_{k=1}^{m'} r'_k,$$

then we have (i) $m' < m$ *and* (ii) $L^S(s, \sigma, \widetilde{r}_j) = L^S(s, \sigma_j, \widetilde{r}'_1)$.

Example. If we go back to our example in the symplectic group, in $G = Sp_{2n} \supset P = MN$ with $M \simeq GL_r \times Sp_{2t}$ then we saw that ${}^L M \simeq GL_r(\mathbb{C}) \times SO_{2t+1}(\mathbb{C})$ and that $m = 2$ with $r_1 = St_{GL_r} \otimes St_{SO_{2t+1}}$ and $r_2 = \Lambda^2(St_{GL_r})$. If we take $G' = SO_{2r} \supset P' = M'N'$ with P' the Siegel parabolic with $M' \simeq GL_r$, then ${}^L M' \simeq GL_r(\mathbb{C})$ and in fact

$${}^L M' = \left\{ \begin{pmatrix} A & 0 \\ 0 & A^* \end{pmatrix} \middle| A \in GL_r \right\} \quad \text{and} \quad \mathrm{Lie}({}^L N') = \left\{ \begin{pmatrix} 0 & S \\ 0 & 0 \end{pmatrix} \middle| {}^t S = -S \right\}$$

so that $m' = 1$ and $r'_1 = \Lambda^2(St_{GL_r})$.

If we apply this proposition inductively we can understand *all*

- $m = 1$ cases;

- $m = 2$ cases, since $L^S(s, \sigma, \widetilde{r}_2) = L^S(s, \sigma', \widetilde{r}'_1)$, which we would now understand;

- $m = 3$ cases, etc ...

Note. The $G \supset P = MN$ fall into a finite number of infinite families plus a finite number of exceptional cases. So we could prove this proposition by "checking the list".

As an application, we have the following result.

Proposition 4.2. *Each $L^S(s, \sigma, \widetilde{r}_j)$ has a meromorphic continuation to $\mathbb{C}$.*

To see this, since the Eisenstein series $E(g, s, f_s)$ has a meromorphic continuation iff its constant term does, we see that $M(s, \sigma)f_s$ must have a meromorphic continuation. But by the explicit computation, this implies that

$$\prod_{j=1}^{m} \frac{L^S(js, \sigma, \widetilde{r}_j)}{L^S(js+1, \sigma, \widetilde{r}_j)}$$

has a meromorphic continuation. By the above proposition, we can inductively strip off factors and can conclude that

$$\frac{L^S(js, \sigma, \widetilde{r}_j)}{L^S(js+1, \sigma, \widetilde{r}_j)}$$

has a meromorphic continuation for each j. But now we utilize the shift in the numerator and denominator to set up the shift equation

$$L^S(s, \sigma, \tilde{r}_j) = \frac{L^S(js, \sigma, \widetilde{r}_j)}{L^S(js+1, \sigma, \widetilde{r}_j)} L^S(s+1, \sigma, \widetilde{r}_j)$$

and inductively shift from the original half-plane of convergence to all of $\mathbb{C}$.

4.3 Resolution of the Problem of the Missing Local Factors

There are two main tools here.

(i) A Local/Global Principle. Since we hope to use the Eisenstein, which are global in nature, we will need some type of a local/global principal. Here it is.

Proposition 4.3 ([10, Proposition 7.3]). *Let F be a local field, σ' a χ_F-generic supercuspidal representation of $M'(F)$. Then there exists a number field k and a group M/k and a globally χ-generic cuspidal representation σ of $M(\mathbb{A})$ such that*

(a) *for some place v_0 of k we have $F = k_{v_0}$, $M(k_{v_0}) = M'(F)$, $\chi_{v_0} = \chi_F$, and $\sigma_{v_0} = \sigma'$.*

(b) *for all $v \neq v_0$, σ_v is unramified (or v is archimedean).*

So, not only can we embed any local generic supercuspidal representation into a globally generic cuspidal representation, but at all the other places we completely understand the local theory!

(ii) Shahidi's Local Coefficient. One of the tools we will exploit is the functional equation of the Eisenstein series

$$E(g, s, f_s) = E(g, -s, M(s, \sigma)f_s)$$

which involves the global intertwining operator. We will also use the global Fourier coefficient $E_\chi(g, s, f_s)$. Shahidi's local coefficient combines these.

Begin with $I(s, \sigma_v)$ acting on $V(s, \sigma_v)$. We have local intertwining operators and local Whittaker functionals. If we combine them, we get the following diagram.

$$\begin{array}{ccc} V(s, \sigma_v) & \xrightarrow{A(s,\sigma_v,w_0)} & V(-s, w_0(\sigma_v)) \\ \lambda_{\chi_v} \Big\downarrow & & \Big\downarrow \lambda_{\chi_v} \\ \mathbb{C} & \xleftarrow[C_{\chi_v}(s,\sigma_v)]{} & \mathbb{C} \end{array}$$

By uniqueness of the Whittaker functionals or models, the functionals λ_{χ_v} and $\lambda_{\chi_v} \circ A(s, \sigma_v, w_0)$ differ by a complex scalar. This scalar is *Shahidi's local coefficient* and denoted by $C_{\chi_v}(s, \sigma_v)$, i.e., for each $f_{s,v} \in V(s, \sigma_v)$ we have

$$\lambda_{\chi_v}(f_{s,v}) = C_{\chi_v}(s, \sigma_v) \cdot \lambda_{\chi_v}(A(s, \sigma_v, w_0)(f_{s,v})).$$

Remark 4.2. The analytic properties of $C_{\chi_v}(s, \sigma_v)$ will be related to those of the local intertwining operator $A(s, \sigma_v, w_0)$ just as those of $E(g, s, f_s)$ are related to $M(s, \sigma)$.

How should we think of these local coefficients? Look at the places where we understand the local L-function, i.e., $v \notin S$.

Proposition 4.4 ([10, Proposition 7.4]). *Let $v|\infty$ or $v < \infty$ and σ_v unramified. Then*

$$C_{\chi_v}(s, \sigma_v) = \prod_{j=1}^{m} \gamma(js, \sigma_v, \widetilde{r}_j, \psi_v^{-1}),$$

where

$$\gamma(s, \sigma_v, \widetilde{r}_j, \psi_v^{-1}) = \frac{\varepsilon(s, \sigma_v, \widetilde{r}_j, \psi_v^{-1})L(1 - s, \sigma_v, r_j)}{L(s, \sigma_v, \widetilde{r}_j)}.$$

Note. This expression for the localγ-factor is the same as for GL_n given in Part I.

Remark 4.3. Since in these cases, we have the either the local Langlands conjecture at the archimedean places or Langlands' interpretation of the Satake parameters in terms of the L-group, the L-functions that appear can all be interpreted as Artin L-functions.

As an application of Shahidi's local coefficients, we obtain the *crude functional equation* for the partial global L-functions. We begin with the functional equation of the Eisenstein series

$$E(g,s,f_s) = E(g,-s,M(s,\sigma)f_s),$$

take the χ-Fourier coefficient

$$E_\chi(g,s,f_s) = E_\chi(g,-s,M(s,\sigma)f_s),$$

then use the calculation of the local Fourier coefficient at the unramified places $v \notin S$ and the definition of the local coefficient at $v \in S$ we obtain a crude functional equation.

Theorem 4.2 ([10, Theorem 6.3]).

$$\prod_{j=1}^{m} L^S(js,\sigma,r_j) = \prod_{v\in S} C_{\chi_v^{-1}}(s,\widetilde{\sigma}_v) \prod_{j=1}^{m} L^S(1-js,\sigma,\widetilde{r}_j).$$

5 The Results: Shahidi

5.1 The Local γ-factor

We have seen that at the unramified places and the archimedean places, there is a relation between Shahidi's local coefficient and the local γ-factor. Shahidi now uses *all* our tools plus the crude functional equation to establish a consistent theory of local γ-factors for all places.

Theorem 5.1 ([10, Theorem 7.1]). *Let G be a split reductive algebraic group over a local field F of characteristic 0. Let $P = MN$ be a (self associate) maximal parabolic subgroup of G, σ an irreducible admissible χ-generic representation of $M = M(F)$. Then there exist m complex functions*

$$\gamma(s,\sigma,r_j,\psi), \quad 1 \leqslant j \leqslant m$$

such that

(1) *if F is archimedean or F is non-archimedean and σ is unramified, then*

$$\gamma(s,\sigma,r_j,\psi) = \frac{\varepsilon(s,\sigma,r_j,\psi)L(1-s,\sigma,\widetilde{r}_j)}{L(s,\sigma,r_j)};$$

(2)
$$C_\chi(s,\sigma) = \prod_{j=1}^{m} \gamma(js,\sigma,\widetilde{r}_j,\psi^{-1});$$

(3) $\gamma(s,\sigma,r_j,\psi)$ *is "multiplicative" with respect to induction;*

(4) *if σ_v is the local component of a globally generic representation σ of $M(\mathbb{A})$ then*

$$L^S(s, \sigma, r_j) = \prod_{v \in S} \gamma(s, \sigma_v, r_j, \psi_v) L^S(1 - s, \sigma, \widetilde{r}_j).$$

Moreover, (1), (3) *and* (4) *characterize the γ-factors uniquely.*

Example. We will illustrate what is meant by "multiplicativity" in part (3) in our example. So we return, once again, to $G = Sp_{2n}$ with $n = r + t$. We have our maximal parabolic subgroup $P = MN$ with $M \simeq GL_r \times Sp_{2t}$. Our representation σ of $M(F)$ will then decompose as $\sigma \simeq \sigma_1 \otimes \tau$ with σ_1 an irreducible admissible representation of $GL_r(F)$ and τ a similar representation of $Sp_{2t}(F)$. Now the Levi subgroup M is itself reductive and within M we take a parabolic subgroup $P' = M'N'$ with

$$M' \simeq (GL_{r_1} \times GL_{r_2}) \times (GL_a \times Sp_{2b}) \subset GL_r \times Sp_{2t} \simeq M$$

with $r_1 + r_2 = r$ and $2a + 2b = 2t$. Let

$$\sigma' = (\sigma_1' \otimes \sigma_2') \otimes (\sigma'' \otimes \tau')$$

be a representation of $M'(F)$ with the tensor decomposition of the representation corresponding to the decomposition of M'. Suppose that in the decomposition $\sigma = \sigma_1 \otimes \tau$ we have

$$\sigma_1 \subset \mathrm{Ind}(\sigma_1' \otimes \sigma_2') \quad \text{and} \quad \tau \subset \mathrm{Ind}(\sigma'' \otimes \tau')$$

so that $\sigma \subset \mathrm{Ind}(\sigma')$.

We also have

$$\begin{aligned} {}^L M' &\simeq (GL_{r_1}(\mathbb{C}) \times GL_{r_2}(\mathbb{C})) \times (GL_a(\mathbb{C}) \times SO_{2b+1}(\mathbb{C})) \\ &\subset GL_r(\mathbb{C}) \times SO_{2t+1}(\mathbb{C}) \simeq {}^L M. \end{aligned}$$

Recall that the action r of ${}^L M$ on $\mathrm{Lie}({}^L N)$ decomposes as $r = r_1 \oplus r_2$ with $r_1 = St_{GL_4} \otimes St_{SO_{2t+1}}$ and $r_2 = \Lambda^2(St_{GL_r})$. Both of these will further decompose when we restrict to ${}^L M' \subset {}^L M$. This decomposition then gives the following decomposition of the resulting γ-factors, which is what is referred to as "multiplicativity".

$$\begin{aligned} \gamma(s, \sigma_1 \otimes \tau, r_1, \psi) &= \gamma(s, \sigma_1 \times \tau, \psi) \\ &= \prod_{k=1}^{2} [\gamma(s, \sigma_k' \times \sigma'', \psi) \gamma(s, \sigma_k' \times \widetilde{\sigma}'', \psi)] \prod_{k=1}^{2} \gamma(s, \sigma_k' \times \tau', \psi) \end{aligned}$$

and

$$\begin{aligned} \gamma(s, \sigma_1 \times \tau, r_2, \psi) &= \gamma(s, \sigma_1, \Lambda^2, \psi) \\ &= \gamma(s, \sigma_1' \times \sigma_2', \psi) \prod_{k=1}^{2} \gamma(s, \sigma_k', \Lambda^2, \psi). \end{aligned}$$

5.2 The Definition of Local L- and ε-factors at the Ramified Places

F is again a local field. Recall that if $v|\infty$ or $v < \infty$ and σ is unramified, then

$$\gamma(s, \sigma, r_j, \psi) = \frac{\varepsilon(s, \sigma, r_j, \psi) L(1 - s, \sigma, \widetilde{r}_j)}{L(s, \sigma, r_j)}.$$

For the other places, we take this relation as the paradigm for defining the L- and ε-factors.

(i) Tempered Representations. For tempered representations we can make the following definition.

Definition 5.1. If σ is χ-generic and tempered, set

$$L(s, \sigma, r_j) = P_{\sigma,j}(q^{-s})^{-1},$$

where $P_{\sigma,j}(X) \in \mathbb{C}[X]$ with $P(0) = 1$, and $P_{\sigma,j}(q^{-s})$ is the normalized polynomial in q^{-s} in the numerator of $\gamma(s, \sigma, r_j, \psi)$. Then we define $\varepsilon(s, \sigma, r_j, \psi)$ by

$$\gamma(s, \sigma, r_j, \psi) = \frac{\varepsilon(s, \sigma, r_j, \psi) L(1 - s, \sigma, \widetilde{r}_j)}{L(s, \sigma, r_j)}.$$

This determines L- and ε-uniquely for σ tempered.

From the behavior of the local intertwining operator $A(s, \sigma, w_0)$ for σ tempered, we can draw the following conclusion.

Theorem 5.2 ([10, Theorem 7.2]). *Suppose σ is χ-generic and tempered. Then $L(s, \sigma, r_j)$ is holomorphic for* $\mathrm{Re}(s) > 0$.

Corollary 5.1. *$L(s, \sigma, r_j)$ is multiplicative for σ tempered.*

(ii) General Representations. If σ is χ-generic but not necessarily tempered we appeal to the (representation theoretic) Langlands classification. This allows us to write σ uniquely as a sub-representation

$$\sigma \subset \mathrm{Ind}_{M'N'}^{M}(\sigma'_\nu)$$

with σ' a tempered representation of $M' = M'(F)$ and σ'_ν a quasi-tempered (or deformed) representation with ν in the negative Weyl chamber. Then $L(s, \sigma, r_j)$ is defined in terms of σ'_ν by formal multiplicativity, as for γ and for L- in the tempered case, and the $\varepsilon(s, \sigma, r_j, \psi)$ is derived from $L(s, \sigma, r_j)$ and $\gamma(s, \sigma, r_j, \psi)$ so the standard relation holds.

This now defines $L(s, \sigma, r_j)$ and $\varepsilon(s, \sigma, r_j)$ for *all* irreducible admissible χ-generic representations of $M = M(F)$.

5.3 Global L-functions

We now return to k to be a number field and $\sigma \simeq \otimes' \sigma_v$ a globally χ-generic cuspidal representation of $M(\mathbb{A})$. Let r_j be one of the representations of ${}^L M$ occurring in the action of ${}^L M$ on $\mathrm{Lie}({}^L N)$.

Definition 5.2.

$$L(s, \sigma, r_j) = \prod_v L(s, \sigma_v, r_j),$$

$$\varepsilon(s, \sigma, r_j) = \prod_v \varepsilon(s, \sigma_v, r_j, \psi_v).$$

Then from property (4) of the γ-factor (the local/global principal for γ) and the definition of L- and ε for $v \in S$ we obtain the true global functional equation.

Theorem 5.3 ([10, Theorem 7.7]). *For each $j = 1, \ldots, m$,*

$$L(s, \sigma, r_j) = \varepsilon(s, \sigma, r_j) L(1 - s, \sigma, \widetilde{r}_j).$$

5.4 Holomorphy and BVS

In general, the $L(s, \sigma, r_j)$ can have poles coming from the analytic properties of the Eisenstein series. If an Eisenstein series has a pole in $\mathrm{Re}(s) \geqslant 0$, its residual representation lies in the residual L^2-spectrum. If you "twist" the Eisenstein series, the residue can't be in L^2, i.e., there can be no pole. Transferring this to the L-functions gives the following result.

Theorem 5.4 ([10, Theorem 8.2]). *Let ξ be the k-rational character of M given by*

$$\xi(m) = \det(\mathrm{Ad}(m) | \mathrm{Lie}(N)).$$

Let S be a non-empty set of finite places of k. For every globally χ-generic cuspidal representation σ of $M(\mathbb{A})$ there exists non-negative integers f_v for $v \in S$ such that for every idele class character $\eta = \otimes \eta_v$ with $\mathrm{cond}(\eta_v) \geqslant f_v$ for all $v \in S$, if we set

$$\sigma_\eta = \sigma \otimes (\eta \circ \xi)$$

then $L(s, \sigma_\eta, r_j)$ is entire for all $1 \leqslant j \leqslant m$.

The above remarks on the Eisenstein series will rule out the poles in $\mathrm{Re}(s) \geqslant \frac{1}{2}$. Then use the global functional equation.

Also, although this requires other analytic techniques, we have the following result.

Theorem 5.5 ([10, Theorem 8.3]). *With the assumptions of the previous theorem, so η_v sufficiently highly ramified at $v \in S$, we have $L(s, \sigma_\eta, r_j)$ is bounded in vertical strips (BVS) for all $1 \leqslant j \leqslant m$.*

5.5 Summary

Let G be a (quasi)split connected reductive algebraic group over a number field k. Let $P = MN$ be a maximal parabolic subgroup defined over k. Let σ be a globally χ-generic cuspidal representation of $M(\mathbb{A})$. Let r be the representation of ${}^LM$ on Lie$({}^LN)$, decomposed as

$$r = \bigoplus_{j=1}^{m} r_j.$$

Then each global L-function $L(s,\sigma,r_j)$ is nice in the sense that

(i) $L(s,\sigma,r_j)$ converges for $\mathrm{Re}(s) \gg 0$ and has a meromorphic continuation to all $\mathbb{C}$;

(ii) for every sufficiently ramified idele class character η, $L(s,\sigma_\eta,r_j)$ is entire;

(iii) for every sufficiently ramified idele class character η, $L(s,\sigma_\eta,r_j)$ is BVS;

(iv) $L(s,\sigma,r_j) = \varepsilon(s,\sigma,r_j)L(1-s,\sigma,\widetilde{r}_j)$.

Concluding Remark. While this method covers a large family of Langlands L-functions $L(s,\sigma,\rho)$, it has two limitations.

(a) It requires σ to be globally generic.

(b) The representations σ are restricted to representations of a group M which must occur as the Levi subgroup of a maximal parabolic $P = MN \subset G$ for some larger group G. Moreover, the representations ρ of ${}^LM$ must be one of the r_j occurring in the action of ${}^LM$ on Lie$({}^LN)$.

However, other than these restrictions that are forced by the method, the analytic results on the $L(s,\sigma,r_j)$ are quite complete. As we will see, they suffice, when combined with the Converse Theorem of Part I, to establish many cases of Langlands' conjecture by the method of L-functions.

III Functoriality

6 Langlands Conjectures and Functoriality

As we indicated in Section 3, the Langlands Conjectures and Functoriality have their origin in the theory of Eisenstein series. Here I want to present another way to think about these ideas and what they entail. Recall that we ended Part I with a Moral Theorem, which we repeat.

Moral Theorem. *Any suitably nice L-function of degree n must be automorphic (or modular), i.e., associated to an automorphic representation of GL_n.*

What is an interesting source of degree n L-functions?

6.1 Global Class Field Theory

The problem of a global class field theory is to

- understand the global Galois group $\mathrm{Gal}(\overline{\mathbb{Q}}/\mathbb{Q})$ or more generally $\mathrm{Gal}(\overline{k}/k)$.

Of course one way to understand the structure of a complicated group is through its linear representations. So our question becomes

- understand the representations

$$\rho : \mathrm{Gal}(\overline{\mathbb{Q}}/\mathbb{Q}) \to GL_n(\mathbb{C}) \quad (\text{ or } GL_n(\widehat{\overline{\mathbb{Q}_\ell}})).$$

Artin attached to every such complex representation ρ a degree n L-function. If $\rho : \mathrm{Gal}(\overline{\mathbb{Q}}/\mathbb{Q}) \to GL(V)$ with V an n-dimensional complex vector space, then

$$L(s,\rho) = \prod_v L_v(s,\rho) = \prod_v L(s,\rho_v),$$

where

$$L(s,\rho_v) = \begin{cases} \det(1 - p^{-s}\rho(Frob_p)|V^{I_p})^{-1}, & \text{if } v = p, \\ \Gamma(s,\rho), & \text{if } v = \infty \end{cases}$$

with $\Gamma(s,\rho)$ the appropriate product of Γ-functions as defined by Artin. He also defined an ε-factor $\varepsilon(s,\rho)$. So then our problem can possibly be reformulated as

- for each ρ as above, understand its L-function $L(s,\rho)$.

We know the following facts about the Artin L-functions.

(i) $L(s,\rho)$ converges for $\mathrm{Re}(s) \gg 0$;

(ii) $L(s,\rho)$ has a meromorphic continuation to $\mathbb{C}$;

(iii) $L(s,\rho)$ is bounded in vertical strips;

(iv) we have the functional equation $L(s,\rho) = \varepsilon(s,\rho)L(1-s,\widetilde{\rho})$.

We also have the

Artin Conjecture. *$L(s,\rho)$ is entire if ρ is irreducible and not the trivial representation.*

So, morally, this should imply that $L(s,\rho)$ is automorphic. (The twists in our Moral Theorem are here given by the tensor product of Galois representations.) It is this automorphy that is the substance of the Langlands Conjectures.

6.2 The Langlands Conjectures for GL_n

Based on our Moral Theorem, we are led to a first naive formulation of the Langlands conjecture.

Naive Global Langlands Conjecture. *There exist bijections*

$$\left\{ \begin{matrix} \rho : \mathrm{Gal}(\overline{\mathbb{Q}}/\mathbb{Q}) \to GL_n(\mathbb{C}) \\ \textit{irreducible} \end{matrix} \right\} \leftrightarrow \left\{ \begin{matrix} \textit{cuspidal automorphic} \\ \textit{representations } \pi \textit{ of } GL_n(\mathbb{A}) \end{matrix} \right\}$$

such that

$$L(s,\rho) = L(s,\pi),$$
$$L(s,\rho_1 \otimes \rho_2) = L(s,\pi_1 \times \pi_2)$$

with similar equalities for ε-factors.

One advantage of the adelic theory of automorphic representations is a natural formulation of a local version of the conjecture.

Naive Local Langlands Conjecture. *There exist bijections*

$$\left\{ \rho_v : \mathrm{Gal}(\overline{\mathbb{Q}_v}/\mathbb{Q}_v) \to GL_n(\mathbb{C}) \right\} \leftrightarrow \left\{ \begin{matrix} \textit{irreducible admissible} \\ \textit{representations } \pi_v \textit{ of } GL_n(\mathbb{Q}_v) \end{matrix} \right\}$$

such that

$$L(s,\rho_v) = L(s,\pi_v),$$
$$L(s,\rho_{1,v} \otimes \rho_{2,v}) = L(s,\pi_{1,v} \times \pi_{2,v})$$

with similar equalities for ε-factors.

Note that these local and global Langlands Conjectures are compatible. This local/global compatibility principal will be important in our later formulations.

Why did we call this naive? It is not even true for $n = 1$. On the left hand side, we would have only finite order Galois characters, while on the right hand side we have all Hecke characters (or all characters of $\mathbb{Q}_v^\times$ in the local case). Weil realized this and remedied it by introducing what we now refer to as the Weil group $W_\mathbb{Q}$ or $W_{\mathbb{Q}_v}$. Locally at a finite place we have

$$I_p \hookrightarrow W_{\mathbb{Q}_p} \to \mathrm{Gal}(\overline{\mathbb{Q}}_p/\mathbb{Q}_p),$$

where the Weil group has dense image in the Galois group; at the infinite place we have a short exact sequence

$$1 \to \mathbb{Q}^\times \to W_\mathbb{R} \to \mathrm{Gal}(\mathbb{C}/\mathbb{R}) \to 1.$$

The global Weil group $W_\mathbb{Q}$ has only a cohomological definition, but there is a compatibility between the local and global Weil groups. If we replace $\mathrm{Gal}(\overline{\mathbb{Q}}/\mathbb{Q})$ by $W_\mathbb{Q}$ and $\mathrm{Gal}(\overline{\mathbb{Q}}_v/\mathbb{Q}_v)$ by $W_{\mathbb{Q}_v}$, then the Langlands conjectures for $n = 1$ become the theorems of global and local Class Field Theory.

Unfortunately, the replacement of the Galois group by the Weil group did not completely solve our problem. Deligne pointed out that even this modification of the local Langlands conjecture was not true for $GL_2(\mathbb{Q}_p)$ for finite primes p. There were still too many representations on the right hand side. Deligne observed that if one used ℓ-adic coefficients rather than complex coefficients in the representations, one had an extra structure, as shown by Grothendieck. ℓ-adic Galois representations come with a monodromy operator, an extra nilpotent endomorphism that is normalized by the Galois representation. So Deligne imposed this structure in the complex case by introducing what we now call the (local) Weil-Deligne group $W'_{\mathbb{Q}_p}$. It has the structure of a semi-direct product $W'_{\mathbb{Q}_p} = W_{\mathbb{Q}_p} \ltimes \mathbb{G}_a$, where in any representation, the generator of the additive group acts by a nilpotent endomorphism N. So a representation of the Weil-Deligne group is a pair (ρ, N) with ρ a representation of the Weil group on a complex vector space V and N a nilpotent endomorphism of V, with certain compatibilities. (The archimedean Weil-Deligne group is just the Weil group; there is no need to change the theory at these places.) Deligne also extended the definition of the L- and ε-factors to these representations. This resolved the problem of the missing representations for $GL_2(\mathbb{Q}_p)$ and gives us a good formulation of a local Langlands conjecture.

Local Langlands Conjecture. *There exist bijections*

$$\left\{ \begin{array}{c} \rho_v : W'_{\mathbb{Q}_v} \to GL_n(\mathbb{C}) \\ \textit{Frobenius semi-simple} \end{array} \right\} \leftrightarrow \left\{ \begin{array}{c} \textit{irreducible admissible} \\ \textit{representations } \pi_v \textit{ of } GL_n(\mathbb{Q}_v) \end{array} \right\}$$

such that

$$L(s, \rho_v) = L(s, \pi_v),$$
$$L(s, \rho_{1,v} \otimes \rho_{2,v}) = L(s, \pi_{1,v} \times \pi_{2,v})$$

with similar equalities for ε-factors.

This is indeed the correct formulation, for this is now a Theorem, due to Langlands for $\mathbb{Q}_v = \mathbb{R}$ and by Harris-Taylor, followed by Henniart, for $\mathbb{Q}_v = \mathbb{Q}_p$. In fact, their proofs work for any local field of characteristic 0. It is also known for local fields of characteristic p by Laumon-Rapoport-Stuhler.

Unfortunately, there is no known construction of a global Weil-Deligne group. (This is often formulated as the conjectural Langlands group.) Instead we rely on a formulation of a local/global compatibility at least for representations of the Weil group. If we begin with a global representation $\rho : W_{\mathbb{Q}} \to GL_n(\mathbb{C})$ then, from the local/global compatibility of the Weil groups, ρ induces local representations ρ_v of $W_{\mathbb{Q}_v}$ for each v. Then we have the following diagram

$$\begin{array}{ccc} \rho & \dashrightarrow & \pi = \otimes' \pi_v \\ \downarrow & & \uparrow \\ \{\rho_v\} & \xrightarrow[LLC]{} & \{\pi_v\} \end{array}$$

where the top arrow is defined by the composition of the other three.

Global Langlands Conjecture. *With the above formalism, $\pi = \otimes' \pi_v$ is automorphic.*

There are two ways of thinking about the information flow in these conjectures and theorems. For non-abelian Class Field Theory, we think of the information flow as going from the automorphic side (where we can control the analytic properties of L-functions) to the Galois side. However, for the formulation of Functoriality, we view the flow of information as going from the Galois side to the automorphic side, thus giving us *an arithmetic parametrization* of admissible or automorphic representations.

6.3 Langlands Conjectures for Other Groups

Another powerful aspect of the adelic theory of automorphic representations is that it can be formulated for any reductive algebraic group H over $\mathbb{Q}$ or k, as we have seen in Part II. Thinking of the Langlands Conjectures as arithmetic parametrizations of automorphic representations of $H(\mathbb{A})$ or admissible representations of $H(\mathbb{Q}_v)$, what do we replace the $GL_n(\mathbb{C})$ by in our Galois representations? This is where the formulation of the dual group $\widehat{H}$ or the L-group ${}^L H$ that we saw in Part II comes into play.

Local Langlands Conjecture for H. *There is a finite-to-one surjective map*

$$\left\{ \begin{matrix} \textit{irreducible admissible} \\ \textit{representations } \pi_v \textit{ of } H(\mathbb{Q}_v) \end{matrix} \right\} \to \left\{ \begin{matrix} \textit{admissible homomorphisms} \\ \phi_v : W'_{\mathbb{Q}_v} \to {}^L H \end{matrix} \right\}$$

which satisfies ... a list of representation theoretic desiderata.

Part of the desiderata should be an equality of L-functions. But Langlands used this formulation to define the L-functions, namely if $r : {}^L H \to GL_n(\mathbb{C})$ is an L-homomorphism of the type we saw in Part II (although now not restricted to come from the theory of induced representations as there) then Langlands *defined*

$$L(s, \pi_v, r) = L(s, r \circ \phi_v),$$

where now the L-function on the right is a generalized Artin L-function. So we should think of the right hand side of this local Langlands conjecture as giving special families of Galois representations that parametrize representations of H and lead to a formulations of L-functions for H as Artin L-functions. The fibres of the parametrization are called *L-packets*, because under this formalism, the representations in the fibres will have the same L-functions for any representation r of the dual group ${}^L H$.

What to we know in the direction of this conjecture? We know the conjecture in the following instances. (Lets talk over a number field k and its completions.)

(i) $k_v = \mathbb{R}$, $\mathbb{C}$ and any H, by Langlands.

(ii) k_v non-archimedean and π_v unramified, by Satake.

(iii) k_v non-archimedean and $H = GL_n$, the characteristic 0 case by Harris-Taylor, Henniart.

(iv) k_v non-archimedean and $H = GSp_4$ or Sp_4, by Gan and Takeda.

(v) k_v non-archimedean, $H = SO_{2n+1}$, and π_v generic, by Jiang and Soudry.

(vi) Possible other miscellaneous cases that I am unaware of.

As for GL_n, we formulate a version of the **Global Langlands Conjecture for** H through a local/global compatibility.

6.4 Functoriality

Thinking of the local or global Langlands conjectures in terms of arithmetic parametrizations of admissible or automorphic representations, these Langlands conjectures lead to a formulation of transferring admissible/automorphic representations from H to G. We will concentrate on the case where $G = GL_N$, as this is most compatible with out methods, but any (quasi) split target group will do.

There is a new ingredient necessary for this transfer. This is an L-**homomorphism**, which is a complex analytic map

$$u : {}^L H \to {}^L G = {}^L GL_N = GL_N(\mathbb{C}) \times \mathrm{Gal}(\bar{k}/k)$$

which is compatible with the projections to the Galois factors. (In the local formulation, we use $\mathrm{Gal}(\bar{k}_v/k_v)$.) In our example from Part II, where $H = Sp_{2n}$, then $\widehat{H} = SO_{2n+1}$ and we can take for u the map coming from the natural embedding

$$u : SO_{2n+1}(\mathbb{C}) \hookrightarrow GL_{2n+1}(\mathbb{C}).$$

There are similar embeddings for the other split and even quasi-split classical groups.

A formulation of local functoriality from H to GL_N relies on assuming the local Langlands conjecture for H and recalling that we know the local Langlands conjecture for GL_N by Harris-Taylor/Henniart.

Local Functoriality. *Assume we know the local Langlands conjecture for $H(k_v)$. Let π_v be an irreducible admissible representation of $H(k_v)$ with associated Langlands parameter ϕ_v. Then we simply follow the diagram*

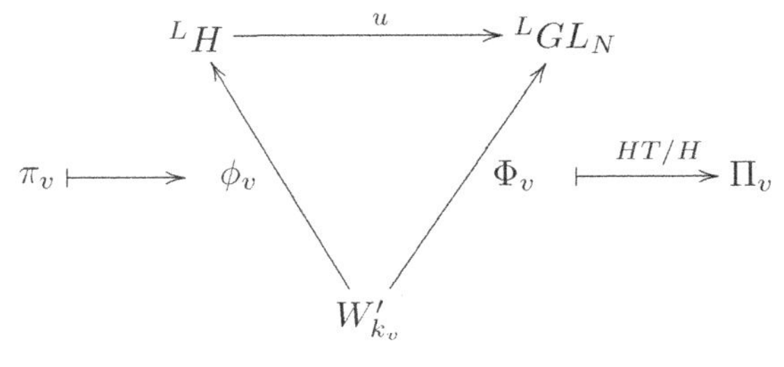

to obtain an irreducible admissible representation Π_v of $GL_n(k_v)$.

This formalism comes complete with an equality of L- and ε-factors, for example

$$L(s, \pi_v, u) = L(s, u \circ \phi_v) = L(s, \Phi_v) = L(s, \Pi_v).$$

Note that we know how to carry this out whenever we know the local Langlands parametrization for π_v, i.e., when k_v is archimedean or when k_v is non-archimedean and π_v is unramified. So in fact, if we are given a global automorphic representation $\pi = \otimes' \pi_v$ of $H(\mathbb{A})$, we can transfer the local representation π_v for all v outside a finite set S of finite places.

We can formulate a global version of this using the local/global principal and the local functoriality diagram. It relies on knowing the local Langlands conjecture for H at all places.

Global Functoriality Conjecture. *Assume the local Langlands conjecture for H at all places of k. If $\pi = \otimes' \pi_v$ is a cuspidal automorphic representation of $H(\mathbb{A})$ then the representation $\Pi = \otimes' \Pi_v$ of $GL_N(\mathbb{A})$ obtained by following the local functoriality diagrams*

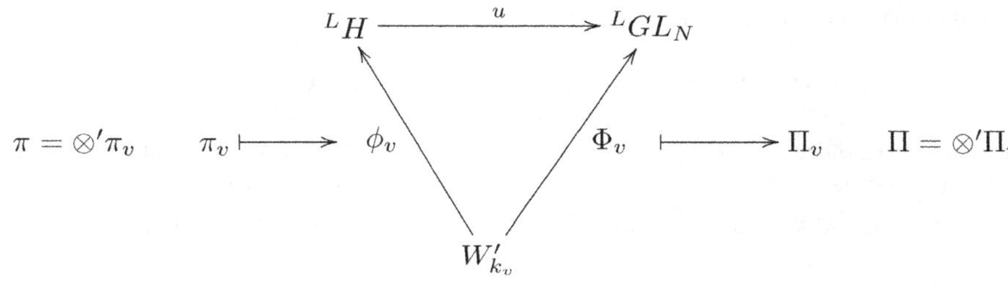

is automorphic.

This global formalism also comes complete with an equality of L- and ε-factors

$$L(s, \pi, u) = \prod_v L(s, \pi_v, u) = \prod_v L(s, \Pi_v) = L(s, \Pi)$$

along with similar equalities for twisted versions.

7 The Converse Theorem and Functoriality

In the global functoriality diagram, one begins with a cuspidal automorphic representation $\pi = \otimes' \pi_v$ of $H(\mathbb{A})$ and produce an irreducible admissible representation $\Pi = \otimes' \Pi_v$ of $GL_N(\mathbb{A})$. Such a representation of GL_N could be the input to the Converse Theorem for GL_N. Since the diagram comes with an identity of twisted L-functions, one could verify the conditions of the Converse Theorem for Π by controlling the analytic properties of the twisted L-functions for H. But this is exactly what came out of the process in Part II. So we are now in the position of combining the L-function techniques of Part I, which come from the theory of integral representations for GL_N, and the L-functions techniques of Part II, which come from the relation with Eisenstein series, to attack the question of functoriality via L-function techniques.

7.1 Functoriality for the Classical Groups

The first instances where we could do this were the families of (quasi)split classical groups. For now, we will content ourselves with the split cases. As we have seen in our example that we carried through in Part II, for Sp_{2n}, we can control the twisted L-functions $L(s, \pi \times \tau)$ for π a globally generic cuspidal representation of $Sp_{2n}(\mathbb{A})$ and τ a cuspidal representation of $GL_m(\mathbb{A})$. The twisted L-functions for the other split classical groups occur in a similar fashion. So we will be able to consider functoriality for globally generic cuspidal representations in the following split situations.

H	${}^LH$	$u : {}^LH \to {}^LGL_N$	${}^LGL_N$	GL_N
SO_{2n+1}	$Sp_{2n}(\mathbb{C})$	$\hookrightarrow$	$GL_{2n}(\mathbb{C})$	GL_{2n}
SO_{2n}	$SO_{2n}(\mathbb{C})$	$\hookrightarrow$	$GL_{2n}(\mathbb{C})$	GL_{2n}
Sp_{2n}	$SO_{2n+1}(\mathbb{C})$	$\hookrightarrow$	$GL_{2n+1}(\mathbb{C})$	GL_{2n+1}

Take an H from this table and let $\pi = \otimes'\pi_v$ be a globally generic cuspidal representation of $H(\mathbb{A})$. For almost all places v of k we know the local Langlands conjecture for $H(k_v)$, namely for the archimedean places and the finite places for which π_v is unramified. So for these places we have our local functoriality diagram

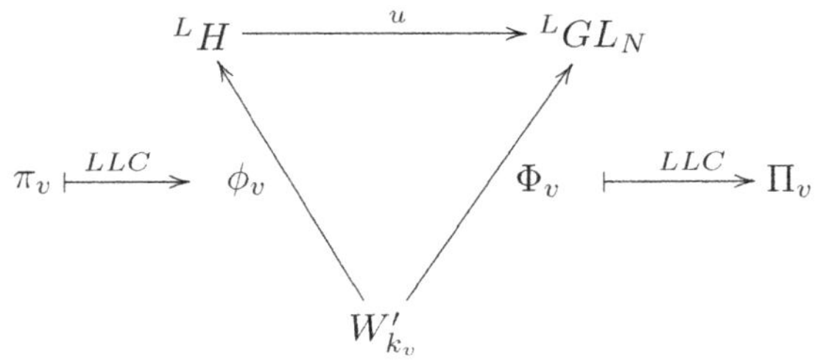

which preserves L-functions. However, there remains a finite set of S of non-archimedean places for which we do not know the local Langlands parameters for π_v. In spite of this we can still prove the following result.

Theorem 7.1. *Let H be a (quasi)split classical group, $\pi = \otimes'\pi_v$ a globally generic cuspidal representation of $H(\mathbb{A})$, and $u : {}^LH \to {}^LGL_N$ the natural L-homomorphism. Then there exists an automorphic representation $\Pi = \otimes'\Pi_v$ of $GL_n(\mathbb{A})$ such that for all $v \notin S$, Π_v is the local Langlands lift of π_v as in the above diagram.*

The proof has three main steps, but three lingering questions about their implementation.

Step 1. Construct a candidate lift $\Pi = \otimes'\Pi_v$ as an irreducible admissible representation of $GL_N(\mathbb{A})$ with the property that for all cuspidal automorphic representations τ of $GL_m(\mathbb{A})$ in an appropriate twisting set we have

$$L(s,\pi\times\tau)=L(s,\Pi\times\tau),$$
$$\varepsilon(s,\pi\times\tau)=\varepsilon(s,\Pi\times\tau).$$

For the places $v\notin S$, we can obtain this locally from the local functoriality diagram.

Question 1. What to do about Π_v for $v\in S$?

Step 2. Control the analytic properties of $L(s,\Pi\times\tau)$ through those of $L(s,\pi\times\tau)$. We do this using the Langlands-Shahidi method, as in the Summary in Section 5, where we are guaranteed that these twisted L-functions are meromorphic, BVS and satisfy an appropriate functional equation. But the Converse Theorem requires that all L-functions be entire, which is more that is guaranteed.

Question 2. How to uniformly control the poles of the L-functions for all the twists simultaneously?

Step 3. Apply the Converse Theorem to $\Pi=\otimes'\Pi_v$.

Question 3. Which one, i.e., what is the appropriate twisting set?

7.2 Resolution of the Difficulties

We will take our difficulties in order, since the resolution of Questions 1 and 2 will force the answer to Question 3.

Question 1. What to do about the $v\in S$ where we do not know the local Langlands parametrization for π_v. We finesse the lack of the local Langlands parametrization by a purely local result called the "stability of L and ε under highly ramified twists". The proof of this is quite involved and is established in the context of the "stability of γ under highly ramified twists". The statement of stability is the following.

Theorem 7.2. *Suppose that η_v is a suitably highly ramified character of $k_v^\times$. Then for irreducible admissible representations π_v of $H(k_v)$ or Π_v of $GL_N(k_v)$ we have*

$$L(s,\pi_v\times\eta_v)\equiv 1\equiv L(s,\Pi_v\times\eta_v)$$

and both $\varepsilon(s,\pi_v\times\eta_v,\psi_v)$ and $\varepsilon(s,\Pi_v\times\eta_v,\psi_v)$ stabilize, with the stable form only depending on the central character of the local representations.

The degree of ramification depends on π_v or Π_v respectively. Once you have stability, one can replace π_v and Π_v by a full induced representation with the same central character and then compute these stable form of the ε-factors (from the stable form of the γ-factors) by using multiplicativity of the γ-factors. When one does this, one obtains the following corollary.

Corollary 7.1. *Assume H is split. Then for any irreducible admissible generic representation π_v of $H(k_v)$ and any irreducible admissible representation Π_v of*

$GL_N(k_v)$ with trivial central character, for sufficiently highly ramified character η_v, depending on both representations π_v of $H(k_v)$ and Π_v of $GL_N(k_v)$, we have

$$L(s, \pi_v \times \eta_v) = L(s, \Pi_v \times \eta_v),$$
$$\varepsilon(s, \pi_v \times \eta_v, \psi_v) = \varepsilon(s, \Pi_v \times \eta_v, \psi_v).$$

So if we have the freedom to twist, we can take as our "local lift" any Π_v with trivial central character. We can extend this to certain twists by $GL_m(k_v)$, again by multiplicativity of L- and ε- for certain representations.

Corollary 7.2. *Assume H is split. Let π_v be an irreducible admissible generic representation of $H(k_v)$ and Π_v irreducible admissible representation of $GL_N(k_v)$ with trivial central character. Let η_v be a sufficiently highly ramified character of $k_v^\times$ as in the above corollary. Then for every unramified representation τ_v° of $GL_m(k_v)$ if we set $\tau_v = \tau_v^\circ \otimes \eta_v$ then*

$$L(s, \pi_v \times \tau_v) = L(s, \Pi_v \times \tau_v),$$
$$\varepsilon(s, \pi_v \times \tau_v, \psi_v) = \varepsilon(s, \Pi_v \times \tau_v, \psi_v).$$

If we combine this last corollary at the places $v \in S$ with the local functoriality diagram for $v \notin S$ we arrive at the end of Step 1. Let $\pi = \otimes' \pi_v$ be a generic cuspidal automorphic representation of $H(\mathbb{A})$. For $v \notin S$, let Π_v be the local lift of π_v obtained through the local functoriality diagram. For $v \in S$ take Π_v any irreducible admissible representation of $GL_n(k_v)$ with trivial central character. Consider the candidate lift $\Pi = \otimes \Pi_v$ of $GL_N(\mathbb{A})$.

Theorem 7.3. *Let $\eta = \otimes \eta_v$ be a fixed idele class character such that at all $v \in S$, η_v is sufficiently highly ramified so that the previous corollaries hold for π_v and Π_v. Then for every $\tau \in \mathcal{T}_0^S(N-2) \otimes \eta$ we have*

$$L(s, \pi \times \tau) = L(s, \Pi \times \tau),$$
$$\varepsilon(s, \pi \times \tau) = \varepsilon(s, \Pi \times \tau).$$

For the definition of $\mathcal{T}_0^S(N-2)$ see the paragraph before Theorem 2.4. This completes Step 1.

Question 2. The answer to our second question, controlling the poles of $L(s, \Pi \times \tau)$ and hence $L(s, \Pi \times \tau)$ we can now find in the Summary in Section 5 and the previous theorem. In Section 5 we saw that if we twisted by a sufficiently ramified character η then the resulting L-function had no poles. When one unravels the twisting there in terms of the Rankin-Selberg L-functions we are considering here, then we have the following result.

Theorem 7.4. *Let H be a split classical group and $\pi = \otimes' \pi$ a globally generic cuspidal representation of $H(\mathbb{A})$. Let S be a non-empty finite set of finite places. Let $\eta = \otimes \eta_v$ be a idele class character which is sufficiently highly ramified at the places $v \in S$ so that Theorems 5.4 and 5.5 are satisfied. Then for every $\tau \in \mathcal{T}_0^S(N-2) \otimes \eta$ we have that $L(s, \pi \times \tau)$ is nice, that is*

(i) *$L(s, \pi \times \tau)$ is an entire function of s;*

(ii) *$L(s, \pi \times \tau$ is bounded in vertical strips;*

(iii) *$L(s, \pi \times \tau) = \varepsilon(s, \pi \times \tau) L(1-s, \widetilde{\pi} \times \widetilde{\tau})$.*

So a similar twisting by a ramified character gives us the the twisted L-functions on the classical group are nice as desired in the Converse Theorem. This completes Step 2.

Question 3. We are now in a position to apply a Converse Theorem. We just need to combine Theorems 7.3 and 7.4. So we take our globally generic cuspidal representation $\pi = \otimes \pi_v$ of $H(\mathbb{A})$. We construct a candidate lift $\Pi = \otimes' \Pi_v$ as in Step 1. We take S to be a finite set of finite places such that (i) S contains all finite places where we do not have the local Langlands parametrization for π_v and (ii) S is not empty. We fix an idele class character $\eta = \otimes \eta_v$ such that the local components at $v \in S$ are sufficiently ramified that both Theorems 7.3 and 7.4 hold. Then combining these with the flexible Converse Theorem given in Theorem 2.4, we obtain Theorem 7.1, the functorial lifting from classical groups to GL_N for globally generic cuspidal representations. This completes Step 3.

7.3 The Image of Functoriality

We have two remaining issues.

(1) When is the functorial lift Π cuspidal?

(2) Can we characterize the image of functoriality?

These questions are answered by yet another set of techniques. This is the method of *automorphic descent* of Ginzburg, Rallis, and Soudry. It was

- motivated by integral representations of twisted L-functions for classical groups,
- implemented in terms of residues of Eisenstein series.

So it represents a nice synthesis of ideas from the two methods for studying L-functions. The descent allows you to begin with a self-dual representation of GL_N, local or global, and descend back to a representation of the classical group H (or a metaplectic group).

If we combine our approach to functoriality via the Converse Theorem with the Ginzburg-Rallis-Soudry theory of descent we obtain the following more complete result.

Theorem 7.5. *Let H be a split classical group, π a globally generic cuspidal representation of $H(\mathbb{A})$. Then there is a representation $R = R_H$ of ${}^L GL_n$ such that the functorial lift of π to an automorphic representation Π of $GL_N(\mathbb{A})$ as above has trivial central character and is of the form*

$$\Pi = \mathrm{Ind}(\Pi_1 \otimes \cdots \otimes \Pi_d) = \Pi_1 \boxplus \cdots \boxplus \Pi_d,$$

where each Π_i *is a unitary self-dual representation of* $GL_{N_i}(\mathbb{A})$ *such that the* L*-function* $L(s,\Pi_i,R)$ *has a pole at* $s=1$ *and* $\Pi_i \not\simeq \Pi_j$ *for* $i \neq j$*. Moreover any such* Π *is the functorial lift of some globally generic cuspidal representation* π *of* $H(\mathbb{A})$*.*

To finish, we only need to identify $R = R_H$. For $H = Sp_{2n}$ or SO_{2n}, so that ${}^LH$ is an orthogonal group, $R = \mathrm{Sym}^2$, while for $H = SO_{2n+1}$, where ${}^LH = Sp_{2n}$, we have $R = \Lambda^2$.

For the case of the (quasi)split classical groups, the result is similar, but there is also a central character condition.

8 Symmetric Powers and Applications

There are a number of other functorialities that one can obtain by this method that have striking arithmetic applications. As the method is the same as before, here I will describe the L-homomorphisms and how one controls the twisted L-functions needed for the Converse Theorem within the Langlands-Shahidi method.

8.1 The Tensor Product Lifting From $GL_2 \times GL_3$ to GL_6

The L-group of GL_n is $GL_n(\mathbb{C})$. The tensor product isomorphism $\mathbb{C}^2 \otimes \mathbb{C}^3 \simeq \mathbb{C}^6$ induces a map

$$GL_2(\mathbb{C}) \times GL_3(\mathbb{C}) \xrightarrow{\ \otimes\ } GL_6(\mathbb{C})$$

and there should be an associated map of automorphic representations given by the following functoriality diagram.

Let $\pi_1 = \otimes' \pi_{1,v}$ be a cuspidal automorphic representation of $GL_2(\mathbb{A})$ and $\pi_2 = \otimes' \pi_{2,v}$ a cuspidal automorphic representation of $GL_3(\mathbb{A})$. Write $\pi_1 \times \pi_2$ for the corresponding representation of $GL_2(\mathbb{A}) \times GL_3(\mathbb{A})$. Then for each place v of k, since we know local Langlands for GL_n, we have

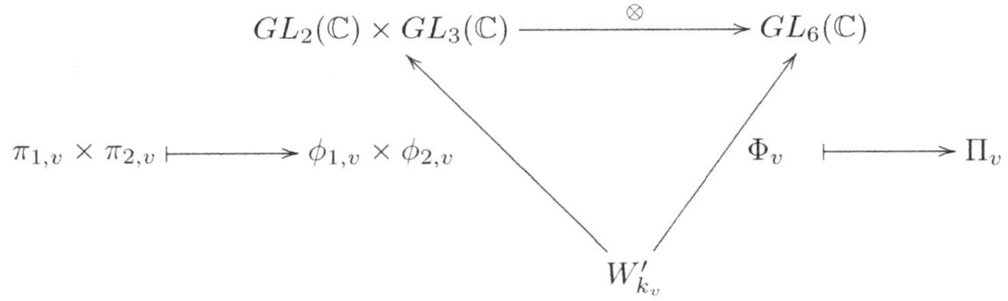

which then gives us an irreducible admissible representation $\Pi = \otimes \Pi_v$ of $GL_6(\mathbb{A})$. Let us denote $\Pi_v = \pi_{1,v} \boxtimes \pi_{2,v}$ and $\Pi = \pi_1 \boxtimes \pi_2$.

Theorem 8.1 ([10, Theorem 9.1])**.** $\pi_1 \boxtimes \pi_2$ *is an automorphic representation of* $GL_6(\mathbb{A})$*.*

To apply the Converse Theorem for GL_6 we must now control the triple product L-functions $L(s, \pi_1 \times \pi_2 \times \tau)$ for τ a cuspidal representation of $GL_m(\mathbb{A})$ with $m = 1, 2, 3, 4$. Where do these occur in the Langlands-Shahidi list?

GL_1 **twists.** We take $G = GL_5$. Its Dynkin diagram is

and if we remove the second vertex

we obtain a maximal parabolic subgroup $P = MN$ whose Levi has derived group $M \simeq GL_2 \times GL_3$. If we take the representation of M to be $\sigma \simeq \pi_1 \otimes (\pi_2 \otimes \tau)$, then this is an $m = 1$ situation and we have, in the notation of Part II,

$$L(s, \sigma, r_1) = L(s, \pi_1 \times \pi_2 \times \tau).$$

GL_2 **twists.** We take $G = D_5^{sc} = Spin_{10}$. Its Dynkin diagram is

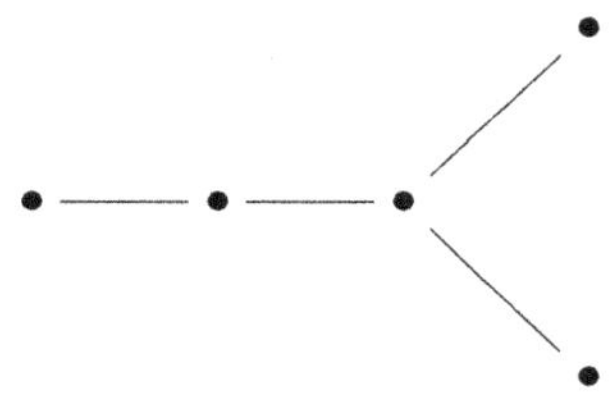

and if we remove the third vertex

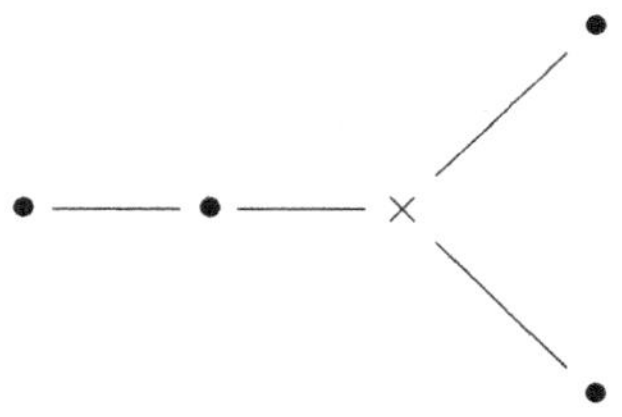

we obtain a maximal parabolic subgroup $P = MN$ whose Levi subgroup has derived group $M_{der} \simeq SL_3 \times SL_2 \times SL_2$. This is the same as the derived group of $GL_3 \times GL_2 \times GL_2$ and there is a k-rational morphism from M to $GL_3 \times GL_2 \times GL_2$ which is the identity on the derived groups. There is then a way to transfer the representation $\pi_2 \otimes \pi_1 \otimes \tau$ of $GL_3 \times GL_2 \times GL_2$ to a representation σ of M such that

$$L(s, \sigma, r_1) = L(s, \pi_1 \times \pi_2 \times \tau).$$

We will denote this (and similar) process by writing $\sigma \sim \pi_2 \otimes \pi_1 \otimes \tau$. This is an $m = 2$ situation.

GL_3 **twists.** We take G to be the exceptional group E_6^{sc}. Its Dynkin diagram is

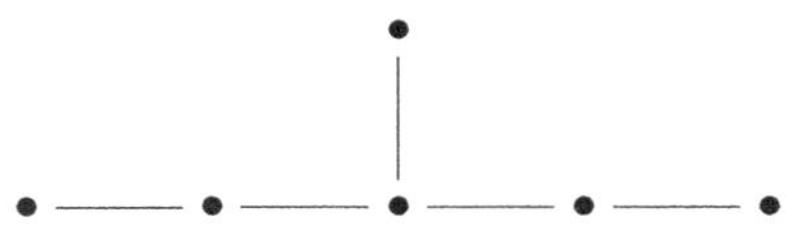

and if we remove the third vertex

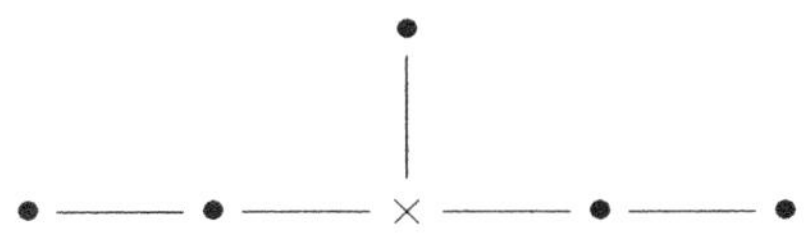

we obtain a maximal parabolic subgroup $P = MN$ whose Levi subgroup has derived group $M_{der} \simeq SL_3 \times SL_2 \times SL_3$. If we take the representation of M so that $\sigma \sim \pi_2 \otimes \pi_1 \otimes \tau$, then this is an $m = 3$ situation with

$$L(s, \sigma, r_1) = L(s, \pi_1 \times \pi_2 \times \tau).$$

GL_4 **twists.** We take G to be the exceptional group E_7^{sc}. Its Dynkin diagram is

and if we remove the fourth vertex

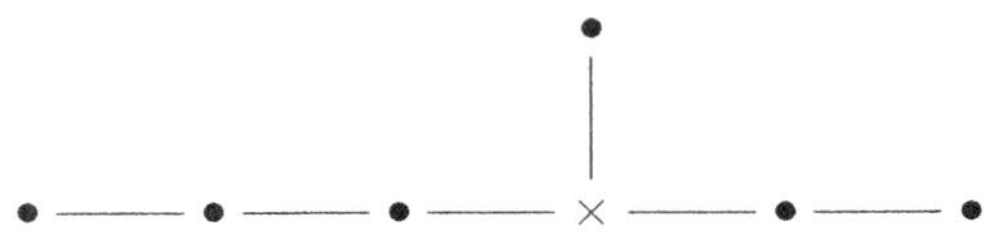

we obtain a maximal parabolic subgroup $P = MN$ whose Levi subgroup has derived group $M_{der} \simeq SL_4 \times SL_2 \times SL_3$. If we take the representation of M so that $\sigma \sim \tau \otimes \pi_1 \otimes \pi_2$, then this is an $m = 4$ situation with

$$L(s, \sigma, r_1) = L(s, \pi_1 \times \pi_2 \times \tau).$$

Even in this relatively simple case of a tensor product functoriality, we need Eisenstein series on exceptional groups! As usual, the Converse Theorem gives an automorphic representation Π' with $\Pi'_v \simeq \pi_{1,v} \boxtimes \pi_{2,v}$ for almost all places v. To fill in the correct lift at the $v \in S$ we need other techniques: base change and the theory of types.

8.2 Symmetric Power Functorialities

$GL_2(\mathbb{C})$, the L-group of GL_2, has an irreducible $n+1$ dimensional representation on the space of symmetric n-tensors $\mathrm{Sym}^n(\mathbb{C}^2)$. This gives a map

$$GL_2(\mathbb{C}) \xrightarrow{\mathrm{Sym}^n} GL_{n+1}(\mathbb{C})$$

and there should be an associated lifting of automorphic representations from $GL_2(\mathbb{A})$ to $GL_{n+1}(\mathbb{A})$. The relevant functoriality diagram is

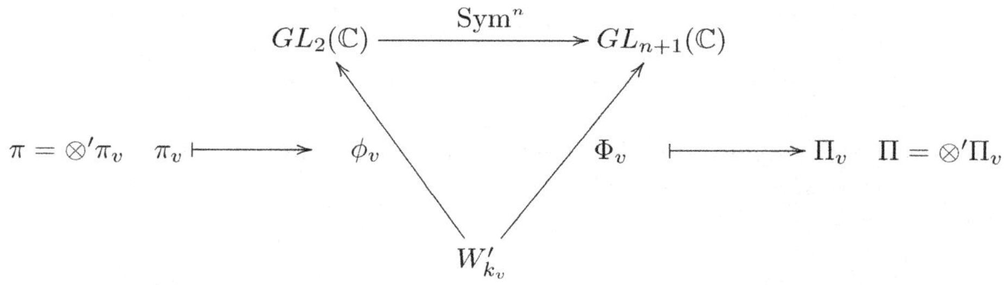

Let us then write $\Pi_v = \mathrm{Sym}^n(\pi_v)$, a well defined irreducible admissible representation of $GL_{n+1}(k_v)$, and $\Pi = \mathrm{Sym}^n(\pi) = \prod \mathrm{Sym}^n(\pi_v)$ an irreducible admissible representation of $GL_{n+1}(\mathbb{A})$. The question of Sym^n functoriality for GL_2 is then whether $\mathrm{Sym}^n(\pi)$ is automorphic.

As a consequence of the $GL_2 \times GL_3 \to GL_6$ tensor product functoriality, we have that $\mathrm{Sym}^3(\pi)$ is indeed automorphic.

Theorem 8.2 ([10, Theorem 9.2]). *If π is a cuspidal automorphic representation of $GL_2(\mathbb{A})$ then $\mathrm{Sym}^3(\pi)$ is an automorphic representation of $GL_4(\mathbb{A})$. It is cuspidal unless π is of dihedral or tetrahedral type.*

This result is finessed from the tensor product functoriality as follows. Begin with π, our cuspidal automorphic representation of $GL_2(\mathbb{A})$. Gelbart and Jacquet had proved that $\mathrm{Sym}^2(\pi)$ is an automorphic representation of $GL_3(\mathbb{A})$ via the Converse Theorem for GL_3 in 1978. Then by the tensor product functoriality for $GL_2 \times GL_3$ above we have that $\pi \boxtimes \mathrm{Sym}^2(\pi)$ is an automorphic representation of $GL_6(\mathbb{A})$. Moreover, we have a decomposition

$$\pi \boxtimes (\mathrm{Sym}^2(\pi) \otimes \omega_\pi^{-1}) = (\mathrm{Sym}^3(\pi) \otimes \omega_\pi^{-1}) \boxplus \pi$$

from which we deduce that $\mathrm{Sym}^3(\pi)$ is automorphic.

8.3 The Exterior Square Lifting for GL_4

From the six dimensional irreducible representation of $GL_4(\mathbb{C})$ on the space of antisymmetric two tensors $\Lambda^2(\mathbb{C}^4)$ we obtain an L-homomorphism

$$GL_4(\mathbb{C}) \xrightarrow{\Lambda^2} GL_6(\mathbb{C})$$

and from it local and global lifts. If $\pi = \otimes \pi_v$ is a cuspidal automorphic representation of $GL_4(\mathbb{C})$ then $\Lambda^2(\pi_v)$ is an irreducible admissible representation of $GL_6(k_v)$ and $\Lambda^2(\pi) = \otimes' \Lambda^2(\pi_v)$ is an irreducible admissible representation of $GL_6(\mathbb{A})$. Henry Kim established the corresponding functoriality in the following form.

Theorem 8.3 ([10, Theorem 9.3]). *Let π be a cuspidal automorphic representation of $GL_4(\mathbb{A})$. Then there exists an automorphic representation Π' of $GL_6(\mathbb{A})$ such that for almost all places v, $\Pi'_v \simeq \Lambda^2(\pi_v)$.*

Again, we use the Converse Theorem for GL_6 and thereby must control the twisted L-functions

$$L(s, \pi \otimes \tau, \Lambda^2 \otimes St_m)$$

for τ cuspidal automorphic representations of $GL_m(\mathbb{A})$ for $1 \leqslant m \leqslant 4$, where St_m is the standard representation of $GL_m(\mathbb{C})$. These are controlled as in Part II, by the Langlands-Shahidi method. In that notation, we take $G = D^{sc}_{k+4} = Spin_{2k+8}$ for $k = 0, 1, 2, 3$. The Dynkin diagram for G is

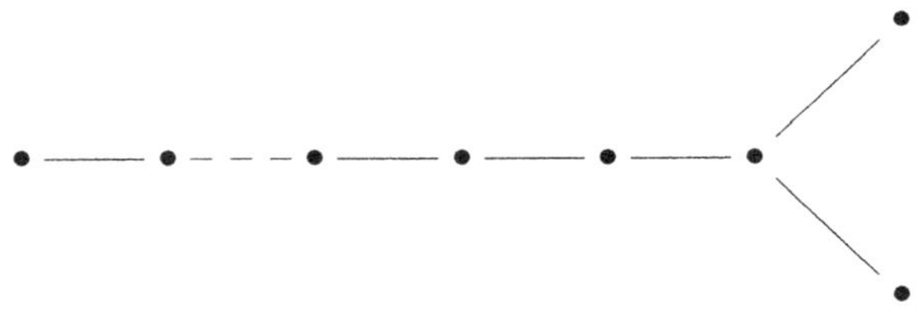

and if we remove the appropriate vertex

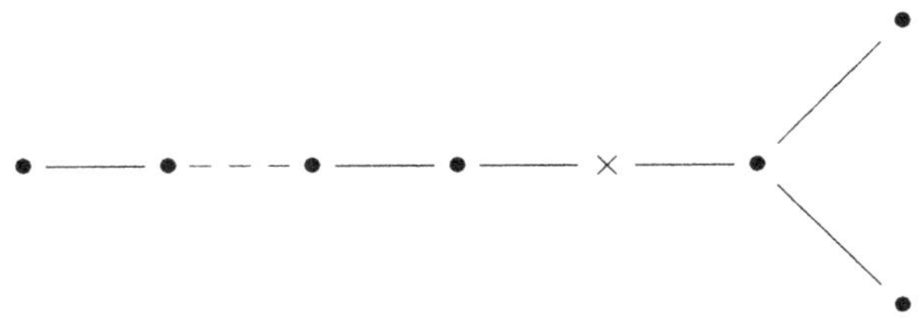

we obtain a maximal parabolic subgroup $P = MN$ whose Levi subgroup has derived group $M_{der} \simeq SL_{k+1} \times SL_4$. If we take the representation of M so that $\sigma \sim \tau \otimes \pi$, then this is an $m = 2$ situation with

$$L(s, \sigma, r_1) = L(s, \pi \otimes \tau, \Lambda^2 \otimes St_{k+1}).$$

This then gives the theorem.

From this result we can deduce another symmetric power functoriality for GL_2.

Corollary 8.1 ([10, Corollary 9.1]). *If π is a cuspidal automorphic representation of $GL_2(\mathbb{A})$ then $\mathrm{Sym}^4(\pi)$ is an automorphic representation of $GL_5(\mathbb{A})$.*

To derive this, we begin with our cuspidal representation π of $GL_2(\mathbb{A})$. We know that $\mathrm{Sym}^3(\pi)$ is an automorphic representation of $GL_4(\mathbb{A})$, so we can now form $\Lambda^2(\mathrm{Sym}^3(\pi))$ which is an automorphic representation of $GL_6(\mathbb{A})$. This is not irreducible, but decomposes as

$$\Lambda^2(\mathrm{Sym}^3(\pi)) = (\mathrm{Sym}^4(\pi) \otimes \omega_\pi) \boxplus \omega_\pi^3$$

from which we deduce the automorphy of $\mathrm{Sym}^4(\pi)$.

8.4 Applications

The new symmetric power functorialities give us improved bounds towards the Ramanujan and Selberg conjectures for GL_2 over an arbitrary number field.

Let $\pi = \otimes' \pi_v$ be a unitary cuspidal representation of $GL_2(\mathbb{A})$. If v is a finite place where π_v is unramified and $\phi_v : W'_{k_v} \to GL_2(\mathbb{C})$ the arithmetic Langlands parameter for π_v, then ϕ_v is also unramified and

$$\phi_v(Frob_v) = \begin{pmatrix} \alpha_v & 0 \\ 0 & \beta_v \end{pmatrix} = t_v \in GL_2(\mathbb{C}).$$

These are the Satake parameters for π_v. The Ramanujan conjecture states that

$$q_v^{-\epsilon} < |\alpha_v|, |\beta_v| < q_v^{\epsilon}$$

for all $\epsilon > 0$.

Theorem 8.4 ([10, Theorem 9.4]). *For π a unitary cuspidal representation of $GL_2(\mathbb{A})$, the Satake parameters for π_v satisfy*

$$q_v^{-1/9} < |\alpha_v|, |\beta_v| < q_v^{1/9}.$$

The first idea is to use the automorphy of the symmetric powers. Suppose that $\mathrm{Sym}^4(\pi)$ is cuspidal. (The non-cuspidal case is in fact easier.) Then the Satake parameters for $\mathrm{Sym}^4(\pi)$ are given by

$$\mathrm{Sym}^4(t_v) = \mathrm{Sym}^4\left(\begin{pmatrix} \alpha_v & 0 \\ 0 & \beta_v \end{pmatrix}\right) = \begin{pmatrix} \alpha_v^4 & & & & \\ & \alpha_v^3\beta_v & & & \\ & & \ddots & & \\ & & & \alpha_v\beta_v^3 & \\ & & & & \beta_v^4 \end{pmatrix}.$$

Now there is a general first nontrivial bound towards Ramanujan valid for any GL_n due to Jacquet and Shalika. It says that the Satake parameters for any unitary cuspidal representation for $GL_n(\mathbb{A})$ satisfy

$$q_v^{-1/2} \leqslant |\alpha_{v,i}| \leqslant q_v^{1/2}.$$

Applying this to the Satake parameters for $\mathrm{Sym}^4(\pi)$ gives

$$q_v^{-1/2} \leqslant |\alpha_v^4|, \beta_v^4| \leqslant q_v^{1/2}$$

or

$$q_v^{-1/8} \leqslant |\alpha_v|, \beta_v| \leqslant q_v^{1/8}.$$

To achieve the "1/9 bound" one has to invoke other techniques. What Kim and Shahidi prove is that $L(s, \pi_v, \mathrm{Sym}^9)$ is holomorphic for $\mathrm{Re}(s) \geqslant 1$ and this gives the stated 1/9 bound.

There is an archimedean analogue giving a 1/9 bound towards the general Selberg conjecture for GL_2.

References

[1] J. Arthur, The Endoscopic Classification of Representations, AMS Colloquium Publications, Vol. 61, AMS, Providence, 2013.

[2] J. W. Cogdell, Analytic theory of L-functions for GL_n; Langlands conjectures for GL_n; Dual groups and Langlands functoriality, in: An Introduction to the Langlands Program, Eds. J. Bernstein and S. Gelbart, Birkhäuser, Boston, 2003, 197–268.

[3] J. W. Cogdell, Lectures on L-functions, converse theorems, and functoriality for $GL(n)$, in: Lectures on Automorphic L-functions, Fields Institute Monographs No. 20, Eds. J. W. Cogdell, H. Kim and R. Murty, AMS, Providence, 2004, 1–96.

[4] J. W. Cogdell, L-functions and converse theorems for GL_n, in: Automorphic Forms and Applications, IAS/Park City Mathematics Series, Vol. 12, AMS, Providence, 2007, 97–177.

[5] J. W. Cogdell, I. I. Piatetski-Shapiro and F. Shahidi, Functoriality for quasisplit classical groups, in: On Certain L-functions, Clay Math. Proc., Vol. 13, AMS, Providence, 2011, 117–140.

[6] J. W. Cogdell, et.al., Ilya Piatetski-Shapiro, In Memoriam, Notices of the AMS **57** (2010), No. 10, 1260–1275.

[7] H. Kim, Automorphic L-functions, in: Lectures on Automorphic L-functions, Fields Institute Monographs No. 20, Eds. J. W. Cogdell, H. Kim and R. Murty, AMS, Providence, 2004, 97–201.

[8] F. Shahidi, Langlands-Shahidi method, in: Automorphic Forms and Applications, IAS/Park City Mathematics Series, Vol. 12, AMS, Providence, 2007, 299–330.

[9] F. Shahidi, Eisenstein Series and Automorphic L-functions, AMS Colloquium Publications, Vol. 58, AMS, Providence, 2010.

[10] F. Shahidi, Eisenstein series, L-functions, and representation theory, in: Automorphic Forms and L-functions, ALM 30, 2014, 149–178.

Automorphic Forms and
L-functions
ALM 30, pp. 49–67

Twisted Hilbert Modular L-functions and Spectral Theory

Gergely Harcos*

Abstract

These are notes for four lectures given at the 2010 CIMPA Research School "Automorphic Forms and L-functions" in Weihai, China. The lectures focused on a Burgess-like subconvex bound for twisted Hilbert modular L-functions published jointly with Valentin Blomer in the same year. They discussed the proof in some detail, especially how spectral theory can be used to estimate the relevant shifted convolution sums efficiently. They also discussed briefly an application for the number of representations by a totally positive ternary quadratic form over a totally real number field. The notes below follow the leisurely style of the lectures, hence they do not constitute a comprehensive survey of the subject.

2010 Mathematics Subject Classification: Primary 11F41; Secondary 11E12, 11E20, 11E45, 11F67, 11F70, 11F72, 11M41.

Keywords and Phrases: Hilbert modular forms, Ternary quadratic forms, Subconvexity, Shifted convolution sums, Spectral decomposition, Kirillov model.

1 Lecture One: Some Quadratic Forms

In the lectures we shall discuss a state-of-the-art bound for twisted Hilbert modular L-functions, namely an analogue of Burgess' bound [7] for Dirichlet L-functions. For motivation and context, we start with an application to quadratic forms.

For a positive integral ternary quadratic form Q let us denote by $r^*(n,Q)$ the number of *primitive* integral representations of n by Q, that is

$$r^*(n,Q) := \#\left\{(x,y,z)\in\mathbb{Z}^3 \;\middle|\; Q(x,y,z)=n \text{ and } \gcd(x,y,z)=1\right\}.$$

*Alfréd Rényi Institute of Mathematics, Hungarian Academy of Sciences, POB 127, Budapest H-1364, Hungary, gharcos@renyi.hu; Central European University, Nador u. 9, Budapest H-1051, Hungary, harcosg@ceu.hu. The author was supported by OTKA grants K101855 and K104183 and ERC Advanced Grant 228005.

Geometrically, this is the number of *visible* lattice points on an ellipsoid of size $\sqrt{n}$. We would like to understand this quantity, in particular determine which positive integers n are primitively represented by Q over $\mathbb{Z}$. Note that when n is square-free, all integral representations are primitive.

1.1 Sums of Three Squares

Let us look at the simplest example

$$Q(x, y, z) := x^2 + y^2 + z^2.$$

Clearly, for $n \equiv 0, 4, 7 \pmod 8$ there are no primitive representations. However, for $n \equiv 1, 2, 3, 5, 6 \pmod 8$ there are, and we have the following elegant formula for their number:

$$r^*(n, Q) = \frac{24}{\pi} \sqrt{n}\, L\left(1, \left(\frac{D}{\cdot}\right)\right), \tag{1.1}$$

where

$$D = \begin{cases} -4n, & n \equiv 1, 2, 5, 6 \pmod 8, \\ -n, & n \equiv 3 \pmod 8). \end{cases}$$

This formula arises naturally from the work of Gauss (1801) and Dirichlet (1839) on the class number. For a negative discriminant D let $h(D)$ denote the number of equivalence classes of positive *primitive* binary quadratic forms of discriminant D, and let w denote the number of automorphs[1] of such a form:

$$w = \begin{cases} 6, & D = -3, \\ 4, & D = -4, \\ 2, & D < -4. \end{cases}$$

In this notation a fundamental result of Gauss [17] states that

$$r^*(n, Q) = \begin{cases} \dfrac{24}{w} h(-4n), & n \equiv 1, 2, 5, 6 \pmod 8, \\ \dfrac{48}{w} h(-n), & n \equiv 3 \pmod 8, \end{cases} \tag{1.2}$$

while the class number formula of Dirichlet [12] reads

$$h(D) = \frac{w}{2\pi} |D|^{1/2} L\left(1, \left(\frac{D}{\cdot}\right)\right). \tag{1.3}$$

For a modern account of these results we refer the reader to [19, §4] and [47, §8], including the exercises. Combining (1.2) and (1.3), we obtain (1.1).

It is not hard to see that the special L-value involved in (1.1) is not too large, while a deeper result of Siegel [39] shows that it is not too small either:

$$L\left(1, \left(\frac{D}{\cdot}\right)\right) = |D|^{o(1)}. \tag{1.4}$$

[1] In these notes, equivalence classes and automorphs are meant in the narrow/strict/proper sense, i.e. they are defined in terms of the action of SL_n on n-ary quadratic forms.

We infer that $r^*(n,Q)$ is either zero or about $\sqrt{n}$:

$$r^*(n,Q) = n^{\frac{1}{2}+o(1)}, \qquad n \equiv 1,2,3,5,6 \pmod 8. \tag{1.5}$$

That is, for the case of $Q(x,y,z) = x^2+y^2+z^2$, we know precisely when there is at least one visible lattice point on our ellipsoid (the sphere of radius $\sqrt{n}$ centered at the origin), and we also know that one visible lattice point implies many others. We can look at (1.5) as a quantitative local-to-global principle:

$$\boxed{n \text{ is primitively represented by } Q \text{ over any } \mathbb{Z}_p \quad \Longrightarrow \quad r^*(n,Q) \text{ is large.}}$$

Here we could restrict the left hand side to $p=2$, because for $p>2$ the condition is automatically met. This formulation, of course, does not reveal that $r^*(n,Q)$ has a nice analytic description as in (1.1). The two viewpoints are unified by the celebrated mass formula[2] of Siegel [40]. Namely, let us assume that $n \equiv 1,2,3,5,6 \pmod 8$ so that $x^2+y^2+z^2=n$ has a solution in $\mathbb{R}$ and a primitive solution in each $\mathbb{Z}_p$. Then we can calculate the right hand side of (1.1) as a product of densities of the primitive local solutions over $\mathbb{R}$ and the various completions $\mathbb{Z}_p$:

$$\beta_\infty = 2\pi\sqrt{n}, \qquad \beta_2 = \frac{\frac{3}{2}}{1-(\frac{D}{2})\frac{1}{2}}, \qquad \beta_p = \frac{1-\frac{1}{p^2}}{1-(\frac{D}{p})\frac{1}{p}} \quad (p \neq 2),$$

whose product is indeed

$$\beta_\infty \prod_p \beta_p = 2\pi\sqrt{n}\,\frac{\frac{3}{2}}{1-\frac{1}{2^2}}\,\frac{1}{\zeta(2)}\,L\left(1,\left(\frac{D}{\cdot}\right)\right) = \frac{24}{\pi}\sqrt{n}\,L\left(1,\left(\frac{D}{\cdot}\right)\right).$$

1.2 Ramanujan's Ternary Quadratic Form

What about more general ternary quadratic forms? As a second example let us consider[3]

$$Q(x,y,z) := x^2+y^2+10z^2.$$

Is there a similar elegant formula as (1.1) for the number of primitive representations? Well, almost. What happens is that there is another ternary quadratic form which is equivalent to Q over $\mathbb{R}$ and all the completions $\mathbb{Z}_p$, but not over $\mathbb{Z}$:

$$Q'(x,y,z) := 2x^2+2y^2+3z^2-2xz.$$

So Q and Q' produce the same primitive local densities of representations over the reals and the p-adic integers, yet they are really different over the integers. A quadratic form that is locally equivalent to Q is globally equivalent to Q or Q', hence perhaps it is not surprising that the product of primitive local densities

[2] The original formula concerns all integral representations, but it is easy to reformulate it in terms of primitive representations for the cases considered here. For the general case see [25, §6.8].

[3] This section was inspired by a nice paper of Ono and Soundararajan [33].

is related to a combination of $r^*(n,Q)$ and $r^*(n,Q')$. For this case Siegel's mass formula [40] precisely tells us that

$$\frac{1}{3}r^*(n,Q)+\frac{2}{3}r^*(n,Q')=\frac{4\sqrt{10}}{3\pi}\sqrt{n}\,L\left(1,\left(\frac{-10n}{\cdot}\right)\right), \tag{1.6}$$

at least when $\gcd(n,10)=1$ as we assume for simplicity. Here the weight of $r^*(n,Q)$ is one-half of the weight of $r^*(n,Q')$, because Q has twice as many automorphs as Q' (eight vs. four). The right hand side manifests as the product of the common primitive local densities of Q and Q',

$$\beta_\infty=\frac{2\pi\sqrt{n}}{\sqrt{10}},\qquad \beta_2=1,\qquad \beta_5=\frac{4}{5},\qquad \beta_p=\frac{1-\frac{1}{p^2}}{1-(\frac{-10n}{p})\frac{1}{p}}\quad (p\neq 2,5).$$

Indeed, the product of these quantities equals

$$\frac{2\pi\sqrt{n}}{\sqrt{10}}\,\frac{1}{1-\frac{1}{2^2}}\,\frac{\frac{4}{5}}{1-\frac{1}{5^2}}\,\frac{1}{\zeta(2)}\,L\left(1,\left(\frac{-10n}{\cdot}\right)\right)=\frac{4\sqrt{10}}{3\pi}\sqrt{n}\,L\left(1,\left(\frac{-10n}{\cdot}\right)\right).$$

We are still assuming that $\gcd(n,10)=1$. Comparing (1.6) with (1.4), we see that n has many primitive representations by Q or Q'. However, we would like to know if n has any or many primitive representations by Q alone! Using automorphic forms and L-functions one can show that $r^*(n,Q)\approx r^*(n,Q')$ with great precision, so that both $r^*(n,Q)$ and $r^*(n,Q')$ are close to the right hand side of (1.6). Specifically, by the work of Schulze-Pillot [34] and Waldspurger [43] based on the Shimura [38] correspondence, there exists a primitive holomorphic cusp form f of weight 2, level 20, and trivial nebentypus such that

$$\left(r^*(n,Q)-r^*(n,Q')\right)^2\ll_\varepsilon n^{\frac{1}{2}+\varepsilon}\,L\left(\frac{1}{2},f\otimes\left(\frac{-10n}{\cdot}\right)\right). \tag{1.7}$$

If we assume the Grand Riemann Hypothesis (GRH) for automorphic L-functions, then the central L-value on the right hand side of (1.7) is $\ll_\varepsilon n^\varepsilon$, whence from (1.6) we infer that

$$r^*(n,Q)=\frac{4\sqrt{10}}{3\pi}\sqrt{n}\,L\left(1,\left(\frac{-10n}{\cdot}\right)\right)+O_\varepsilon\left(n^{\frac{1}{4}+\varepsilon}\right). \tag{1.8}$$

The main term here is $\gg_\varepsilon n^{\frac{1}{2}-\varepsilon}$ by (1.4), and all the implied constants are effective under GRH. Roughly, this means that

GRH and $\gcd(n,10)=1$ $\Longrightarrow$ $r^*(n,Q)$ is large.

This is another instance (albeit conditionally) of a quantitative local-to-global principle, because the integers coprime with 10 are primitively represented by Q over any $\mathbb{Z}_p$. In particular, under GRH, there are only finitely many positive integers such that $\gcd(n,10)=1$ and $r^*(n,Q)=0$, and GRH even provides a *theoretical* algorithm to determine them. It is a challenging task to turn GRH into a *practical* algorithm for this problem, solved for n square-free by Ono and Soundararajan [33].

We can also estimate the right hand side of (1.7) without GRH. The functional equation for the twisted L-functions coupled with the Phragmén-Lindelöf convexity principle shows that the L-value in (1.7) is $\ll_\varepsilon n^{\frac{1}{2}+\varepsilon}$. This *convexity bound*, however, yields the error term $O_\varepsilon(n^{\frac{1}{2}+\varepsilon})$ in (1.8), which is too weak to imply $r^*(n,Q)>0$. If, instead, we employ the *subconvexity bound*

$$L\left(\frac{1}{2}, f\otimes\left(\frac{-10n}{\cdot}\right)\right) \ll_\varepsilon n^{\frac{1}{2}-\delta+\varepsilon}$$

for some $\delta>0$, then we arrive at a useful variant of (1.8) for $\gcd(n,10)=1$:

$$r^*(n,Q)=\frac{4\sqrt{10}}{3\pi}\sqrt{n}\,L\left(1,\left(\frac{-10n}{\cdot}\right)\right)+O_\varepsilon\left(n^{\frac{1-\delta}{2}+\varepsilon}\right).$$

Such a conclusion was first established unconditionally with $\delta=\dfrac{1}{14}$ by Duke [13] using a method of Iwaniec [24] (see also [15]), and then with $\delta=\dfrac{1}{8}$ by Bykovskiĭ [8] using a method of Duke, Friedlander and Iwaniec [14] (see also [4] and [22]). Furthermore, it seems likely that the technical assumptions in a deep result of Conrey and Iwaniec [11] can be relaxed so as to yield the conclusion with $\delta=\dfrac{1}{6}$. To summarize,

Subconvexity and $\gcd(n,10)=1$ $\implies$ $r^*(n,Q)$ is large.

2 Lecture Two: More Quadratic Forms

The examples presented in the first lecture can be generalized to a large extent. Let K be a totally real number field with discriminant D and ring of integers $\mathfrak{o}$. Let (a_{ij}) be a 3×3 matrix with $a_{ij}=a_{ji}\in\mathfrak{o}$ and $a_{ii}\in 2\mathfrak{o}$ such that the corresponding integral quadratic form

$$Q(x_1,x_2,x_3):=\frac{1}{2}\sum_{i,j}a_{ij}x_ix_j$$

is totally positive. Then the determinant $d:=\det(a_{ij})$ is totally positive and lies in $2\mathfrak{o}$ by [20, Lemma 2.1]. We are interested in the number of *primitive* integral representations of a totally positive integer $n\in\mathfrak{o}$ by Q, that is

$$r^*(n,Q):=\#\left\{(x,y,z)\in\mathfrak{o}^3\mid Q(x,y,z)=n \text{ and } \gcd(x,y,z)=\mathfrak{o}\right\}.$$

An obvious necessary condition for $r^*(n,Q)>0$ is that n is primitively represented by Q over any non-archimedean completion $\mathfrak{o}_\mathfrak{p}$ of $\mathfrak{o}$. This condition is invariant under replacing Q by any form in its genus, that is, by any totally positive ternary quadratic form over $\mathfrak{o}$ which is locally equivalent to Q over any $\mathfrak{o}_\mathfrak{p}$. In fact the primitive local densities of the representations only depend on the genus, and by Siegel's mass formula [40] their product equals a weighted average over the finitely many equivalence classes contained in the genus:

$$r^*(n, \operatorname{gen} Q) := \left(\sum_{[Q'] \in \operatorname{gen} Q} \frac{r^*(n, Q')}{\operatorname{aut}(Q')} \right) \left(\sum_{[Q'] \in \operatorname{gen} Q} \frac{1}{\operatorname{aut}(Q')} \right)^{-1} = \beta_\infty \prod_{\mathfrak{p}} \beta_{\mathfrak{p}}.$$

Here $r^*(n, Q')$ and $\operatorname{aut}(Q')$, the number of automorphs of Q', only depend on the class $[Q']$ of Q', hence the sums are well-defined. For $\gcd(n, d) = \mathfrak{o}$ we can simplify the right hand side using [40, §7 of Part III] and [20, Lemma 3.2] to obtain

$$r^*(n, \operatorname{gen} Q) = c(n, Q)\, (\mathcal{N} n)^{\frac{1}{2}}\, L\left(1, \left(\frac{-2dn}{\cdot}\right)\right), \tag{2.1}$$

where $\mathcal{N}$ stands for the norm and $c(n, Q)$ equals, up to a positive constant depending on K, the density of primitive solutions of the congruence $Q(x, y, z) \equiv n \pmod{4d}$. In particular, $c(n, Q) > 0$ is equivalent to primitive local representability modulo $4d$, in which case $r^*(n, \operatorname{gen} Q) = (\mathcal{N} n)^{\frac{1}{2} + o(1)}$ by a straightforward extension of (1.4).

Surprisingly, $r^*(n, \operatorname{gen} Q)$ being large does not ensure that n is primitively represented by Q, even when $\mathcal{N} n$ is large and coprime with d. Here are two examples, borrowed from [16] and [37]. When $K = \mathbb{Q}$, the forms $x^2 + 3y^2 + 36z^2$ and $3x^2 + 4y^2 + 9z^2$ are in the same genus, yet any square number coprime with 6 is primitively represented by exactly one of them. When $K = \mathbb{Q}(\sqrt{35})$, an integer of the form $7p^2$ with a rational prime $p \equiv 1 \pmod 7$ is not a sum of three squares in $\mathfrak{o}$, although a sum of three coprime squares in $\mathfrak{o}_{\mathfrak{p}}$ for any prime ideal $\mathfrak{p}$. The proper discussion of this phenomenon would lead us too far as it relies on the theory of spinor genera and theta series, see the recent surveys [21, 37] and the references therein. Let us just say that in the modern theory one considers lattices and their representations in a quadratic space over K, and the restriction to free lattices yields the classical theory of integral quadratic forms and their representations [29, §82]. A class is an orbit of lattices under the group of rotations of the space. A genus is an orbit under the group of adelic rotations, while a spinor genus is an orbit under a certain normal subgroup of adelic rotations [29, §102]. Lattices in the same genus are isomorphic as $\mathfrak{o}$-modules[4], hence the set of free lattices is a disjoint union of genera (resp. spinor genera) in the modern sense.

We shall conveniently avoid the above type of exceptions by restricting to totally positive *non-square* integers $n \in \mathfrak{o}$ coprime with dD. Let $r^*(n, \operatorname{spn} Q)$ be defined similarly as $r^*(n, \operatorname{gen} Q)$, but over the spinor genus of Q. By the work of

[4]This was kindly explained to me by Rainer Schulze-Pillot, here is a variant of his argument. Let L be a lattice in a fixed quadratic space V over K. By [29, §81:3], there is a basis (x_i) of V and fractional ideals $\mathfrak{a}_i$ in K such that $L = \mathfrak{a}_1 x_1 + \cdots + \mathfrak{a}_n x_n$. Writing $\mathfrak{a} = \mathfrak{a}_1 \cdots \mathfrak{a}_n$, the volume of L equals $\mathfrak{a}^2 \operatorname{disc}(x_1, \ldots, x_r)$, which is $\mathfrak{a}^2 \operatorname{disc}(V)$ modulo $(F^\times)^2$. Therefore the volume of L (hence also the genus of L) determines $\mathfrak{a}$ modulo $F^\times$, which is the Steinitz class of L as an $\mathfrak{o}$-module.

Kneser [26] and Hsia [23], the two averages agree now[5], so that the bounds usually proved for $r^*(n,Q)-r^*(n,\operatorname{spn}Q)$ apply for $r^*(n,Q)-r^*(n,\operatorname{gen}Q)$. By the work of Baruch and Mao [1], Blasius [2], Schulze-Pillot [36], Waldspurger [44], there exist finitely many primitive holomorphic Hilbert cusp forms $f_1,\dots,f_r$ over K depending on Q, each of weight $(2,\dots,2)$ and trivial nebentypus, such that

$$\left(r^*(n,Q)-r^*(n,\operatorname{gen}Q)\right)^2 \ll_{Q,K,\varepsilon} (\mathcal{N}n)^{\frac{1}{2}+\varepsilon} \max_{1\leqslant i\leqslant r} L\left(\frac{1}{2}, f_i\otimes\left(\frac{-2dn}{\cdot}\right)\right). \tag{2.2}$$

As before, any subconvex bound for the central L-values on the right hand side yields an asymptotic formula for $r^*(n,Q)$ with a power saving error term. More specifically and generally, let us assume the bound

$$L\left(\frac{1}{2},\pi\otimes\chi\right) \ll_{\pi,\chi_\infty,K,\varepsilon} (\mathcal{N}\mathfrak{q})^{\frac{1}{2}-\delta+\varepsilon}, \tag{2.3}$$

where δ is a positive constant, π is any irreducible unitary cuspidal representation of GL_2 over K, and χ is any Hecke character of K of conductor $\mathfrak{q}$. Then from (2.1) and (2.2) we infer, for totally positive non-square $n\in\mathfrak{o}$ coprime with dD,

$$r^*(n,Q)=c(n,Q)\,(\mathcal{N}n)^{\frac{1}{2}}\,L\left(1,\left(\frac{-2dn}{\cdot}\right)\right)+O_{Q,K,\varepsilon}\left((\mathcal{N}n)^{\frac{1-\delta}{2}+\varepsilon}\right). \tag{2.4}$$

Recall that $c(n,Q)>0$ is equivalent to primitive local representability of n by Q modulo $4d$, in which case the main term dominates the error term for $\mathcal{N}n$ sufficiently large:

$$\boxed{\gcd(n,dD)=\mathfrak{o} \text{ and } n\neq\square \text{ and } c(n,Q)>0 \quad\Longrightarrow\quad r^*(n,Q)\text{ is large.}}$$

The conclusion (2.3) was first established by Cogdell, Piatetski-Shapiro, Sarnak [10] with the value $\delta=\dfrac{1-2\theta}{14+4\theta}$, at least for π induced by a holomorphic Hilbert cusp form, which suffices for the application (2.4). For general π (and arbitrary K), the breakthrough is due to Venkatesh [42] who achieved $\delta=\dfrac{(1-2\theta)^2}{14-2\theta}$. In these results, $0\leqslant\theta\leqslant\dfrac{1}{2}$ is an approximation towards the Ramanujan-Petersson conjecture, the current record being $\theta=\dfrac{7}{64}$ due to Blomer and Brumley [3]. Interestingly, under the Ramanujan-Petersson conjecture $\theta=0$ both expressions become $\delta=\dfrac{1}{14}$, matching the already mentioned results of Duke [13] and Iwaniec [24]. For totally real K and general π, Blomer and Harcos [6] established the Burgess-like subconvexity saving $\delta=\dfrac{1-2\theta}{8}$, and the proof of this result is outlined in the

[5] Let L be a free ternary lattice corresponding to the class of Q. It suffices to show that $r(n,\operatorname{spn}\mathfrak{a}L)=r(n,\operatorname{gen}\mathfrak{a}L)$ for any ideal $\mathfrak{a}$ in $\mathfrak{o}$. Take a prime ideal $\mathfrak{p}\nmid dD$ such that $\operatorname{ord}_{\mathfrak{p}}(n)$ is odd. The quadratic extension $K_{\mathfrak{p}}(\sqrt{-2dn})/K_{\mathfrak{p}}$ is ramified, while $L_{\mathfrak{p}}$ is unimodular, hence local class field theory combined with [29, §92:5] shows that [35, (4)] fails at $\mathfrak{p}$ in the present setting.

rest of these notes. Recently Maga [27, 28] extended the method of [6] to arbitrary number fields. Wu [46] obtained the same $\delta = \frac{1-2\theta}{8}$ in full generality, even uniformly in χ_∞, by a different method based on the deep work of Michel and Venkatesh [30].

Theorem 2.1 ([6]). *Let K be a totally real number field. Let π be an irreducible unitary cuspidal representation of GL_2 over K, and χ a Hecke character of K of conductor $\mathfrak{q}$. Then for any $\varepsilon > 0$ one has*

$$L\left(\frac{1}{2}, \pi \otimes \chi\right) \ll_{\pi,\chi_\infty,K,\varepsilon} (\mathcal{N}\mathfrak{q})^{\frac{3+2\theta}{8}+\varepsilon}.$$

Theorem 2.2 ([1, 2, 6, 36, 40, 44]). *Let K be a totally real number field with discriminant D. Let Q be a totally positive integral ternary quadratic form over K with determinant d. Let n be a totally positive non-square integer in K coprime to dD. Then the number of primitive representations $Q(x, y, z) = n$ equals*

$$r^*(n, Q) = c(n, Q)\,(\mathcal{N}n)^{\frac{1}{2}}\, L\left(1, \left(\frac{-2dn}{\cdot}\right)\right) + O_{Q,K,\varepsilon}\left((\mathcal{N}n)^{\frac{7+2\theta}{16}+\varepsilon}\right),$$

where $c(n, Q)$ is a constant times the density of primitive solutions of the congruence $Q(x, y, z) \equiv n \pmod{4d}$.

3 Lecture Three: Preliminaries from Number Theory

We collect some preliminaries for the proof of Theorem 2.1, to be outlined in the fourth lecture. As explained in the first two lectures, Theorem 2.2 follows from Theorem 2.1. For more detail concerning the preliminaries the reader should consult [6, 18, 32, 45].

3.1 Adeles and Ideles

The adele ring of K is a restricted direct product of the completions K_v at the various places of K:

$$\mathbb{A} := K_\infty \times \mathbb{A}_{\text{fin}} = \prod_{v|\infty} K_v \times \prod_{\mathfrak{p}}{}' K_{\mathfrak{p}},$$

where $\mathfrak{p}$ runs through the prime ideals of $\mathfrak{o}$. The topology of the additive group $(\mathbb{A}, +)$ is determined by the fundamental neighborhoods of zero in $\mathbb{A}_{\text{fin}}$:

$$U^{\mathfrak{m}} := \prod_{\mathfrak{p}} U_{\mathfrak{p}}^{(\text{ord}_{\mathfrak{p}}\, \mathfrak{m})},$$

where $\mathfrak{m} \subseteq \mathfrak{o}$ is any nonzero ideal, $\text{ord}_{\mathfrak{p}}\, \mathfrak{m}$ denotes the exponent of $\mathfrak{p}$ in $\mathfrak{m}$, and

$$U_{\mathfrak{p}}^{(n)} := \mathfrak{p}^n \mathfrak{o}_{\mathfrak{p}}, \qquad n \in \mathbb{N}.$$

Then $(\mathbb{A}, +)$ is a locally compact Hausdorff topological group such that

$$K \overset{\text{diag}}{\hookrightarrow} \mathbb{A} \text{ is discrete}, \qquad K\backslash\mathbb{A} \text{ is compact}.$$

Similarly, the group of ideles of K is a restricted direct product

$$\mathbb{A}^\times := K_\infty^\times \times \mathbb{A}_{\text{fin}}^\times = \prod_{v|\infty} K_v^\times \times \prod_{\mathfrak{p}}{}' K_{\mathfrak{p}}^\times,$$

whose topology is determined by the fundamental neighborhoods of one in $\mathbb{A}_{\text{fin}}^\times$:

$$V^{\mathfrak{m}} := \prod_{\mathfrak{p}} V_{\mathfrak{p}}^{(\text{ord}_{\mathfrak{p}} \mathfrak{m})},$$

where $\mathfrak{m} \subseteq \mathfrak{o}$ is any nonzero ideal, $\text{ord}_{\mathfrak{p}} \mathfrak{m}$ denotes the exponent of $\mathfrak{p}$ in $\mathfrak{m}$, and

$$V_{\mathfrak{p}}^{(n)} := \begin{cases} \mathfrak{o}_{\mathfrak{p}}^\times, & n = 0, \\ 1 + \mathfrak{p}^n \mathfrak{o}_{\mathfrak{p}}, & n > 0. \end{cases}$$

Then $(\mathbb{A}^\times, \cdot)$ is a locally compact Hausdorff topological group such that

$$\mathbb{A}^\times = K_{\infty,+}^{\text{diag}} \mathbb{A}^1 \cong \mathbb{R}_{>0} \times \mathbb{A}^1, \qquad K^\times \overset{\text{diag}}{\hookrightarrow} \mathbb{A}^1 \text{ is discrete}, \qquad K^\times\backslash\mathbb{A}^1 \text{ is compact}.$$

Here $\mathbb{A}^1$ is the subgroup of ideles of module 1, and $K_{\infty,+}^{\text{diag}}$ denotes $\mathbb{R}_{>0}$ diagonally embedded into $K_\infty^\times$.

3.2 Hecke Characters and Grössencharacters

A *Hecke character* is a continuous homomorphism $\chi : \mathbb{A}^\times \to S^1$ which is trivial on $K^\times$. The kernel of such a character χ always contains a subgroup of the form $\{(1, \ldots, 1)\} \times V^{\mathfrak{m}}$. Let $I^{\mathfrak{m}}$ denote the group of fractional ideals of K coprime with $\mathfrak{m}$, this is a free abelian group generated by the prime ideals $\mathfrak{p} \nmid \mathfrak{m}$. Choose a prime element $p_{\mathfrak{p}} \in \mathfrak{p} - \mathfrak{p}^2$ for each $\mathfrak{p} \nmid \mathfrak{m}$, and define the character $\tilde{\chi} : I^{\mathfrak{m}} \to S^1$ via

$$\tilde{\chi}(\mathfrak{p}) := \chi(\ldots, 1, 1, p_{\mathfrak{p}}, 1, 1, \ldots), \qquad \mathfrak{p} \nmid \mathfrak{m}.$$

Observe that $\tilde{\chi}$ is independent of the choice of the $p_{\mathfrak{p}}$'s. Moreover, for any *principal* ideal $(a) \in I^{\mathfrak{m}}$ have

$$\begin{aligned} \tilde{\chi}((a)) &= \chi\big(\underbrace{1, \ldots, 1}_{\text{arch}}, \underbrace{\ldots, p_{\mathfrak{p}}^{\text{ord}_{\mathfrak{p}}(a)}, \ldots}_{\text{non-arch}}\big) \\ &= \chi\big(\underbrace{1, \ldots, 1}_{v|\infty}, \underbrace{1, \ldots, 1}_{\mathfrak{p}|\mathfrak{m}}, \underbrace{a, a, \ldots}_{\mathfrak{p}\nmid\mathfrak{m}}\big) \\ &= \chi\big(\underbrace{a^{-1}, \ldots, a^{-1}}_{v|\infty}, \underbrace{a^{-1}, \ldots, a^{-1}}_{\mathfrak{p}|\mathfrak{m}}, \underbrace{1, 1, \ldots}_{\mathfrak{p}\nmid\mathfrak{m}}\big). \end{aligned}$$

If we interpret $a \in K^\times$ as an element of $K_\infty^\times$ by embedding a diagonally, and as an element of $(\mathfrak{o}/\mathfrak{m})^\times$ by reducing a modulo $\mathfrak{m}$, then we infer that

$$\tilde{\chi}((a)) = \tilde{\chi}_\infty(a)\tilde{\chi}_{\text{fin}}(a), \qquad (a) \in I^{\mathfrak{m}},$$

where $\tilde{\chi}_\infty : K_\infty^\times \to S^1$ and $\tilde{\chi}_{\text{fin}} : (\mathfrak{o}/\mathfrak{m})^\times \to S^1$ are uniquely determined characters[6]. A character $\tilde{\chi} : I^{\mathfrak{m}} \to S^1$ with this property is called a *Grössencharacter* modulo $\mathfrak{m}$. So any Hecke character can be regarded as a Grössencharacter.

Conversely, any Grössencharacter $\tilde{\chi} : I^{\mathfrak{m}} \to S^1$ arises in this way from a Hecke character $\chi : \mathbb{A}^\times \to S^1$ trivial on $\{(1, \ldots, 1)\} \times V^{\mathfrak{m}}$. To see this, define

$$V := K_\infty^\times \times \prod_{\mathfrak{p}\mid\mathfrak{m}} V_{\mathfrak{p}}^{(\operatorname{ord}_{\mathfrak{p}} \mathfrak{m})} \prod_{\mathfrak{p}\nmid\mathfrak{m}} K_{\mathfrak{p}}^\times;$$

$$\chi(a) := \tilde{\chi}_\infty^{-1}(a_\infty)\,\tilde{\chi}\Big(\prod_{\mathfrak{p}\nmid\mathfrak{m}} \mathfrak{p}^{\operatorname{ord}_{\mathfrak{p}} a_{\mathfrak{p}}}\Big), \qquad a \in V.$$

Observe that V is a subgroup of $\mathbb{A}^\times$, and $\chi : V \to S^1$ is a character trivial on $\{(1, \ldots, 1)\} \times V^{\mathfrak{m}}$. This character extends uniquely to a character $\chi : K^\times V \to S^1$ trivial on $K^\times$, because for $a \in K^\times \cap V$ we have $a \equiv 1 \pmod{\mathfrak{m}}$, whence

$$\chi(a) = \tilde{\chi}_\infty^{-1}(a)\tilde{\chi}((a)) = \tilde{\chi}_{\text{fin}}(a) = 1, \qquad a \in K^\times \cap V.$$

On the other hand, $K^\times V$ equals $\mathbb{A}^\times$ by weak approximation in K, so we are done.

If, for given χ, we choose the largest $\mathfrak{m}$ (called the *conductor*), then $\tilde{\chi}$ will be primitive, and vice versa. To summarize, we have a natural bijection

$$\boxed{\chi \text{ a Hecke character} \quad \longleftrightarrow \quad \tilde{\chi} \text{ a primitive Grössencharacter.}}$$

In the rest of these notes, χ will always stand for a Hecke character, and $\tilde{\chi}$ for the corresponding primitive Grössencharacter.

3.3 The Automorphic Spectrum of GL_2

Let us restrict to trivial central character for simplicity. Then the corresponding L^2 space of automorphic functions on GL_2 over K decomposes as a direct integral of irreducible automorphic representations,

$$L^2(GL_2(K)\backslash GL_2(\mathbb{A})) = \Big(\bigoplus_\pi V_\pi\Big) \oplus \Big(\bigoplus_{\chi^2=1} V_\chi\Big) \oplus \Big(\int_{\{\chi,\chi^{-1}\}} V_{\chi,\chi^{-1}}\, d\{\chi,\chi^{-1}\}\Big),$$

in the sense that each function in the L^2 space is a convergent integral of functions from each subspace, and a Plancherel formula holds. In this decomposition, which is compatible with the right action of $GL_2(\mathbb{A})$,

[6] In these notes, all characters are continuous.

- V_π is an irreducible subspace of the cuspidal space defined by

$$\int_{K\backslash\mathbb{A}} \phi\left(\begin{pmatrix}1 & x\\ 0 & 1\end{pmatrix} g\right) dx = 0 \quad \text{for almost all } g \in GL_2(\mathbb{A}).$$

- V_χ is the one-dimensional subspace spanned by the function $g \mapsto \chi(\det g)$.

- $V_{\chi,\chi^{-1}}$ consists of the Eisenstein series

$$E(\varphi, g) \doteq \sum_{\gamma\in P(K)\backslash GL_2(K)} \varphi(\gamma g),$$

where P stands for upper triangular matrices, and $\varphi : GL_2(\mathbb{A}) \to \mathbb{C}$ satisfies

$$\int_{\mathcal{K}} |\varphi(k)|^2 dk < \infty, \qquad \mathcal{K} := SO_2(K_\infty) \times \mathcal{K}(\mathfrak{o});$$

$$\varphi\left(\begin{pmatrix}a & x\\ 0 & b\end{pmatrix} g\right) = \chi\left(\frac{a}{b}\right)\left|\frac{a}{b}\right|^{1/2} \varphi(g), \qquad \begin{pmatrix}a & x\\ 0 & b\end{pmatrix} \in P(\mathbb{A}).$$

Here $\mathcal{K}(\mathfrak{o})$ is a maximal compact subgroup of $GL_2(\mathbb{A}_{\text{fin}})$ to be defined in the next subsection. More precisely, the sum converges only when the exponent $1/2$ is replaced by any $s \in \mathbb{C}$ with $\Re s > 1$. The symbol "$\doteq$" stands for evaluation at $s = 1/2$ of the function obtained by meromorphic continuation to $s \in \mathbb{C}$, keeping the restriction of φ to $\mathcal{K}$ fixed all the way.

3.4 Conductor, L-function, Kirillov Model

Let $\mathfrak{d}$ denote the different ideal of K. For each cuspidal space V_π (and also for each Eisenstein space $V_{\chi,\chi^{-1}}$) there is a largest congruence subgroup

$$\mathcal{K}(\mathfrak{m}) := \prod_{\mathfrak{p}} \left\{\begin{pmatrix}a & b\\ c & d\end{pmatrix} \in GL_2(K_\mathfrak{p}) \;\middle|\; a, d \in \mathfrak{o}_\mathfrak{p},\ b \in \mathfrak{d}_\mathfrak{p}^{-1},\ c \in \mathfrak{m}\mathfrak{d}_\mathfrak{p},\ ad - bc \in \mathfrak{o}_\mathfrak{p}^\times\right\},$$

regarded as a subgroup of $GL_2(\mathbb{A})$ in the obvious way, such that V_π contains a nonzero vector fixed by the right action of $\mathcal{K}(\mathfrak{m})$. The corresponding nonzero ideal $\mathfrak{m}$ is called the *conductor* of π, denoted $\mathfrak{c}_\pi$. It is known that $\mathfrak{c}_{\chi,\chi^{-1}} = \mathfrak{c}_\chi^2$. If $V_\pi(\mathfrak{m})$ denotes the $\mathcal{K}(\mathfrak{m})$-fixed subspace of V_π, then $V_\pi(\mathfrak{c}_\pi)$ (the space of *newforms*) is particularly nice. Namely, for each nonzero ideal $\mathfrak{m}$, the Hecke operator $T(\mathfrak{m})$ acts by some scalar $\lambda_\pi(\mathfrak{m}) \in \mathbb{C}$ on this space. These Hecke eigenvalues satisfy $\lambda_\pi(\mathfrak{m}) \ll_\varepsilon (\mathcal{N}\mathfrak{m})^{\theta+\varepsilon}$ with $\theta = \dfrac{7}{64}$ by [3], and they determine the L-function of π:

$$\begin{aligned} L(s,\pi) &= \sum_{\{0\}\neq\mathfrak{m}\subseteq\mathfrak{o}} \frac{\lambda_\pi(\mathfrak{m})}{(\mathcal{N}\mathfrak{m})^s} \\ &= \prod_{\mathfrak{p}\mid\mathfrak{c}_\pi} \frac{1}{1-\lambda_\pi(\mathfrak{p})(\mathcal{N}\mathfrak{p})^{-s}} \prod_{\mathfrak{p}\nmid\mathfrak{c}_\pi} \frac{1}{1-\lambda_\pi(\mathfrak{p})(\mathcal{N}\mathfrak{p})^{-s}+(\mathcal{N}\mathfrak{p})^{-2s}}. \end{aligned}$$

Fixing a nontrivial character $\psi : K\backslash\mathbb{A} \to S^1$, any newform $\phi \in V_\pi(\mathfrak{c}_\pi)$ has a Fourier expansion

$$\phi\left(\begin{pmatrix} y & x \\ 0 & 1 \end{pmatrix}\right) = \sum_{r\in K^\times} \frac{\lambda_\pi(ry_{\text{fin}})}{\sqrt{\mathcal{N}(ry_{\text{fin}})}} W_\phi(ry_\infty)\psi(rx), \qquad y \in \mathbb{A}^\times, \quad x \in \mathbb{A},$$

where $W_\phi \in L^2(K_\infty^\times, d^\times y)$ is given by

$$W_\phi(y) := \int_{K\backslash\mathbb{A}} \phi\left(\begin{pmatrix} y & x \\ 0 & 1 \end{pmatrix}\right) \psi(-x)\,dx, \qquad y \in K_\infty^\times.$$

So we restricted here the Whittaker model to upper triangular matrices (this is called the Kirillov model), and we separated the archimedean and non-archimedan parts. An important feature is that any Whittaker function $W_\phi \in L^2(K_\infty^\times, d^\times y)$ occurs for some newform $\phi \in V_\pi(\mathfrak{c}_\pi)$, and $\|W_\phi\|$ is proportional to $\|\phi\|$ with a constant depending very mildly on π.

What about "oldforms of level $\mathfrak{c}$", i.e. the elements of $V_\pi(\mathfrak{c})$ for $\mathfrak{c}$ divisible by $\mathfrak{c}_\pi$? For any nonzero ideal $\mathfrak{t} \mid \mathfrak{c}\mathfrak{c}_\pi^{-1}$ we have an isometric embedding of complex vector spaces

$$R_\mathfrak{t} : V_\pi(\mathfrak{c}_\pi) \hookrightarrow V_\pi(\mathfrak{c}), \qquad (R_\mathfrak{t}\phi)(g) := \phi\left(g\begin{pmatrix} t^{-1} & 0 \\ 0 & 1 \end{pmatrix}\right),$$

where $t \in \mathbb{A}_{\text{fin}}^\times$ is any finite idele representing $\mathfrak{t}$. Then it follows from the local result of Casselman [9] or the global result of Miyake [31] that

$$V_\pi(\mathfrak{c}) = \bigoplus_{\mathfrak{t}\mid\mathfrak{c}\mathfrak{c}_\pi^{-1}} R_\mathfrak{t} V_\pi(\mathfrak{c}_\pi), \qquad \mathfrak{c}_\pi \mid \mathfrak{c}.$$

A technical difficulty here is that the spaces $R_\mathfrak{t} V_\pi(\mathfrak{c}_\pi)$ are in general not orthogonal. Nevertheless, using a Gram-Schmidt orthogonalization process, we can obtain an orthogonal decomposition

$$V_\pi(\mathfrak{c}) = \bigoplus_{\mathfrak{t}\mid\mathfrak{c}\mathfrak{c}_\pi^{-1}} R^{(\mathfrak{t})} V_\pi(\mathfrak{c}_\pi), \qquad \mathfrak{c}_\pi \mid \mathfrak{c},$$

and for every $\phi \in R_\mathfrak{t} V_\pi(\mathfrak{c}_\pi)$ a Fourier expansion

$$\phi\left(\begin{pmatrix} y & x \\ 0 & 1 \end{pmatrix}\right) = \sum_{r\in K^\times} \frac{\lambda_\pi^{(\mathfrak{t})}(ry_{\text{fin}})}{\sqrt{\mathcal{N}(ry_{\text{fin}})}} W_\phi(ry_\infty)\psi(rx), \qquad y \in \mathbb{A}^\times, \quad x \in \mathbb{A},$$

$$W_\phi := W_{(R^{(\mathfrak{t})})^{-1}\phi} \quad \text{and} \quad \lambda_\pi^{(\mathfrak{t})}(\mathfrak{m}) := \sum_{\mathfrak{s}\mid\gcd(\mathfrak{t},\mathfrak{m})} \alpha_{\mathfrak{t},\mathfrak{s}} (\mathcal{N}\mathfrak{s})^{1/2} \lambda_\pi(\mathfrak{m}\mathfrak{s}^{-1}).$$

The constants $\alpha_{\mathfrak{t},\mathfrak{s}} \in \mathbb{C}$ are explicit, although difficult to estimate in general.

Similarly, the Eisenstein spaces have an orthogonal decomposition

$$V_{\chi,\chi^{-1}}(\mathfrak{c}) = \bigoplus_{\mathfrak{t}\mid\mathfrak{c}\mathfrak{c}_\chi^{-2}} R^{(\mathfrak{t})} V_{\chi,\chi^{-1}}(\mathfrak{c}_\chi^2), \qquad \mathfrak{c}_\chi^2 \mid \mathfrak{c},$$

such that every $\phi \in R^{(\mathfrak{t})}V_{\chi,\chi^{-1}}(\mathfrak{c}_\chi^2)$ has a Fourier expansion

$$\phi\left(\begin{pmatrix} y & x \\ 0 & 1 \end{pmatrix}\right) = \phi_{\mathrm{const}}(y) + \sum_{r \in K^\times} \frac{\lambda^{(\mathfrak{t})}_{\chi,\chi^{-1}}(ry_{\mathrm{fin}})}{\sqrt{\mathcal{N}(ry_{\mathrm{fin}})}} W_\phi(ry_\infty)\psi(rx), \qquad y \in \mathbb{A}^\times, \ x \in \mathbb{A},$$

$$\lambda^{(\mathfrak{t})}_{\chi,\chi^{-1}}(\mathfrak{m}) \ll_{K,\varepsilon} (\mathcal{N}\gcd(\mathfrak{t},\mathfrak{m}))(\mathcal{N}\mathfrak{m})^\varepsilon.$$

4 Lecture Four: Subconvexity of Twisted L-functions

In this final lecture we highlight the main ideas in the proof of Theorem 2.1, following closely the original source [6]. The proof is inspired by and builds on important earlier work by several researchers: see [5, 6] for references and history. For the sake of readability, we omit some technicalities, and we do not indicate the dependence of implied constants on π, χ_∞, K, ε.

By an approximate functional equation it suffices to bound finite sums

$$\mathcal{L}_{\chi_{\mathrm{fin}}} := \sum_{0 \ll r \in \mathfrak{y}} \frac{\lambda_\pi(r\mathfrak{y}^{-1})\tilde{\chi}_{\mathrm{fin}}(r)}{\sqrt{\mathcal{N}(r\mathfrak{y}^{-1})}} W(r), \tag{4.1}$$

where $W : K^\times_{\infty,+} \to \mathbb{C}$ is a smooth weight function of compact support cutting off at about $\mathcal{N}r \approx (\mathcal{N}\mathfrak{q})^{1+\varepsilon}$, and $\mathfrak{y}$ represents a narrow ideal class. Recall that $\tilde{\chi}_{\mathrm{fin}} : (\mathfrak{o}/\mathfrak{q})^\times \to S^1$ is determined by the Hecke character $\chi : \mathbb{A}^\times \to S^1$. We shall estimate (4.1) by a general principle of analytic number theory: sums in a harmonic family are easier to bound together than individually. To illustrate this point, let us assume that we are back in school, and we need to prove the inequality

$$|\sin x + \cos x| \leqslant \sqrt{2}.$$

Of course, we can accomplish this task in many ways, but a particularly nice way is to "invent the harmonic complement" $|\sin x - \cos x|$ and observe the identity

$$(\sin x + \cos x)^2 + (\sin x - \cos x)^2 = 2.$$

In order to estimate (4.1), we consider *all the sums*

$$\mathcal{L}_\xi := \sum_{0 \ll r \in \mathfrak{y}} \frac{\lambda_\pi(r\mathfrak{y}^{-1})\xi(r)}{\sqrt{\mathcal{N}(r\mathfrak{y}^{-1})}} W(r),$$

where ξ is any character of $(\mathfrak{o}/\mathfrak{q})^\times$, and then the *amplified second moment*

$$\mathcal{S} := \sum_{\xi \in \widehat{(\mathfrak{o}/\mathfrak{q})^\times}} \left| \sum_\ell \xi(\ell)\overline{\tilde{\chi}_{\mathrm{fin}}(\ell)} \right|^2 |\mathcal{L}_\xi|^2,$$

where $\ell \in \mathfrak{o}$ runs through certain elements of norm $\mathcal{N}\ell \in [L, 2L]$ generating prime ideals $(\ell) \nmid \mathfrak{q}$. The quantity $L > 1$ is the *amplifier length*, and the previous sum is

$$\mathcal{S} \gg (\mathcal{N}\mathfrak{q})^{-\varepsilon} L^2 \left|\mathcal{L}_{\tilde{\chi}_{\mathrm{fin}}}\right|^2$$

by positivity. Applying Plancherel and some easy estimates for the diagonal contribution, we arrive at

$$\frac{|\mathcal{L}_{\tilde{\chi}_{\mathrm{fin}}}|^2}{(\mathcal{N}\mathfrak{q})^{1+\varepsilon}} \ll \frac{1}{L} + \sum_{0 \neq q \in \mathfrak{q}\mathfrak{y} \cap \mathcal{B}} \sum_{\substack{\ell_1 r_1 - \ell_2 r_2 = q \\ 0 \neq r_1, r_2 \in \mathfrak{y}}} \frac{\lambda_\pi(r_1 \mathfrak{y}^{-1}) \overline{\lambda_\pi(r_2 \mathfrak{y}^{-1})}}{\sqrt{\mathcal{N}(r_1 r_2 \mathfrak{y}^{-2})}} W(r_1) \overline{W(r_2)}, \quad (4.2)$$

where (ℓ_1), (ℓ_2) are some prime ideals of norm about L, and $\mathcal{B} \subset K_\infty$ is some ball of volume at most $L(\mathcal{N}\mathfrak{q})^{1+\varepsilon}$ centered at the origin.

To handle the *shifted convolution sum* inside (4.2), we write it as

$$\sum_{\substack{r_1 - r_2 = q \\ r_1, r_2 \in K^\times}} \frac{\lambda_\pi(r_1 \ell_1^{-1} \mathfrak{y}^{-1}) \overline{\lambda_\pi(r_2 \ell_2^{-1} \mathfrak{y}^{-1})}}{\sqrt{\mathcal{N}(r_1 \ell_1^{-1} r_2 \ell_2^{-1} \mathfrak{y}^{-2})}} W_1(r_1) \overline{W_2(r_2)}. \quad (4.3)$$

The weight functions $W_1, W_2 : K_\infty^\times \to \mathbb{C}$ are nice, hence they determine vectors $\phi_1, \phi_2 \in V_\pi(\mathfrak{c}_\pi)$ such that

$$\phi_j\left(\begin{pmatrix} y & x \\ 0 & 1 \end{pmatrix}\right) = \sum_{r \in K^\times} \frac{\lambda_\pi(r y_{\mathrm{fin}})}{\sqrt{\mathcal{N}(r y_{\mathrm{fin}})}} W_j(r y_\infty) \psi(r x), \qquad y \in \mathbb{A}^\times, \quad x \in \mathbb{A}.$$

Let us fix $y \in \mathbb{A}^\times$ such that $y_\infty = (1, \ldots, 1)$ and $(y_{\mathrm{fin}}) = \mathfrak{y}^{-1}$. Let us also define $\Phi \in L^2(GL_2(K) \backslash GL_2(\mathbb{A}))$ via

$$\Phi(g) := \phi_1\left(g \begin{pmatrix} \ell_1^{-1} & 0 \\ 0 & 1 \end{pmatrix}\right) \overline{\phi_2}\left(g \begin{pmatrix} \ell_2^{-1} & 0 \\ 0 & 1 \end{pmatrix}\right), \qquad g \in GL_2(\mathbb{A}).$$

It is straightforward to check that Φ is fixed by the right action of $\mathcal{K}(\mathfrak{c})$ for

$$\mathfrak{c} := \mathfrak{c}_\pi \operatorname{lcm}((\ell_1), (\ell_2)).$$

Moreover,

$$\Phi\left(\begin{pmatrix} y & x \\ 0 & 1 \end{pmatrix}\right) = \sum_{r_1, r_2 \in K^\times} \frac{\lambda_\pi(r_1 \ell_1^{-1} \mathfrak{y}^{-1}) \overline{\lambda_\pi(r_2 \ell_2^{-1} \mathfrak{y}^{-1})}}{\sqrt{\mathcal{N}(r_1 \ell_1^{-1} r_2 \ell_2^{-1} \mathfrak{y}^{-2})}} W_1(r_1) \overline{W_2(r_2)} \psi((r_1 - r_2)x),$$

whence (4.3) really equals

$$\int_{K \backslash \mathbb{A}} \Phi\left(\begin{pmatrix} y & x \\ 0 & 1 \end{pmatrix}\right) \psi(-qx)\, dx.$$

We decompose Φ in the level $\mathfrak{c}$ spectrum of $L^2(GL_2(K) \backslash GL_2(\mathbb{A}))$, following the discussion of the third lecture. We obtain

$$\Phi = \Phi_{sp} + \int_{(\mathfrak{c})} \sum_{\mathfrak{t} \mid \mathfrak{c}\mathfrak{c}_\varpi^{-1}} \Phi_{\varpi, \mathfrak{t}}\, d\varpi,$$

where Φ_{sp} lies in the span of the functions $g \mapsto \chi(\det g)$ for quadratic Hecke characters χ, the representations V_ϖ run through the cuspidal spaces V_π and the Eisenstein spaces $V_{\chi,\chi^{-1}}$ of conductor $\mathfrak{c}_\varpi \mid \mathfrak{c}$, and

$$\Phi_{\varpi,\mathfrak{t}} \in R^{(\mathfrak{t})} V_\varpi(\mathfrak{c}_\varpi), \qquad \mathfrak{t} \mid \mathfrak{c}\mathfrak{c}_\varpi^{-1}.$$

Upon defining $W_{\varpi,\mathfrak{t}} := W_{\Phi_{\varpi,\mathfrak{t}}}$, we can now rewrite (4.2) as

$$\frac{|\mathcal{L}_{\tilde{\chi}_{\text{fin}}}|^2}{(\mathcal{N}\mathfrak{q})^{1+\varepsilon}} \ll \frac{1}{L} + \sum_{0 \neq q \in \mathfrak{q}\mathfrak{y} \cap \mathcal{B}} \int_{(\mathfrak{c})} \sum_{\mathfrak{t} \mid \mathfrak{c}\mathfrak{c}_\varpi^{-1}} \frac{\lambda_\varpi^{(\mathfrak{t})}(q\mathfrak{y}^{-1})}{\sqrt{\mathcal{N}(q\mathfrak{y}^{-1})}} W_{\varpi,\mathfrak{t}}(q)\, d\varpi. \tag{4.4}$$

The advantage of this expression is that it is not quadratic but linear in the Hecke eigenvalues. The price to pay is the spectral averaging in ϖ.

The Whittaker functions $W_{\varpi,\mathfrak{t}} : K_\infty^\times \to \mathbb{C}$ depend on the original weight function $W : K_{\infty,+}^\times \to \mathbb{C}$ included in (4.1). In order to proceed further, one needs to understand the "size" of these functions, at least on average over the spectrum. We carry this out via Sobolev type norms, the process being schematized as follows:

$$\|W\|_* \rightsquigarrow \|\phi_1\|_* \|\phi_2\|_* \rightsquigarrow \|\Phi\|_* \rightsquigarrow \|\Phi_{\varpi,\mathfrak{t}}\|_* \rightsquigarrow \|W_{\varpi,\mathfrak{t}}\|_*.$$

For example, over the Eisenstein spectrum we can readily derive the bound

$$\int_{\varpi \in \mathcal{E}(\mathfrak{c})} \sum_{\mathfrak{t} \mid \mathfrak{c}\mathfrak{c}_\varpi^{-1}} |W_{\varpi,\mathfrak{t}}(y)|\, d\varpi \ll (\mathcal{N}(\ell_1\ell_2))^\varepsilon,$$

using that the number of cusps in level $\mathfrak{c}$ is $\ll (\mathcal{N}(\ell_1\ell_2))^\varepsilon$, because $\mathfrak{c}\mathfrak{c}_\pi^{-1}$ is squarefree. This implies that the contribution of the Eisenstein spectrum in (4.4) is

$$\ll \frac{1}{\mathcal{N}\mathfrak{q}} \sqrt{L(\mathcal{N}\mathfrak{q})^{1+\varepsilon}} (\mathcal{N}(\ell_1\ell_2))^\varepsilon \ll \frac{L^{\frac{1}{2}}}{(\mathcal{N}\mathfrak{q})^{\frac{1}{2}-\varepsilon}}.$$

Bounding the cuspidal contribution in (4.4) is much harder, but at least we can initially restrict to very small spectral parameters, namely $\mathcal{N}\tilde{\lambda}_\varpi \leqslant (\mathcal{N}\mathfrak{q})^\varepsilon$, thanks to the bound (valid for any $A > 0$)

$$\int_{\varpi \in \mathcal{C}(\mathfrak{c})} (\mathcal{N}\tilde{\lambda}_\varpi)^A \sum_{\mathfrak{t} \mid \mathfrak{c}\mathfrak{c}_\varpi^{-1}} |W_{\varpi,\mathfrak{t}}(y)|\, d\varpi \ll_A |\mathcal{N}(\ell_1\ell_2)|^{\frac{1}{2}+\varepsilon}.$$

Using this observation we find that the contribution of the cuspidal spectrum in (4.4) is, for some values $f(\mathfrak{a}) \ll (\mathcal{N}\mathfrak{q})^\varepsilon$,

$$\ll (\mathcal{N}\mathfrak{q})^\varepsilon \left(\sum_{\varpi \in \mathcal{C}(\mathfrak{c},\varepsilon)} \sum_{\mathfrak{t} \mid \mathfrak{c}\mathfrak{c}_\varpi^{-1}} \left| \sum_{\mathcal{N}\mathfrak{m} \ll L(\mathcal{N}\mathfrak{q})^\varepsilon} \frac{\lambda_\varpi^{(\mathfrak{t})}(\mathfrak{m}\mathfrak{q})}{\sqrt{\mathcal{N}(\mathfrak{m}\mathfrak{q})}} f(\mathfrak{m}\mathfrak{q}) \right|^2 \right)^{1/2}.$$

Here we essentially factor out $\lambda_\varpi(\mathfrak{q})$ and bound it individually by $(\mathcal{N}\mathfrak{q})^{\theta+\varepsilon}$, which is why the parameter θ appears in Theorems 2.1 and 2.2. Now we arrive at the endgame. We majorize the remaining sum by a smooth spectral sum involving an analogous Eisenstein contribution and rapidly decaying spectral weights. We open the square and apply a variant of the Bruggeman-Kuznetsov formula originally developed by Venkatesh [41] and further extended by Maga [27]. Finally we apply familiar bounds for the resulting Kloosterman sums and Bessel transforms to infer that the contribution of the cuspidal spectrum in (4.4) is

$$\ll (\mathcal{N}\mathfrak{q})^{\varepsilon}\Bigg(\underbrace{\frac{L^2}{(\mathcal{N}\mathfrak{q})^{1-2\theta-\varepsilon}}}_{\text{diagonal}}+\underbrace{\frac{L^{3/2}}{(\mathcal{N}\mathfrak{q})^{1-2\theta-\varepsilon}}}_{\text{off-diagonal}}\Bigg)^{1/2} \ll \frac{L}{(\mathcal{N}\mathfrak{q})^{\frac{1}{2}-\theta-\varepsilon}}.$$

Collecting terms, we deduce from (4.4) that

$$\frac{|\mathcal{L}_{\tilde{\chi}_{\mathrm{fin}}}|^2}{(\mathcal{N}\mathfrak{q})^{1+\varepsilon}} \ll \frac{1}{L}+\frac{L}{(\mathcal{N}\mathfrak{q})^{\frac{1}{2}-\theta}}.$$

The right hand side is smallest when $L := (\mathcal{N}\mathfrak{q})^{\frac{1-2\theta}{4}}$, in which case we get

$$\mathcal{L}_{\chi_{\mathrm{fin}}} \ll (\mathcal{N}\mathfrak{q})^{\frac{3+2\theta}{8}+\varepsilon}.$$

This concludes the proof of Theorem 2.1.

Acknowledgments

I thank my friends Valentin Blomer and Jianya Liu for making these notes possible. Valentin and I worked on the results between 2007 and 2009, and we learnt a lot from each other. Jianya invited me to lecture at a great summer school in 2010, and he showed infinite patience in waiting for my written account. Finally, Rainer Schulze-Pillot explained to me some subtleties of quadratic forms over number fields, for which I am most grateful.

References

[1] E. M. Baruch and Z. Mao, Central value of automorphic L-functions, Geom. Funct. Anal. **17** (2007), 333–384.

[2] D. Blasius, Hilbert modular forms and the Ramanujan conjecture, in: Noncommutative Geometry and Number Theory, Aspects Math. E37, Vieweg, Wiesbaden, 2006, 35–56.

[3] V. Blomer and F. Brumley, On the Ramanujan conjecture over number fields, Ann. Math. **174** (2011), 581–605.

[4] V. Blomer and G. Harcos, Hybrid bounds for twisted L-functions, J. Reine Angew. Math. **621** (2008), 53–79; Addendum, ibid., to appear.

[5] V. Blomer and G. Harcos, The spectral decomposition of shifted convolution sums, Duke Math. J. **144** (2008), 321–339.

[6] V. Blomer and G. Harcos, Twisted L-functions over number fields and Hilbert's eleventh problem, Geom. Funct. Anal. **20** (2010), 1–52; Erratum available at the authors' webpages.

[7] D. A. Burgess, On character sums and L-series. II, Proc. London Math. Soc. **13** (1963), 524–536.

[8] V. A. Bykovskiĭ, A trace formula for the scalar product of Hecke series and its applications, translation in J. Math. Sci. (New York) **89** (1998), 915–932.

[9] W. Casselman, On some results of Atkin and Lehner, Math. Ann. **201** (1973), 301–314.

[10] J. Cogdell, I. Piatetski-Shapiro and P. Sarnak, Estimates on the critical line for Hilbert modular L-functions and applications, preprint.

[11] B. Conrey and H. Iwaniec, The cubic moment of central values of automorphic L-functions, Ann. Math. **151** (2000), 1175–1216.

[12] P. G. L. Dirichlet, Lectures on Number Theory [translated by John Stillwell], History of Mathematics, Vol. 16, Amer. Math. Soc., Providence, RI; London Math. Soc., London, 1999.

[13] W. Duke, Hyperbolic distribution problems and half-integral weight Maass forms, Invent. Math. **92** (1988), 73–90.

[14] W. Duke, J. B. Friedlander and H. Iwaniec, Bounds for automorphic L-functions, Invent. Math. **112** (1993), 1–8.

[15] W. Duke and R. Schulze-Pillot, Representation of integers by positive ternary quadratic forms and equidistribution of lattice points on ellipsoids, Invent. Math. **99** (1990), 49–57.

[16] A. G. Earnest, J. S. Hsia and D. C. Hung, Primitive representations by spinor genera of ternary quadratic forms, J. London Math. Soc. **50** (1994), 222–230.

[17] C. F. Gauss, Disquisitiones Arithmeticae [translated by A. A. Clarke and revised by W. C. Waterhouse, C. Greither and A. W. Grootendorst], Springer-Verlag, New York, 1986.

[18] S. Gelbart and H. Jacquet, Forms of $GL(2)$ from the analytic point of view, in: Automorphic Forms, Representations, and L-functions, Part 1, Eds. A. Borel and W. Casselman, Proc. Sympos. Pure Math., Vol. 33, 1979, 213–251.

[19] E. Grosswald, Representations of Integers as Sums of Squares, Springer-Verlag, New York, 1985.

[20] J. Hanke, Local densities and explicit bounds for representability by a quadratic form, Duke Math. J. **124** (2004), 351–388.

[21] J. Hanke, Some recent results about (ternary) quadratic forms, in: Number Theory, CRM Proc. Lecture Notes, Vol. 36, Amer. Math. Soc., Providence, RI, 2004, 147–164.

[22] J. Hoffstein and T. A. Hulse, Multiple Dirichlet series and shifted convolutions (with an appendix by Andre Reznikov), submitted; preprint available at `http://arxiv.org/abs/1110.4868`.

[23] J. S. Hsia, Representations by spinor genera, Pacific J. Math. **63** (1976), 147–152.

[24] H. Iwaniec, Fourier coefficients of modular forms of half-integral weight, Invent. Math. **87** (1987), 385–401.

[25] Y. Kitaoka, Arithmetic of Quadratic Forms, Cambridge Tracts in Mathematics, Vol. 106, Cambridge University Press, Cambridge, 1993.

[26] M. Kneser, Darstellungsmaße indefiniter quadratischer Formen, Math. Z. **77** (1961), 188–194.

[27] P. Maga, A semi-adelic Kuznetsov formula over number fields, Int. J. Number Theory, **9** (2013), 1649–1681.

[28] P. Maga, Shifted convolution sums and Burgess type subconvexity over number fields, submitted; preprint available at `http://arxiv.org/abs/1312.0553`.

[29] O. T. O'Meara, Introduction to Quadratic Forms, reprint of the 1973 edition, Classics in Mathematics, Springer-Verlag, Berlin, 2000.

[30] P. Michel and A. Venkatesh, The subconvexity problem for GL_2, Publ. Math. Inst. Hautes Études Sci. **111** (2010), 171–271.

[31] T. Miyake, On automorphic forms on GL_2 and Hecke operators, Ann. Math. **94** (1971), 174–189.

[32] J. Neukirch, Algebraic Number Theory, translated from the 1992 German original, Grundlehren der Mathematischen Wissenschaften, Vol. 322, Springer-Verlag, Berlin, 1999.

[33] K. Ono and K. Soundararajan, Ramanujan's ternary quadratic form, Invent. Math. **130** (1997), 415–454.

[34] R. Schulze-Pillot, Thetareihen positiv definiter quadratischer Formen, Invent. Math. **75** (1984), 283–299.

[35] R. Schulze-Pillot, Darstellungsmaße von Spinorgeschlechtern ternärer quadratischer Formen, J. Reine Angew. Math. **352** (1984), 114–132.

[36] R. Schulze-Pillot, Ternary quadratic forms and Brandt matrices, Nagoya Math. J. **102** (1986), 117–126.

[37] R. Schulze-Pillot, Representation by integral quadratic forms – a survey, in: Algebraic and Arithmetic Theory of Quadratic Forms, Contemp. Math., Vol. 344, Amer. Math. Soc., Providence, RI, 2004, 303–321.

[38] G. Shimura, On modular forms of half integral weight, Ann. Math. **97** (1973), 440–481.

[39] C. L. Siegel, Über die Classenzahl quadratischer Zahlkörper, Acta Arith. **1** (1935), 83–86.

[40] C. L. Siegel, Über die analytische Theorie der quadratischen Formen, Ann. Math. **36** (1935), 527–606.; II, ibid. **37** (1936), 230–263.; III, ibid. **38** (1937), 212–291.

[41] A. Venkatesh, Beyond endoscopy and special forms on $GL(2)$, J. Reine Angew. Math. **577** (2004), 23–80.

[42] A. Venkatesh, Sparse equidistribution problems, period bounds, and subconvexity, Ann. Math. **172** (2010), 989–1094.

[43] J.-L. Waldspurger, Sur les coefficients de Fourier des formes modulaires de poids demi-entier, J. Math. Pures Appl. **60** (1981), 375–484.

[44] J.-L. Waldspurger, Correspondances de Shimura et quaternions, Forum Math. **3** (1991), 219–307.

[45] A. Weil, Basic Number Theory, reprint of the second (1973) edition, Classics in Mathematics, Springer-Verlag, Berlin, 1995.

[46] H. Wu, Burgess-like subconvex bounds for $GL_2 \times GL_1$, Geom. Funct. Anal., to appear; preprint available at `http://arxiv.org/abs/1209.5950`.

[47] D. B. Zagier, Zetafunktionen und Quadratische Körper, Springer-Verlag, Berlin-New York, 1981.

Automorphic Forms and
L-functions
ALM 30, pp. 69–90

The Voronoi Formula for the Triple Divisor Function

Xiaoqing Li*

Abstract

In this note, we generalize the method in [6] to derive the Voronoi formula for the triple divisor function twisted by additive characters as well as the Voronoi formula for the minimal Eisenstein series on $GL(3)$.

2010 Mathematics Subject Classification: 11F11, 11F66.

Keywords and Phrases: Voronoi formulas, Eisenstein series, Triple divisors.

1 Introduction

Voronoi formulas for automorphic forms are Poisson summation formulas weighted by Fourier coefficients of automorphic forms with possible twists by characters or other arithmetic weights. They usually serve as tools to study the shifted sum $\sum_{n\leqslant N} \alpha(n)\beta(n+k)$ where $\alpha(n), \beta(n)$ are Fourier coefficients of automorphic forms. Such problems are often encountered when evaluating power moments of L-functions, see, for example, [4, 8, 15]. Voronoi formulas for automorphic forms on $GL(2)$ including the two triple divisor function $\tau(n)$, which is the number of solutions of the equation $d_1d_2 = n$ in natural numbers, were well-established (see [3, 8, 12] for example) and play important roles in the theory of $GL(2)$ L-functions (see the excellent survey papers [9, 13]).

Voronoi formulas for Maass cusp forms on $GL(3)$ twisted by additive characters were first derived by Miller and Schmid [14] using their powerful method of automorphic distributions. In the previous paper with Goldfeld [6], we gave a different proof of Miller and Schmid's formula. In this note, we generalize the method in [5,

*Department of Mathematics, State University of New York-Buffalo, Buffalo, NY 14260, USA, XL29@buffalo.edu.

Chapter VII] and [6] to derive a Voronoi formula for the minimal parabolic Eisenstein series on $GL(3)$. As a consequence, we obtain a Voronoi formula for the triple divisor function $\tau_3(n)$ which is the number of solutions of the equation $d_1d_2d_3 = n$ in natural numbers. Voronoi formulas for the n-tuple divisor functions are known in different forms, see [10] for example. But our formula is more explicit and the method of derivation illustrates the role of the modularity of the Eisenstein series.

To state our main theorem, let

$$\tau(n) = \sum_{d|n} 1, \quad \tau_3(n) = \sum_{\substack{n=d_1d_2d_3\\ d_1,d_2,d_3>0}} 1;$$

let

$$\sigma_{0,0}(k,l) = \sum_{\substack{d_1|l\\ d_1>0}} \sum_{\substack{d_2|\frac{l}{d_1}\\ d_2>0\\ (d_2,k)=1}} 1;$$

let $\zeta(s)$ be the Riemann zeta function,

$$\gamma := \lim_{s\to 1}\left(\zeta(s) - \frac{1}{s-1}\right)$$

be the Euler constant and

$$\gamma_1 := -\frac{d}{ds}\left(\zeta(s) - \frac{1}{s-1}\right)\Big|_{s=1}$$

be a Stieltjes constant;

$$S(0,n;c) = \sum_{\substack{d \ (\mathrm{mod}\ c)\\ (d,c)=1}} e\left(\frac{nd}{c}\right) \tag{1.1}$$

is the Ramanujan sum and

$$S(m,n;c) = \sum_{\substack{d \ (\mathrm{mod}\ c)\\ d\bar{d}\equiv 1 (\mathrm{mod}\ c)}} e\left(\frac{md + n\bar{d}}{c}\right) \tag{1.2}$$

is the clasical Kloosterman sum where $e(x) = e^{2\pi i x}$ throughout the paper; for $\phi(x) \in C_c^\infty(0,\infty), k = 0,1, \sigma > 3$, set

$$\Phi_k(x) = \frac{1}{2\pi i}\int_{\Re w=\sigma} (\pi^3 x)^{-w} \frac{\Gamma\left(\frac{1+w+2k}{2}\right)^3}{\Gamma(\frac{-w}{2})^3} \tilde{\phi}(-w-k)dw$$

with

$$\tilde{\phi}(s) = \int_0^\infty \phi(x)x^s \frac{dx}{x}$$

be the Mellin transform of ϕ, we also set

$$\Phi_{0,1}^0(x) = \Phi_0(x) + \frac{\pi^{-3}q^3}{n^2mi}\Phi_1(x)$$

and

$$\Phi_{0,1}^1(x) = \Phi_0(x) - \frac{\pi^{-3}q^3}{n^2mi}\Phi_1(x),$$

then our main theorem is:

Theorem 1.1 (Main Theorem). *For $\phi(x) \in C_c^\infty(0,\infty)$, $h, \bar{h}, q$ be positive integers with $h\bar{h} \equiv 1(\mathrm{mod}\ q)$, we have*

$$\begin{aligned}
&\sum_{m>0} \tau_3(m) e\left(\frac{m\bar{h}}{q}\right)\phi(m) \\
&= \frac{1}{2}q\pi^{-\frac{3}{2}} \sum_{n|q}\sum_{m>0} \frac{1}{nm} \sum_{n_1|n}\sum_{n_2|\frac{n}{n_1}} \sigma_{0,0}\left(\frac{n}{n_1n_2}, m\right) S(m,h;qn^{-1})\Phi_{0,1}^0\left(\frac{mn^2}{q^3}\right) \\
&+ \frac{1}{2}q\pi^{-\frac{3}{2}} \sum_{n|q}\sum_{m>0} \frac{1}{nm} \sum_{n_1|n}\sum_{n_2|\frac{n}{n_1}} \sigma_{0,0}\left(\frac{n}{n_1n_2}, m\right) S(-m,h;qn^{-1})\Phi_{0,1}^1\left(\frac{mn^2}{q^3}\right) \\
&+ \frac{1}{2}\sum_{n|q} S(0,h;qn^{-1})nq^{-2}\tau(n)\Big\{(\log n)^2 - 5\log q\log n + \frac{9}{2}(\log q)^2 + 3\gamma^2 - 3\gamma_1 \\
&+ \tau(n)^{-1}\Big[\sum_{d|n}\log d(\log n + \log q - 5\gamma) - \frac{3}{2}\sum_{d|n}(\log d)^2\Big] + 7\gamma\log n - 9\gamma\log q\Big\}\tilde{\phi}(1) \\
&+ \frac{1}{2}\sum_{n|q} S(0,h;qn^{-1})nq^{-2}\tau(n)\left[\frac{5}{3}\log n - 3\log q + 3\gamma - \frac{1}{3}\tau(n)^{-1}\sum_{d|n}\log d\right]\tilde{\phi}'(1) \\
&+ \frac{1}{4}\sum_{n|q} S(0,h;qn^{-1})nq^{-2}\tau(n)\tilde{\phi}''(1).
\end{aligned}$$

2 Proof of the Main Theorem

Let $G = GL(3,\mathbb{R}), \Gamma = SL(3,\mathbb{Z})$ and

$$\mathfrak{h}^3 = GL(3,\mathbb{R})/\langle O(3,\mathbb{R})\cdot \mathbb{R}^\times\rangle$$

be the generalized upper half plane. By the Iwasawa decomposition, every element $z \in \mathfrak{h}^3$ has the form $z = xy$ where

$$x = \begin{pmatrix} 1 & x_2 & x_3 \\ & 1 & x_1 \\ & & 1 \end{pmatrix}, \quad y = \begin{pmatrix} y_1y_2 & & \\ & y_1 & \\ & & 1 \end{pmatrix}.$$

Let Γ_∞ be the minimal parabolic subgroup of Γ which consists of upper triangular matrices. For $\Re\nu_1, \Re\nu_2 > \frac{2}{3}, z \in \mathfrak{h}^3$, we define the minimal parabolic Eisenstein series:

$$E(z,\nu_1,\nu_2) := \sum_{\gamma\in\Gamma_\infty\backslash\Gamma} y_1(\gamma z)^{2\nu_1+\nu_2} y_2(\gamma z)^{\nu_1+2\nu_2}. \tag{2.1}$$

We also define its dual form

$$\tilde{E}(z,\nu_1,\nu_2) := E(w_1{}^t z^{-1} w_1, \nu_1, \nu_2),$$

where

$$w_1 = \begin{pmatrix} & & -1 \\ & 1 & \\ 1 & & \end{pmatrix}.$$

Since $E(z,\nu_1,\nu_2)$ is automorphic for Γ, it is easy to check that

$$\tilde{E}(w_2{}^t(Auz)^{-1}, \nu_1, \nu_2) = E(Auz, \nu_1, \nu_2), \tag{2.2}$$

where

$$w_2 = \begin{pmatrix} 0 & 0 & 1 \\ 1 & 0 & 0 \\ 0 & 1 & 0 \end{pmatrix}. \tag{2.3}$$

On the other hand, (2.1) of $E(z,\nu_1,\nu_2)$, we have

$$\tilde{E}(z,\nu_1,\nu_2) = E(z,\nu_2,\nu_1). \tag{2.4}$$

Combining (2.2) and (2.4) we have

$$E(w_2{}^t(Auz)^{-1}, \nu_2, \nu_1) = E(Auz, \nu_1, \nu_2). \tag{2.5}$$

We introduce some notations which will be used in the remaining part of the paper: for $(\nu_1,\nu_2)\in\mathbb{C}^2$, set

$$\nu := (\nu_1,\nu_2), \qquad \tilde{\nu} := (\nu_2,\nu_1); \tag{2.6}$$

$$\alpha := -\nu_1 - 2\nu_2 + 1,\ \ \beta := -\nu_1+\nu_2,\ \ \gamma := 2\nu_1+\nu_2-1; \tag{2.7}$$

$$\begin{aligned} &W^*_{\text{Jacquet}}\left(\begin{pmatrix} y_1y_2 & & \\ & y_1 & \\ & & 1 \end{pmatrix}, \nu\right) \\ &:= 4y_1^{1+\frac{\nu_1-\nu_2}{2}} y_2^{1-\frac{\nu_1-\nu_2}{2}} \\ &\quad \cdot \int_0^\infty K_{\frac{3\nu_1+3\nu_2-2}{2}}(2\pi y_2\sqrt{1+u^{-2}}) K_{\frac{3\nu_1+3\nu_2-2}{2}}(2\pi y_1\sqrt{1+u^2}) u^{\frac{3\nu_1-3\nu_2}{2}} \frac{du}{u}, \end{aligned}$$

where $K_\nu(s)$ is the K-Bessel function; set

$$\xi(s) := \pi^{-\frac{s}{2}}\Gamma\left(\frac{s}{2}\right)\zeta(s), \quad c(s) := \frac{\xi(2s-1)}{\xi(2s)};$$

for $\begin{pmatrix} a & b \\ c & d \end{pmatrix} \in SL(2,\mathbb{Z})$, let

$$u_{c,d} := \Im \frac{az_2+b}{cz_2+d} = \frac{y_2}{|cz_2+d|^2}$$

with $z_2 = x_2 + iy_2$, we also set

$$\sigma_s(n) := \sum_{\substack{d|n \\ d>0}} d^s, \quad \sigma_{\alpha,\beta}(n,m) := \sum_{\substack{d_1|m \\ d_1>0}} \sum_{\substack{d_2|\frac{m}{d_1} \\ d_2>0 \\ (d_2,n)=1}} d_1^\alpha d_2^\beta,$$

we then have

Lemma 2.1 ([17]). *The function $\hat{E}(z,\nu_1,\nu_2)$, defined for $\Re\nu_1, \Re\nu_2 > \frac{2}{3}$ by*

$$\hat{E}(z,\nu_1,\nu_2) = \prod_{j=1}^{3}\left(\nu_j - \frac{2}{3}\right)\zeta(3\nu_j)E(z,\nu_1,\nu_2),$$

is an entire function of the two complex variables ν_1 and ν_2. For all ν_1 and ν_2, except those on the complex lines $\nu_1 = \frac{2}{3}, \frac{1}{3}\rho_1$ or $\nu_2 = \frac{2}{3}, \frac{1}{3}\rho_2$ where ρ_1, ρ_2 run through all nontrivial zeroes of $\zeta(s)$, there is the Fourier expansion of $E(z,\nu_1,\nu_2)$:

$$E(z,\nu_1,\nu_2) = E_0(z,\nu_1,\nu_2) + E_1(z,\nu_1,\nu_2),$$

where

$$\begin{aligned} E_0(z,\nu_1,\nu_2) &= y_1^{2\nu_1+\nu_2} y_2^{\nu_1+2\nu_2} \\ &+ c\left(\frac{3\nu_1}{2}\right) y_1^{1+\nu_2-\nu_1} y_2^{\nu_1+2\nu_2} + c\left(\frac{3\nu_2}{2}\right) y_1^{2\nu_1+\nu_2} y_2^{\nu_1-\nu_2+1} \\ &+ c\left(\frac{3\nu_1}{2}\right) c\left(\frac{3\nu_1+3\nu_2-1}{2}\right) y_1^{1+\nu_2-\nu_1} y_2^{-\nu_2-2\nu_1+3} \\ &+ c\left(\frac{3\nu_2}{2}\right) c\left(\frac{3\nu_1+3\nu_2-1}{2}\right) y_1^{2-\nu_1-2\nu_2} y_2^{1+\nu_1-\nu_2} \\ &+ c\left(\frac{3\nu_1}{2}\right) c\left(\frac{3\nu_2}{2}\right) c\left(\frac{3\nu_1+3\nu_2-1}{2}\right) y_1^{2-\nu_1-2\nu_2} y_2^{2-2\nu_1-\nu_2}, \end{aligned}$$

$$E_1(z,\nu_1,\nu_2) = E_{11}(z,\nu_1,\nu_2) + E_{12}(z,\nu_1,\nu_2) + E_{13}(z,\nu_1,\nu_2)$$

with

$$\begin{aligned} &E_{11}(z,\nu_1,\nu_2) \\ &= \frac{2(y_1^2 y_2)^{1-\frac{\nu_1}{2}-\nu_2} y_2^{\frac{1}{2}}}{\xi(3\nu_1)} c\left(\frac{3\nu_2}{2}\right) c\left(\frac{3\nu_1+3\nu_2-1}{2}\right) \\ &\cdot \sum_{m\neq 0} |m|^{\frac{3\nu_1}{2}-\frac{1}{2}} \sigma_{1-3\nu_1}(|m|) K_{\frac{3\nu_1-1}{2}}(2\pi|m|y_2) e(mx_2) \\ &+ \frac{2(y_1^2 y_2)^{\nu_1+\frac{\nu_2}{2}} y_2^{\frac{1}{2}}}{\xi(3\nu_2)} \sum_{m\neq 0} |m|^{\frac{3\nu_2}{2}-\frac{1}{2}} \sigma_{1-3\nu_2}(|m|) K_{\frac{3\nu_2-1}{2}}(2\pi|m|y_2) e(mx_2) \end{aligned}$$

$$+\frac{2(y_1^2y_2)^{\frac{1}{2}+\frac{3}{4}(\nu_2-\nu_1)}y_2^{\frac{1}{2}}}{\xi(3\nu_1+3\nu_2-1)}c\left(\frac{3\nu_1}{2}\right)\sum_{m\neq 0}|m|^{\frac{3\nu_1+3\nu_2-2}{2}}\sigma_{4-3\nu_1-3\nu_2}(|m|)$$
$$\cdot K_{\frac{3\nu_1+3\nu_2-2}{2}}(2\pi|m|y_2)e(mx_2),$$
$$E_{12}(z,\nu_1,\nu_2)=E_{12}^1(z,\nu_1,\nu_2)+E_{12}^2(z,\nu_1,\nu_2)+E_{12}^3(z,\nu_1,\nu_2)$$

with

$$E_{12}^1(z,\nu_1,\nu_2)=\frac{2(y_1^2y_2)^{\frac{1}{4}+\frac{1}{4}\nu_1+\frac{1}{2}\nu_2}}{\xi(3\nu_1)}\sum_{(c,d)=1}\sum_{n\geqslant 1}n^{\frac{3}{2}\nu_1-\frac{1}{2}}\sigma_{1-3\nu_1}(n)$$
$$\cdot u_{c,d}^{\frac{3}{4}\nu_1+\frac{3}{2}\nu_2-\frac{1}{4}}K_{\frac{3\nu_1-1}{2}}\left(2\pi n\sqrt{\frac{y_1^2y_2}{u_{c,d}}}\right)e\big((cx_3+dx_1)n\big),$$

$$E_{12}^2(z,\nu_1,\nu_2)=\frac{2(y_1^2y_2)^{\frac{3}{4}-\frac{1}{2}\nu_1-\frac{1}{4}\nu_2}}{\xi(3\nu_2)}c\left(\frac{3\nu_1}{2}\right)c\left(\frac{3\nu_1+3\nu_2-1}{2}\right)$$
$$\cdot\sum_{(c,d)=1}\sum_{n\geqslant 1}n^{\frac{3}{2}\nu_2-\frac{1}{2}}\sigma_{1-3\nu_2}(n)u_{c,d}^{\frac{5}{4}-\frac{3}{2}\nu_1-\frac{3}{4}\nu_2}$$
$$\cdot K_{\frac{3\nu_2-1}{2}}\left(2\pi n\sqrt{\frac{y_1^2y_2}{u_{c,d}}}\right)e\big((cx_3+dx_1)n\big),$$

and

$$E_{12}^3(z,\nu_1,\nu_2)$$
$$=\frac{2(y_1^2y_2)^{\frac{1}{2}+\frac{1}{4}(\nu_1-\nu_2)}}{\xi(3\nu_1+3\nu_2-1)}c\left(\frac{3\nu_2}{2}\right)$$
$$\cdot\sum_{(c,d)=1}\sum_{n\geqslant 1}n^{\frac{3\nu_1+3\nu_2-1}{2}}\sigma_{-3\nu_1-3\nu_2+2}(n)u_{c,d}^{\frac{1}{2}+\frac{3}{4}\nu_1-\frac{3}{4}\nu_2}$$
$$\cdot K_{\frac{3\nu_1+3\nu_2-2}{2}}\left(2\pi n\sqrt{\frac{y_1^2y_2}{u_{c,d}}}\right)e\big((cx_3+dx_1)n\big),$$

$$E_{13}(z,\nu_1,\nu_2)$$
$$=\frac{2}{\xi(3\nu_1)\xi(3\nu_2)\xi(3\nu_1+3\nu_2-1)}$$
$$\cdot\sum_{(c,d)=1}\sum_{n\geqslant 1}\sum_{m\neq 0}\sum_{n_1|n}\sum_{n_2|\frac{n}{n_1}}n^{-2+\nu_1+2\nu_2}|m|^{-2+2\nu_1+\nu_2}n_1^{-3\nu_1-3\nu_2+2}n_2^{1-3\nu_2}$$
$$\cdot\sigma_{1-3\nu_1,2-3\nu_1-3\nu_2}\left(\frac{n}{n_1n_2},|m|\right)W^*_{\text{Jacquet}}\left(\begin{pmatrix}\frac{n|m|y_1y_2}{|cz_2+d|} & & \\ & ny_1|cz_2+d| & \\ & & 1\end{pmatrix},\nu\right)$$
$$\cdot e\left(n(cx_3+dx_1)+m\Re\frac{az_2+b}{cz_2+d}\right).$$

The above Fourier expansion of the Eisenstein series is the starting point of our work.

To derive the Voronoi formula, we also need to compute the Fourier coefficients of the Eisenstein series after additive twists. For that purpose, let

$$A = \begin{pmatrix} 1 & 0 & 0 \\ \frac{h}{q} & 1 & 0 \\ 0 & 0 & 1 \end{pmatrix},$$

where h, q are positive integers with $(h, q) = 1$,

$$u = \begin{pmatrix} 1 & 0 & u_3 \\ & 1 & u_1 \\ & & 1 \end{pmatrix} \in SL(3, \mathbb{R}),$$

and

$$z = \begin{pmatrix} 1 & x_2 & x_3 \\ & 1 & x_1 \\ & & 1 \end{pmatrix} \begin{pmatrix} y_1 y_2 & & \\ & y_1 & \\ & & 1 \end{pmatrix} \in \mathfrak{h}^3,$$

recall

$$w_2 = \begin{pmatrix} 0 & 0 & 1 \\ 1 & 0 & 0 \\ 0 & 1 & 0 \end{pmatrix}$$

and all the other notations are as in Lemma 2.1, then we have the following lemma:

Lemma 2.2.

$$\int_0^1 \int_0^1 E_0(Auz, \nu_1, \nu_2) e(-qu_1) du_1 du_3 = 0, \tag{2.8}$$

$$\int_0^1 \int_0^1 E_0(w_2 {}^t(Auz)^{-1}, \nu_2, \nu_1) e(-qu_1) du_1 du_3 = 0, \tag{2.9}$$

$$\int_0^1 \int_0^1 E_{11}(Auz, \nu_1, \nu_2) e(-qu_1) du_1 du_3 = 0, \tag{2.10}$$

$$\int_0^1 \int_0^1 E_{11}(w_2 {}^t(Auz)^{-1}, \nu_2, \nu_1) e(-qu_1) du_1 du_3 = 0, \tag{2.11}$$

$$\begin{aligned} &\int_0^1 \int_0^1 E_{12}^1(Auz, \nu_1, \nu_2) e(-qu_1) du_1 du_3 \\ &= \frac{2}{\xi(3\nu_1)} y_1^{\frac{1}{2}+\frac{1}{2}\nu_1+\nu_2} y_2^{\nu_1+2\nu_2} q^{-\frac{3}{2}\nu_1-3\nu_2+\frac{1}{2}} K_{\frac{3\nu_1-1}{2}}(2\pi q y_1) e(q x_1), \end{aligned} \tag{2.12}$$

$$\begin{aligned} &\int_0^1 \int_0^1 E_{12}^2(Auz, \nu_1, \nu_2) e(-qu_1) du_1 du_3 \\ &= \frac{2}{\xi(3\nu_2)} c\left(\frac{3\nu_1}{2}\right) c\left(\frac{3\nu_1+3\nu_2-1}{2}\right) \\ &\cdot y_1^{\frac{3}{2}-\nu_1-\frac{\nu_2}{2}} y_2^{2-2\nu_1-\nu_2} q^{-\frac{5}{2}+3\nu_1+\frac{3\nu_2}{2}} K_{\frac{3\nu_2-1}{2}}(2\pi q y_1) e(q x_1), \end{aligned} \tag{2.13}$$

$$\begin{aligned}
&\int_0^1 \int_0^1 E_{12}^3(Auz, \nu_1, \nu_2) e(-qu_1) du_1 du_3 \\
&= \frac{2}{\xi(3\nu_1 + 3\nu_2 - 1)} c\left(\frac{3\nu_2}{2}\right) \\
&\quad \cdot y_1^{1+\frac{1}{2}(\nu_1-\nu_2)} y_2^{1+\nu_1-\nu_2} q^{-1-\frac{3}{2}\nu_1+\frac{3\nu_2}{2}} K_{\frac{3\nu_1+3\nu_2-2}{2}}(2\pi q y_1) e(qx_1),
\end{aligned} \tag{2.14}$$

$$\begin{aligned}
&\int_0^1 \int_0^1 E_{12}^1({w_2}^t (Auz)^{-1}, \nu_2, \nu_1) e(-qu_1) du_1 du_3 \\
&= \frac{2}{\xi(3\nu_2)} y_1^{-\nu_1-\frac{\nu_2}{2}+\frac{3}{2}} y_2^{\frac{1}{2}+\frac{\nu_2}{2}+\nu_1} (x_2^2+y_2^2)^{-\frac{3}{4}\nu_2-\frac{3}{2}\nu_1+\frac{1}{4}} q^{-\frac{3}{2}\nu_2-3\nu_1+\frac{1}{2}} \\
&\quad \cdot \sum_{n|q} n^{3\nu_1+3\nu_2-1} \sigma_{1-3\nu_2}(n) S(0,h;qn^{-1}) e(qx_1) \\
&\quad \cdot \int_{-\infty}^{\infty} (u_3^2+1)^{-\frac{3}{4}\nu_2-\frac{3}{2}\nu_1+\frac{1}{4}} K_{\frac{3\nu_2-1}{2}}\left(\frac{2\pi q y_1 y_2 \sqrt{u_3^2+1}}{\sqrt{x_2^2+y_2^2}}\right) \\
&\quad \cdot e\left(\frac{-qx_2u_3y_1}{\sqrt{x_2^2+y_2^2}}\right) du_3,
\end{aligned} \tag{2.15}$$

$$\begin{aligned}
&\int_0^1 \int_0^1 E_{12}^2({w_2}^t (Auz)^{-1}, \nu_2, \nu_1) e(-qu_1) du_1 du_3 \\
&= \frac{2c(\frac{3\nu_2}{2})c(\frac{3\nu_1+3\nu_2-1}{2})}{\xi(3\nu_1)} y_1^{\nu_2+\frac{\nu_1}{2}+\frac{1}{2}} y_2^{\frac{3}{2}-\nu_2-\frac{\nu_1}{2}} (x_2^2+y_2^2)^{\frac{3}{2}\nu_2+\frac{3}{4}\nu_1-\frac{5}{4}} q^{-\frac{5}{2}+3\nu_2+\frac{3}{2}\nu_1} \\
&\quad \cdot \sum_{n|q} n^{-3\nu_2+2} \sigma_{1-3\nu_1}(n) S(0,h;qn^{-1}) e(qx_1) \\
&\quad \cdot \int_{-\infty}^{\infty} (u_3^2+1)^{\frac{3}{2}\nu_2+\frac{3}{4}\nu_1-\frac{5}{4}} K_{\frac{3\nu_1-1}{2}}\left(\frac{2\pi q y_1 y_2 \sqrt{u_3^2+1}}{\sqrt{x_2^2+y_2^2}}\right) e\left(\frac{-qx_2u_3y_1}{\sqrt{x_2^2+y_2^2}}\right) du_3,
\end{aligned} \tag{2.16}$$

$$\begin{aligned}
&\int_0^1 \int_0^1 E_{12}^3({w_2}^t (Auz)^{-1}, \nu_2, \nu_1) e(-qu_1) du_1 du_3 \\
&= \frac{2c(\frac{3\nu_1}{2})}{\xi(3\nu_1+3\nu_2-1)} y_1^{1-\frac{\nu_2}{2}+\frac{\nu_1}{2}} y_2^{1+\frac{\nu_2}{2}-\frac{\nu_1}{2}} (x_2^2+y_2^2)^{\frac{3}{4}\nu_1-\frac{3}{4}\nu_2-\frac{1}{2}} q^{-1-\frac{3}{2}\nu_2+\frac{3}{2}\nu_1} \\
&\quad \cdot \sum_{n|q} n^{3\nu_2} \sigma_{-3\nu_1-3\nu_2+2}(n) S(0,h;qn^{-1}) e(qx_1) \\
&\quad \cdot \int_{-\infty}^{\infty} (u_3^2+1)^{-\frac{3}{4}\nu_2+\frac{3}{4}\nu_1-\frac{1}{2}} K_{\frac{3\nu_1+3\nu_2-2}{2}}\left(\frac{2\pi q y_1 y_2 \sqrt{u_3^2+1}}{\sqrt{x_2^2+y_2^2}}\right) \\
&\quad \cdot e\left(\frac{-qx_2u_3y_1}{\sqrt{x_2^2+y_2^2}}\right) du_3.
\end{aligned} \tag{2.17}$$

Proof. Since

$$Auz = \begin{pmatrix} y_1'y_2' & y_1'x_2' & x_3' \\ & y_1' & x_1' \\ & & 1 \end{pmatrix} \quad \mathrm{mod}(O(3,\mathbb{R})\mathbb{R}^\times) \tag{2.18}$$

with

$$z_2' = x_2' + iy_2' = \frac{z_2}{hq^{-1}z_2 + 1},$$

$$y_1' = |hq^{-1}z_2 + 1|y_1, \quad x_1' = hq^{-1}(x_3 + u_3) + x_1 + u_1,$$

and

$$x_3' = x_3 + u_3,$$

together with a simple fact

$$\int_0^1 e(-qu_1)du_1 = 0, \tag{2.19}$$

(2.8) and (2.10) follow easily.

Similarly, due to

$$w_2 \cdot \left({}^t(Auz)^{-1}\right) = \begin{pmatrix} 1 & -u_3 - x_3 & -u_1 - x_1 + \dfrac{x_2(u_3+x_3)}{x_2^2+y_2^2} \\ & 1 & -\dfrac{h}{q} - \dfrac{x_2}{x_2^2+y_2^2} \\ & & 1 \end{pmatrix} \tag{2.20}$$

$$\cdot \begin{pmatrix} \dfrac{y_1y_2}{\sqrt{x_2^2+y_2^2}} & & \\ & \dfrac{y_2}{x_2^2+y_2^2} & \\ & & 1 \end{pmatrix} \mathrm{mod}(O(3,\mathbb{R})\mathbb{R}^\times),$$

and (2.19), (2.9) and (2.11) are obviously true.

Now we are going to prove (2.12): since

$$\int_0^1 e(-qu_1 + du_1n)du_1 = 0$$

unless $q = dn$, i.e., $d = qn^{-1}$. So

$$\int_0^1 e\big((cn + dnhq^{-1})u_3\big)du_3 = \int_0^1 e\big((cn+h)u_3\big)du_3$$

which is 0 unless $cn = -h$. (2.12) follows easily.

(2.13) and (2.14) hold for the similar reasons.

Due to (2.20),

$$\int_0^1\int_0^1 E_{12}^1(w_2{}^t(Auz)^{-1},\nu_2,\nu_1)e(-qu_1)du_1du_3$$

$$= \int_0^1\int_0^1 \frac{2\big(y_1y_2^2(x_2^2+y_2^2)^{-\frac{3}{2}}\big)^{\frac{1}{4}+\frac{1}{4}\nu_2+\frac{\nu_1}{2}}}{\xi(3\nu_2)} \sum_{(c,d)=1}\sum_{n\geqslant 1} n^{\frac{3}{2}\nu_2-\frac{1}{2}}\sigma_{1-3\nu_2}(n) \tag{2.21}$$

$$\cdot\left(\frac{y_1\sqrt{x_2^2+y_2^2}}{|cz_2^*+d|^2}\right)^{\frac{3}{4}\nu_2+\frac{3}{2}\nu_1-\frac{1}{4}} K_{\frac{3\nu_2-1}{2}}\left(\frac{2\pi n y_2}{x_2^2+y_2^2}|cz_2^*+d|^2\right)e\left(nc(-u_1-x_1)\right)$$

$$\cdot e\left(\frac{ncx_2(u_3+x_3)}{x_2^2+y_2^2}+nd\left(-hq^{-1}-\frac{x_2}{x_2^2+y_2^2}\right)-qu_1\right)du_1du_3,$$

where

$$z_2^* = x_2^*+iy_2^* = -u_3-x_3+iy_1\sqrt{x_2^2+y_2^2}.$$

Since

$$\int_0^1 e\big((-nc-q)u_1\big)du_1 = 0$$

unless $nc=-q$, i.e., $c=-qn^{-1}$. Write $d=lc+r$ with $l\in\mathbb{Z}$ and $1\leqslant r<|c|$ with $(r,c)=1$. Next we change variables $u_3| \to u_3+l-\dfrac{rn}{q}$, then (2.21) becomes

$$\frac{2}{\xi(3\nu_2)}y_1^{\nu_2+2\nu_1}y_2^{\frac{1}{2}+\frac{\nu_2}{2}+\nu_1}(x_2^2+y_2^2)^{-\frac{1}{2}}\sum_{n|q} n^{\frac{3}{2}\nu_2-\frac{1}{2}}\sigma_{1-3\nu_2}(n)$$

$$\cdot e(qx_1)\sum_{l\in\mathbb{Z}}\sum_{\substack{r(\mathrm{mod}\ qn^{-1})\\ (r,qn^{-1})=1}} e\left(\frac{-rh}{qn^{-1}}\right)$$

$$\cdot\int_{-l+\frac{rn}{q}}^{1-l+\frac{rn}{q}} e\left(\frac{-qx_2(u_3+x_3)}{x_2^2+y_2^2}\right)K_{\frac{3\nu_2-1}{2}}\left(\frac{2\pi n y_2}{x_2^2+y_2^2}|cz_2^*|\right)|cz_2^*|^{-\frac{3}{2}\nu_2-3\nu_1+\frac{1}{2}}du_3$$

with

$$z_2^* = -u_3-x_3+iy_1\sqrt{x_2^2+y_2^2}.$$

Changing variables $u_3| \to u_3y_1\sqrt{x_2^2+y_2^2}-x_3$, the above becomes

$$\frac{2}{\xi(3\nu_2)}y_1^{\nu_2+2\nu_1}y_2^{\frac{1}{2}+\frac{\nu_2}{2}+\nu_1}(x_2^2+y_2^2)^{-\frac{1}{2}}\sum_{n|q} n^{\frac{3}{2}\nu_2-\frac{1}{2}}\sigma_{1-3\nu_2}(n)e(qx_1)$$

$$\cdot S(0,h;qn^{-1})y_1\sqrt{x_2^2+y_2^2}\int_{-\infty}^{\infty} e\left(\frac{-qx_2u_3y_1}{\sqrt{x_2^2+y_2^2}}\right)K_{\frac{3\nu_2-1}{2}}\left(\frac{2\pi q y_1 y_2\sqrt{u_3^2+1}}{\sqrt{x_2^2+y_2^2}}\right)$$

$$\cdot(qn^{-1}y_1\sqrt{x_2^2+y_2^2}\sqrt{u_3^2+1})^{-\frac{3}{2}\nu_2-3\nu_1+\frac{1}{2}}du_3$$

which implies (2.15).

(2.16) and (2.17) can be proved in the same manner. This finishes the proof of the lemma. □

Now for $k = 0, 1$, we define

$$F_k(y, \nu_1, \nu_2) := \left(\frac{\partial}{\partial x_2}\right)^k \int_0^1 \int_0^1 E_{13}(Auz, \nu_1, \nu_2) e(-qu_1) du_1 du_3 \Bigg|_{x_1=x_2=0}.$$

It is easy to see from the definition of $E_{13}(Auz, \nu_1, \nu_2)$ that $F_k(y, \nu_1, \nu_2)$ has rapid decay as $y_2 \to \infty$, so for $\Re w_2$ large, we can define

$$L_k(y_1, w_2, \nu) := \int_0^1 F_k(y, \nu_1, \nu_2) y_2^{w_2 - 1} \frac{dy_2}{y_2}.$$

They have the following properties:

Lemma 2.3. *As a function of w_2, $L_1(y_1, w_2, \nu)$ is entire; $L_0(y_1, w_2, \nu)$ has a meromorphic continuation to the whole plane with simple poles at $w_2 = 2 - \nu_1 - 2\nu_2, 2\nu_1 + \nu_2, 1 + \nu_2 - \nu_1, 1 - \nu_1 - 2\nu_2, 2\nu_1 + \nu_2 - 1$ and $\nu_2 - \nu_1$.*

Proof. Notice that

$$L_k(y_1, w_2, \nu) = \int_0^1 F_k(y, \nu_1, \nu_2) y_2^{w_2 - 1} \frac{dy_2}{y_2} + \int_1^\infty F_k(y, \nu_1, \nu_2) y_2^{w_2 - 1} \frac{dy_2}{y_2}$$

and $\displaystyle\int_1^\infty F_k(y, \nu_1, \nu_2) y_2^{w_2 - 1} \frac{dy_2}{y_2}$ is entire in w_2 it is enough to consider

$$\int_0^1 F_k(y, \nu_1, \nu_2) y_2^{w_2 - 1} \frac{dy_2}{y_2}.$$

By (2.5), we can write it as

$$\int_0^1 F_k(y, \nu_1, \nu_2) y_2^{w_2 - 1} \frac{dy_2}{y_2} = \int_0^1 \left(F_{k,1}(y, \nu_1, \nu_2) + F_{k,2}(y, \nu_1, \nu_2)\right) y_2^{w_2 - 1} \frac{dy_2}{y_2},$$

where

$$\begin{aligned} &F_{k,1}(y, \nu_1, \nu_2) \\ &:= \left(\frac{\partial}{\partial x_2}\right)^k \int_0^1 \int_0^1 \Big[E(w_2{}^t(Auz)^{-1}, \tilde{\nu}) - E_0(w_2{}^t(Auz)^{-1}, \tilde{\nu}) \\ &-E_{11}(w_2{}^t(Auz)^{-1}, \tilde{\nu}) - E_{12}(w_2{}^t(Auz)^{-1}, \tilde{\nu})\Big] e(-qu_1) du_1 du_3 \Bigg|_{x_1=x_2=0} \end{aligned}$$

and

$$\begin{aligned} &F_{k,2}(y, \nu_1, \nu_2) \\ &:= \left(\frac{\partial}{\partial x_2}\right)^k \int_0^1 \int_0^1 \Big[E_0(w_2{}^t(Auz)^{-1}, \tilde{\nu}) + E_{11}(w_2{}^t(Auz)^{-1}, \tilde{\nu}) \\ &+E_{12}(w_2{}^t(Auz)^{-1}, \tilde{\nu}) - E_0(Auz, \nu) \\ &-E_{11}(Auz, \nu) - E_{12}(Auz, \nu)\Big] e(-qu_1) du_1 du_3 \Bigg|_{x_1=x_2=0} \end{aligned}$$

which is equal to

$$\left(\frac{\partial}{\partial x_2}\right)^k \int_0^1\int_0^1 \left[E_{12}({w_2}^t(Auz)^{-1},\tilde{\nu}) - E_{12}(Auz,\nu)\right] e(-qu_1)du_1du_3\Bigg|_{x_1=x_2=0}$$

by Lemma 2.2.

Obviously

$$F_{k,1}(y,\nu_1,\nu_2) = \left(\frac{\partial}{\partial x_2}\right)^k \int_0^1\int_0^1 E_{13}({w_2}^t(Auz)^{-1},\tilde{\nu})e(-qu_1)du_1du_3\Bigg|_{x_1=x_2=0},$$

so it has rapid decay as $y_2 \to 0$. It follows that $\int_0^1 F_{k,1}(y,\nu_1,\nu_2)y_2^{w_2-1}\frac{dy_2}{y_2}$ is holomorphic for all $w_2 \in \mathbb{C}$.

For $\Re w_2$ large, set

$$L_{0,2}(y_1,w_2,\nu) := \int_0^1 F_{0,2}(y,\nu_1,\nu_2)y_2^{w_2-1}\frac{dy_2}{y_2}.$$

Due to Lemma 2.2, we have

$$\begin{aligned} &L_{0,2}(y_1,w_2,\nu)\\ &= G_1(y_1,w_2,\nu)\frac{1}{w_2+2\nu_2+\nu_1-2} + G_2(y_1,w_2,\nu)\frac{1}{w_2-\nu_2-2\nu_1}\\ &+G_3(y_1,w_2,\nu)\frac{1}{w_2+\nu_1-\nu_2-1} + G_4(y_1,w_2,\nu)\frac{1}{w_2+\nu_1+2\nu_2-1}\\ &+G_5(y_1,w_2,\nu)\frac{1}{w_2-2\nu_1-\nu_2+1} + G_6(y_1,w_2,\nu)\frac{1}{w_2+\nu_1-\nu_2}, \end{aligned} \tag{2.22}$$

where

$$\begin{aligned} G_1(y_1,w_2,\nu) = {} & \frac{2}{\xi(3\nu_1)}y_1^{\nu_2+\frac{\nu_1}{2}+\frac{1}{2}}q^{3\nu_2+\frac{3}{2}\nu_1-\frac{5}{2}}c\left(\frac{3\nu_2}{2}\right)c\left(\frac{3\nu_1+3\nu_2-1}{2}\right)\\ &\cdot\sum_{n|q} n^{-3\nu_2+2}\sigma_{1-3\nu_1}(n)S(0,h;qn^{-1})\\ &\cdot\int_{-\infty}^{\infty}(u_3^2+1)^{\frac{3}{2}\nu_2+\frac{3}{4}\nu_1-\frac{5}{4}}K_{\frac{3\nu_1-1}{2}}\left(2\pi q y_1\sqrt{u_3^2+1}\right)du_3, \end{aligned}$$

$$\begin{aligned} G_2(y_1,w_2,\nu) = {} & \frac{2}{\xi(3\nu_2)}y_1^{-\nu_1-\frac{\nu_2}{2}+\frac{3}{2}}q^{-3\nu_1-\frac{3}{2}\nu_2+\frac{1}{2}}\\ &\cdot\sum_{n|q} n^{3\nu_1+3\nu_2-1}\sigma_{1-3\nu_2}(n)S(0,h;qn^{-1})\\ &\cdot\int_{-\infty}^{\infty}(u_3^2+1)^{-\frac{3}{4}\nu_2-\frac{3}{2}\nu_1+\frac{1}{4}}K_{\frac{3\nu_2-1}{2}}\left(2\pi q y_1\sqrt{u_3^2+1}\right)du_3, \end{aligned}$$

$$G_3(y_1,w_2,\nu) = \frac{2c(\frac{3\nu_1}{2})}{\xi(3\nu_1+3\nu_2-1)} y_1^{\frac{\nu_1}{2}-\frac{\nu_2}{2}+1} q^{-1-\frac{3}{2}\nu_2+\frac{3\nu_1}{2}}$$
$$\cdot \sum_{n|q} n^{3\nu_2}\sigma_{-3\nu_1-3\nu_2+2}(n) S(0,h;qn^{-1})$$
$$\cdot \int_{-\infty}^{\infty} (u_3^2+1)^{-\frac{3}{4}\nu_2+\frac{3\nu_1}{4}-\frac{1}{2}} K_{\frac{3\nu_1+3\nu_2-2}{2}}\left(2\pi q y_1\sqrt{u_3^2+1}\right) du_3,$$

$$G_4(y_1,w_2,\nu) = -\frac{2}{\xi(3\nu_1)} y_1^{\frac{1}{2}+\frac{\nu_1}{2}+\nu_2} q^{-\frac{3}{2}\nu_1-3\nu_2+\frac{1}{2}} K_{\frac{3\nu_1-1}{2}}(2\pi q y_1),$$

$$G_5(y_1,w_2,\nu) = -\frac{2c(\frac{3\nu_1}{2})c\left(\frac{3\nu_1+3\nu_2-1}{2}\right)}{\xi(3\nu_2)} y_1^{\frac{3}{2}-\nu_1-\frac{\nu_2}{2}} q^{-\frac{5}{2}+3\nu_1+\frac{3}{2}\nu_2} K_{\frac{3\nu_2-1}{2}}(2\pi q y_1),$$

$$G_6(y_1,w_2,\nu) = -\frac{2c\left(\frac{3\nu_2}{2}\right)}{\xi(3\nu_1+3\nu_2-1)} y_1^{1+\frac{\nu_1}{2}-\frac{\nu_2}{2}} q^{-1-\frac{3}{2}\nu_1+\frac{3}{2}\nu_2} K_{\frac{3\nu_1+3\nu_2-2}{2}}(2\pi q y_1).$$

The meromorphic continuation of $L_0(y_1,w_2,\nu)$ with poles at $w_2 = 2\nu_1+\nu_2, 2-\nu_1-2\nu_2, 1+\nu_2-\nu_1, 1-\nu_1-2\nu_2, 2\nu_1+\nu_2-1$ and $\nu_2-\nu_1$ follows from the above expression.

By (2.15), we have

$$\frac{\partial}{\partial x_2}\int_0^1\int_0^1 E_{12}^1(w_2{}^t(Auz)^{-1},\tilde{\nu})e(-qu_1)du_1du_3\Bigg|_{x_1=x_2=0}$$
$$= \frac{2}{\xi(3\nu_2)} y_1^{-\nu_1-\frac{\nu_2}{2}+\frac{3}{2}} y_2^{-\nu_2-2\nu_1+1} q^{-3\nu_1-\frac{3}{2}\nu_2+\frac{1}{2}} \tag{2.23}$$
$$\cdot \sum_{n|q} n^{3\nu_1+3\nu_2-1}\sigma_{1-3\nu_2}(n)S(0,h;qn^{-1})$$
$$\cdot \int_{-\infty}^{\infty}(u_3^2+1)^{-\frac{3}{4}\nu_2-\frac{3}{2}\nu_1+\frac{1}{4}} K_{\frac{3\nu_2-1}{2}}\left(\frac{2\pi q y_1 y_2\sqrt{u_3^2+1}}{\sqrt{x_2^2+y_2^2}}\right)$$
$$\cdot(-q)u_3y_1y_2^{-1}du_3$$

which is 0 because the integrated function in the above integral is an odd function of u_3.

Similarly one can show that

$$\frac{\partial}{\partial x_2}\int_0^1\int_0^1 E_{12}^2(w_2{}^t(Auz)^{-1},\tilde{\nu})e(-qu_1)du_1du_3\Bigg|_{x_1=x_2=0} = 0 \tag{2.24}$$

and

$$\frac{\partial}{\partial x_2}\int_0^1\int_0^1 E_{12}^3(w_2{}^t(Auz)^{-1},\tilde{\nu})e(-qu_1)du_1du_3\Bigg|_{x_1=x_2=0} = 0. \tag{2.25}$$

It is obvious from (2.12), (2.13) and (2.14) that

$$\frac{\partial}{\partial x_2}\int_0^1\int_0^1 E_{12}^1(Auz,\nu)e(-qu_1)du_1du_3 = 0, \tag{2.26}$$

$$\frac{\partial}{\partial x_2}\int_0^1\int_0^1 E_{12}^2(Auz,\nu)e(-qu_1)du_1du_3=0, \tag{2.27}$$

$$\frac{\partial}{\partial x_2}\int_0^1\int_0^1 E_{12}^3(Auz,\nu)e(-qu_1)du_1du_3=0. \tag{2.28}$$

Combining (2.26), (2.27), (2.28), (2.29), (2.30) and (2.31), one concludes that

$$L_{12}(y_1,w_2,\nu):=\int_0^1 F_{1,2}(y,\nu)y_2^{w_2-1}\frac{dy_2}{y_2}=0. \tag{2.29}$$

Hence

$$L_1(y,w_2,\nu)=\int_0^1 F_{1,1}(y,\nu)y_2^{w_2-1}\frac{dy_2}{y_2}+\int_1^\infty F_1(y,\nu)y_2^{w_2-1}\frac{dy_2}{y_2}$$

is an entire function of w_2. This finishes the proof of the lemma. □

Now for $\Re w_1$ large and fixed, set

$$\Lambda_k(w_1,w_2,\nu):=\int_0^\infty L_k(y_1,w_2,\nu)y_1^{w_1-1}\frac{dy_1}{y_1}$$

which inherits the same analytic properties of $L_k(y_1,w_2,\nu)$.

Lemma 2.4. *For $k=0,1$, $\Re w_2$ large, we have*

$$\Lambda_k(w_1,w_2,\nu)=\frac{1}{q^{w_1-2w_2+1}}\sum_{m\neq 0}\frac{A(1,m,\nu)}{|m|^{w_2}}\left(\frac{2\pi i m}{q^2}\right)^k e\left(\frac{m\bar{h}}{q}\right)G_1(w_1,w_2,\nu)$$

with $h\bar{h}\equiv 1(\mathrm{mod}\ q)$,

$$A(n,m,\nu)=\frac{2}{\xi(3\nu_1)\xi(3\nu_2)\xi(3\nu_1+3\nu_2-1)}\sum_{n_1|n}\sum_{n_2|\frac{n}{n_1}}n^{-1+\nu_1+2\nu_2}$$
$$\cdot|m|^{-1+2\nu_1+\nu_2}n_1^{-3\nu_1-3\nu_2+2}n_2^{1-3\nu_2}\sigma_{1-3\nu_1,2-3\nu_1-3\nu_2}\left(\frac{n}{n_1n_2},|m|\right)$$

and

$$G_1(w_1,w_2,\nu)=\int_0^\infty\int_0^\infty W_{\text{Jacquet}}^*(\text{diag}(y_1y_2,y_1,1),\nu)y_1^{w_1-1}y_2^{w_2-1}\frac{dy_1}{y_1}\frac{dy_2}{y_2}.$$

Remark 2.1. By [1, (10.1)], we know $G_1(w_1,w_2,\nu)$ has the following simple formula:

$$G_1(w_1,w_2,\nu)=\frac{\pi^{-w_1-w_2}}{4}\frac{\Gamma(\frac{w_1+\alpha}{2})\Gamma(\frac{w_1+\beta}{2})\Gamma(\frac{w_1+\gamma}{2})\Gamma(\frac{w_2-\alpha}{2})\Gamma(\frac{w_2-\beta}{2})\Gamma(\frac{w_2-\gamma}{2})}{\Gamma(\frac{w_1+w_2}{2})},$$

for $\Re w_1,\Re w_2$ being large, recall α,β,γ are defined by (2.7).

Proof. A simple computation shows that

$$\begin{pmatrix} a & b \\ c & d \end{pmatrix}\begin{pmatrix} 1 & 0 \\ h/q & 1 \end{pmatrix} = \begin{pmatrix} a+bh/q & b \\ c+dh/q & d \end{pmatrix} = \begin{pmatrix} a' & b \\ c' & d \end{pmatrix}.$$

Applying this to $E_{13}(z,\nu)$ in Lemma 2.1, we have

$$\begin{aligned} &\int_0^1\int_0^1 E_{13}(Auz,\nu)e(-qu_1)du_1du_3 \\ &= \sum_{(c,d)=1}\sum_{n=1}^{\infty}\sum_{m\neq 0}\frac{A(n,m,\nu_1,\nu_2)}{n|m|} \qquad (2.30) \\ &\cdot\int_0^1\int_0^1 e\big(n(c'(x_3+u_3)+d'(x_1+u_1))\big)e(-qu_1)e\left(m\Re\frac{a'z_2+b}{c'z_2+d}\right) \\ &\cdot W^*_{\text{Jacquet}}\big(\text{diag}(n|m|y_1y_2/|c'z_2+d|, ny_1|c'z_2+d|, 1),\nu\big)du_1du_3. \end{aligned}$$

Because of the simple fact

$$\int_0^1 e(x\alpha)dx = \begin{cases} 1, & \text{if } \alpha = 0, \\ 0, & \text{otherwise,} \end{cases}$$

the integrals over u_1, u_3 are zero unless $c = -\dfrac{dh}{q}$, $nd = q$. It follows that $n = 1, d = q$ and $c = -h$. The conclusion of the lemma is a direct consequence of the above fact. □

Lemma 2.5. *For $k = 0, 1$, $-\Re w_2$ large, we have*

$$\begin{aligned} \Lambda_k(w_1,w_2,\nu) = &\frac{(-2\pi i q)^k \pi^{\frac{w_1+w_2}{2}}}{q^{w_1+w_2}\Gamma(\frac{w_1+w_2}{2})}\sum_{n|q}\sum_{m\neq 0}\frac{A(n,m,\nu_2,\nu_1)}{n^{2k+1-2w_2}|m_2|^{k+1-w_2}}\left(\frac{im}{|m|}\right)^k \\ &\cdot S(h,m;qn^{-1})G_2(w_2,\nu,k) \end{aligned}$$

with

$$\begin{aligned} G_2(w_1,w_2,\tilde{\nu}) = &\int_0^{\infty}\int_0^{\infty} W^*_{\text{Jacquet}}(y,\tilde{\nu})K_{\frac{w_1+w_2-1-2k}{2}}(2\pi y_2) \\ &\cdot y_1^{2k+w_1-w_2} y_2^{\frac{2k+w_1-w_2-1}{2}}\frac{dy_1}{y_1}\frac{dy_2}{y_2}, \end{aligned}$$

and $S(h,m;c)$ is the Kloosterman sum as in (1.2).

Remark 2.2. By [16, p.357] (or [2]), we have

$$\begin{aligned} &G_2(w_1,w_2,\tilde{\nu}) \\ &= \frac{1}{4}\pi^{\frac{-3w_1-6k+3w_2-3}{2}}\Gamma\left(\frac{w_1+\alpha}{2}\right)\Gamma\left(\frac{w_1+\beta}{2}\right)\Gamma\left(\frac{w_1+\gamma}{2}\right) \\ &\cdot\Gamma\left(\frac{1-w_2+2k+\alpha}{2}\right)\Gamma\left(\frac{1-w_2+2k+\beta}{2}\right)\Gamma\left(\frac{1-w_2+2k+\gamma}{2}\right). \end{aligned}$$

Proof. For $k = 0, 1, -\Re w_2$ large, making use of Lemma 2.2 as in the proof of Lemma 2.3, the computation shows that

$$\int_1^\infty F_{k,2}(y,\nu_1,\nu_2)y_2^{w_2-1}\frac{dy_2}{y_2} = -L_{k,2}(y_1,w_2,\nu),$$

recall that $F_{k,2}(y,\nu_1,\nu_2)$ and $L_{k,2}(y_1,w_2,\nu)$ are both defined in the proof of Lemma 2.3. It follows that for $-\Re w_2$ large,

$$\begin{aligned}\Lambda_k(w_1,w_2,\nu) = &\int_0^\infty\int_0^1 F_{k,1}(y,\nu_1,\nu_2)y_1^{w_1-1}y_2^{w_2-1}\frac{dy_1}{y_1}\frac{dy_2}{y_2}\\ &+\int_0^\infty L_{k,2}(y_1,w_2,\nu)y_1^{w_1-1}\frac{dy_1}{y_1}\\ &+\int_0^\infty\int_1^\infty F_{k,1}(y,\nu_1,\nu_2)y_1^{w_1-1}y_2^{w_2-1}\frac{dy_1}{y_1}\frac{dy_2}{y_2}\\ &+\int_0^\infty\int_1^\infty F_{k,2}(y,\nu_1,\nu_2)y_1^{w_1-1}y_2^{w_2-1}\frac{dy_1}{y_1}\frac{dy_2}{y_2}\\ = &\int_0^\infty\int_0^\infty F_{k,1}(y,\nu_1,\nu_2)y_1^{w_1-1}y_2^{w_2-1}\frac{dy_1}{y_1}\frac{dy_2}{y_2}.\end{aligned}$$

Recall that

$$F_{k,1}(y,\nu_1,\nu_2) = \left(\frac{\partial}{\partial x_2}\right)^k\int_0^1\int_0^1 E_{13}(w_2{}^t(Auz)^{-1},\tilde\nu)e(-qu_1)du_1du_3\Bigg|_{x_1=x_2=0}.$$

By the definition of $E_{13}(z,\tilde\nu)$ in Lemma 2.1 and the indentity (2.20), we have

$$\begin{aligned}&\int_0^1\int_0^1 E_{13}(w_2{}^t(Auz)^{-1},\tilde\nu)e(-qu_1)du_1du_3\\ =&\sum_{(c,d)=1}\sum_{n=1}^\infty\sum_{m\neq0}\frac{A(n,m,\tilde\nu)}{n|m|}\\ &\cdot\int_0^1\int_0^1 e\left(nc\left(-u_1-x_1+\frac{x_2(u_3+x_3)}{x_2^2+y_2^2}\right)\right)\\ &\cdot e\left(-nd\left(\frac{h}{q}+\frac{x_2}{x_2^2+y_2^2}\right)-qu_1+m\Re\frac{az_2'+b}{cz_2'+d}\right)\\ &\cdot W^*_{\text{Jacquet}}\left(\operatorname{diag}\left(\frac{n|m|y_1y_2}{|cz_2'+d|},\frac{ny_2|cz_2'+d|}{x_2^2+y_2^2},1\right),\tilde\nu\right)du_1du_3.\end{aligned}\tag{2.31}$$

Here

$$z_2' = -u_3-x_3+iy_1\sqrt{x_2^2+y_2^2}.\tag{2.32}$$

For the same reason as before, the u_1 intgeral disappears unless $nc = -q$. Note that $\dfrac{az+b}{cz+d} = \dfrac{a}{c}-\dfrac{1}{c(cz+d)}$. Let $d = lc+r$ with $l\in\mathbb{Z}$ and $1\leqslant r<|c|$ with

$(r,c)=1$. After changing variables $u_3 \to u_3 + l - \dfrac{rn}{q}$ in formula (2.36), it follows that the right side of (2.36) becomes

$$e(qx_1)\sum_{n|q}\sum_{m\neq 0}\frac{A(n,m,\tilde{\nu})}{n|m|}S(h,m;qn^{-1}) \\ \cdot\sum_{l\in\mathbb{Z}}\int_{l-\frac{rn}{q}}^{1+l-\frac{rn}{q}} e\left(\frac{-qx_2(u_3+x_3)}{x_2^2+y_2^2}+m\Re\left(\frac{-1}{c^2z_2'}\right)\right) \\ \cdot W^*_{\text{Jacquet}}\left(\operatorname{diag}\left(\frac{n|m|y_1y_2}{\sqrt{x_2^2+y_2^2}|cz_2'|},\frac{ny_2|cz_2'|}{x_2^2+y_2^2},1\right),\tilde{\nu}\right)du_3. \tag{2.33}$$

Making succesive transformtaions

$$u_3 \to u_3 - x_3, \qquad u_3 \to u_3y_1\sqrt{x_2^2+y_2^2},$$

the above becomes

$$e(qx_1)\sum_{n|q}\sum_{m\neq 0}\frac{A(n,m,\tilde{\nu})}{n|m|}S(h,m;qn^{-1}) \\ \cdot\int_{-\infty}^{\infty} e\left(\frac{-qx_2y_1u_3}{\sqrt{x_2^2+y_2^2}}+\frac{n^2mu_3}{q^2y_1\sqrt{x_2^2+y_2^2}(u_3^2+1)}\right) \\ \cdot W^*_{\text{Jacquet}}\left(\operatorname{diag}\left(\frac{n^2|m|y_2}{q(x_2^2+y_2^2)\sqrt{u_3^2+1}},\frac{qy_1y_2\sqrt{u_3^2+1}}{\sqrt{x_2^2+y_2^2}},1\right),\tilde{\nu}\right)y_1\sqrt{x_2^2+y_2^2}du_3.$$

Taking partial derivatives with respect to x_2 and setting $x_1 = x_2 = 0$, we have

$$F_{k,1}(y,\nu_1,\nu_2) = \sum_{n|q}\sum_{m\neq 0}\frac{A(n,m,\tilde{\nu})}{n|m|}S(h,m;qn^{-1}) \\ \cdot\int_{-\infty}^{\infty}\left(\frac{-2\pi iqy_1u_3}{y_2}\right)^k e\left(\frac{n^2mu_3}{q^2y_1y_2(u_3^2+1)}\right) \\ \cdot W^*_{\text{Jacquet}}\left(\operatorname{diag}\left(\frac{n^2|m|}{qy_2\sqrt{u_3^2+1}},qy_1\sqrt{u_3^2+1},1\right),\tilde{\nu}\right)y_1y_2du_3. \tag{2.34}$$

By making the transformations

$$y_1 \to \frac{y_1}{q\sqrt{u_3^2+1}}, \qquad y_2 \to \frac{n^2|m|}{q\sqrt{u_3^2+1}y_1y_2},$$

and invoking the following formulas for $\Re s > 0$ (see [18, p.172, p.79]):

$$\int_{-\infty}^{\infty} e(uy_2)(u^2+1)^{-s}du = \frac{2\pi^s}{\Gamma(s)}|y_2|^{s-\frac{1}{2}}K_{s-\frac{1}{2}}(2\pi|y_2|)$$

and

$$\int_{-\infty}^{\infty} e(uy_2)(u^2+1)^{-s}udu = 2\left(i\frac{y_2}{|y_2|}\right)\frac{\pi^s}{\Gamma(s)}|y_2|^{s-\frac{1}{2}}K_{s-\frac{3}{2}}(2\pi|y_2|),$$

we obtain our conclusion as claimed. □

For $k=0,1$, $\Re w_2$ being large, we define

$$\begin{aligned} L_k(\bar{h},q,w_2) &:= q^{w_1-2w_2+1+2k}(2\pi i)^{-k}G_1^{-1}(w_1,w_2,\nu)\Lambda_k(w_1,w_2,\nu) \\ &= \sum_{m>0}\frac{A(1,m,\nu)}{m^{w_2-k}}\left[e\left(\frac{m\bar{h}}{q}\right)+(-1)^k e\left(\frac{-m\bar{h}}{q}\right)\right]. \end{aligned}$$

Then from the above arguments, $L_1(\bar{h},q,w_2)$ has analytic continuation to the whole plane while $L_0(\bar{h},q,w_2)$ has meromorphic continuation to the whole plane with poles at $w_2=-\nu_1-2\nu_2+2, 2\nu_1+\nu_2, \nu_2-\nu_1+1, -\nu_1-2\nu_2+1, 2\nu_1+\nu_2-1$ and $\nu_2-\nu_1$. We assume that all the poles are different, otherwise we can use continuity argument. Next we will compute the residues of $L_0(\bar{h},q,w_2)$ at the above poles. For that purpose, we need the following formulas:

$$\int_0^{\infty} y^s K_\nu(y)\frac{dy}{y} = 2^{s-2}\Gamma\left(\frac{s+\nu}{2}\right)\Gamma\left(\frac{s-\nu}{2}\right)$$

if $\Re s > |\Re\nu|$ (see [7, 6.561, 16]) and

$$\int_{-\infty}^{\infty}(u_3^2+1)^{-w}dw = B\left(\frac{1}{2},w-\frac{1}{2}\right) = \frac{\Gamma(\frac{1}{2})\Gamma(w-\frac{1}{2})}{\Gamma(w)}$$

if $\Re w > \dfrac{1}{2}$ (see [7, 3.194, 3; 8.384, 1]).

Using the above two formulas and (2.22), one can easily obtain the following (all the notations are the same as in Theorem 1.1 and in Lemma 2.1):

I) the residue of $L_0(\bar{h},q,w_2)$ at $w_2=-\nu_1-2\nu_2+2$ is:

$$\begin{aligned} &\frac{2}{\xi(3\nu_1)}q^{6\nu_2+3\nu_1-5}\pi^{-3\nu_2-\frac{3\nu_1}{2}+\frac{5}{2}} \\ &\cdot\sum_{n|q} n^{-3\nu_2+2}S(0,h;qn^{-1})\sigma_{1-3\nu_1}(n) \\ &\cdot c\left(\frac{3\nu_2}{2}\right)c\left(\frac{3\nu_1+3\nu_2-1}{2}\right)\Gamma^{-1}\left(\frac{-3\nu_2+2}{2}\right)\Gamma^{-1}\left(\frac{-3\nu_1-3\nu_2+3}{2}\right); \end{aligned}$$

II) the residue of $L_0(\bar{h},q,w_2)$ at $w_2=2\nu_1+\nu_2$ is:

$$\begin{aligned} &\frac{2}{\xi(3\nu_2)}q^{-6\nu_1-3\nu_2+1}\pi^{3\nu_1+\frac{3\nu_2}{2}-\frac{1}{2}} \\ &\cdot\sum_{n|q} n^{3\nu_1+3\nu_2-1}S(0,h;qn^{-1})\sigma_{1-3\nu_2}(n)\Gamma^{-1}\left(\frac{3\nu_1}{2}\right)\Gamma^{-1}\left(\frac{3\nu_1+3\nu_2-1}{2}\right); \end{aligned}$$

III) the residue of $L_0(\bar{h}, q, w_2)$ at $w_2 = \nu_2 - \nu_1 + 1$ is:

$$\frac{2}{\xi(3\nu_1 + 3\nu_2 - 1)} q^{3\nu_1 - 3\nu_2 - 2} \pi^{\frac{3\nu_2}{2} - \frac{3\nu_1}{2} + 1}$$

$$\cdot \sum_{n|q} n^{3\nu_2} S(0, h; qn^{-1}) \sigma_{2-3\nu_1-3\nu_2}(n) c\left(\frac{3\nu_1}{2}\right) \Gamma^{-1}\left(\frac{3\nu_2}{2}\right) \Gamma^{-1}\left(\frac{-3\nu_1 + 2}{2}\right);$$

IV) the residue of $L_0(\bar{h}, q, w_2)$ at $w_2 = -\nu_1 - 2\nu_2 + 1 = \alpha$ is:

$$\frac{-2\pi^{\frac{3}{2} - \frac{3}{2}\nu_1 - 3\nu_2}}{\xi(3\nu_1)} \frac{1}{\Gamma\left(\frac{\alpha-\beta}{2}\right)\Gamma\left(\frac{\alpha-\gamma}{2}\right)} \frac{1}{\Gamma\left(\frac{w_2-\alpha}{2}\right)}\Bigg|_{w_2=\alpha} = 0.$$

Similarly one can show that the residues of $L_0(\bar{h}, q, w_2)$ at $w_2 = 2\nu_1 + \nu_2 - 1$ and at $w_2 = \nu_2 - \nu_1$ are all 0.

To derive the Voronoi formula, take $\phi(x) \in C_c^\infty(0, \infty)$ with $\tilde{\phi}(s) = \int_0^\infty \phi(x) x^s \frac{dx}{x}$ being its Mellin transform, then for $\sigma > 3 + |\Re\alpha| + |\Re\beta| + |\Re\gamma|$, by Mellin inversion, we have

$$\sum_{m_2>0} A(1, m_2, \nu_1, \nu_2) \left[e\left(\frac{m_2\bar{h}}{q}\right) + (-1)^k e\left(\frac{-m_2\bar{h}}{q}\right)\right] \phi(m_2)$$

$$= \frac{1}{2\pi i} \int_{\Re w_2 = \sigma} \tilde{\phi}(w_2 - k) L_k(\bar{h}, q, w_2) dw_2.$$

Moving the line of integration to $\Re w_2 = -\sigma + 1$, picking up the poles at $w_2 = \nu_2 + 2\nu_1, -2\nu_2 - \nu_1 + 2$ and $-\nu_1 + \nu_2 + 1$ when $k = 0$ and applying Lemma 2.5, we then end up with the following Voronoi formula for the Eisenstein series $E(z, \nu_1, \nu_2)$. Before we state it, we introduce some notations.

For $k = 0, 1, \phi(x) \in C_c^\infty(0, \infty)$, set

$$V_1 = q^{1+3k} \pi^{-3k-\frac{3}{2}} (-i)^k \sum_{n|q} \sum_{m>0} \frac{n^{\nu_2+2\nu_1-2k-2} m^{2\nu_2+2\nu_1-k-2}}{\xi(3\nu_1)\xi(3\nu_2)\xi(3\nu_1 + 3\nu_2 - 1)}$$

$$\cdot \sum_{n_1|n} \sum_{n_2|\frac{n}{n_1}} n_1^{-3\nu_1-3\nu_2+2} n_2^{1-3\nu_1} \sigma_{1-3\nu_2, 2-3\nu_1-3\nu_2}\left(\frac{n}{n_1 n_2}, m\right)$$

$$\cdot \left[S(m, h; qn^{-1}) + (-1)^k S(-m, h; qn^{-1})\right] \Phi_k^*\left(\frac{mn^2}{q^3}\right),$$

where

$$\Phi_k^*(x) = \frac{1}{2\pi i} \int_{\Re w = \sigma} (\pi^3 x)^{-w} \frac{\Gamma(\frac{1+w+2k+\alpha}{2})\Gamma(\frac{1+w+2k+\beta}{2})\Gamma(\frac{1+w+2k+\gamma}{2})}{\Gamma(\frac{-w-\alpha}{2})\Gamma(\frac{-w-\beta}{2})\Gamma(\frac{-w-\gamma}{2})} \tilde{\phi}(-w - k) dw,$$

$$V_2 = \frac{\delta_{k,0}}{\xi(3\nu_2)} q^{-6\nu_1 - 3\nu_2 + 1} \pi^{3\nu_1 + \frac{3}{2}\nu_2 - \frac{1}{2}} \sum_{n|q} n^{3\nu_1+3\nu_2-1} S(0, h; qn^{-1}) \sigma_{1-3\nu_2}(n)$$

$$\cdot \Gamma^{-1}\left(\frac{3\nu_1}{2}\right) \Gamma^{-1}\left(\frac{3\nu_1 + 3\nu_2 - 1}{2}\right) \tilde{\phi}(\nu_2 + 2\nu_1),$$

$$V_3 = \frac{\delta_{k,0}}{\xi(3\nu_1)} q^{6\nu_2+3\nu_1-5}\pi^{-3\nu_2-\frac{3\nu_1}{2}+\frac{5}{2}} \sum_{n|q} n^{-3\nu_2+2} S(0,h;qn^{-1})\sigma_{1-3\nu_1}(n)$$
$$\cdot c\left(\frac{3\nu_2}{2}\right) c\left(\frac{3\nu_1+3\nu_2-1}{2}\right)\Gamma^{-1}\left(\frac{-3\nu_2+2}{2}\right)\Gamma^{-1}\left(\frac{-3\nu_1-3\nu_2+3}{2}\right)$$
$$\cdot\tilde{\phi}(-2\nu_2-\nu_1+2),$$

$$V_4 = \frac{\delta_{k,0}}{\xi(3\nu_1+3\nu_2-1)} q^{3\nu_1-3\nu_2-2}\pi^{\frac{3\nu_2}{2}-\frac{3\nu_1}{2}+1} \sum_{n|q} n^{3\nu_2} S(0,h;qn^{-1})$$
$$\cdot\sigma_{-3\nu_1-3\nu_2+2}(n) c\left(\frac{3\nu_1}{2}\right)\Gamma^{-1}\left(\frac{3\nu_2}{2}\right)\Gamma^{-1}\left(\frac{-3\nu_1+2}{2}\right)\tilde{\phi}(\nu_2-\nu_1+1),$$

in all of the above

$$\delta_{k,0} = \begin{cases} 1, & \text{if } k=0, \\ 0, & \text{otherwise.} \end{cases}$$

Proposition 2.1. *Let $k = 0,1$, and $\phi(x) \in C_c^\infty(0,\infty)$. Let $h,\bar{h},q$ be positive integers with $h\bar{h} \equiv 1 (\mathrm{mod}\ q)$. For all $(\nu_1,\nu_2) \in \mathbb{C}^2$, except those on the complex lines $\nu_1 = \frac{2}{3}, \frac{1}{3}\rho_1$ or $\nu_2 = \frac{2}{3}, \frac{1}{3}\rho_2$ with ρ_i running through the nontrivial zeroes of $\zeta(s)$, we have*

$$\sum_{m>0} \frac{m^{-1+2\nu_1+\nu_2}\sigma_{1-3\nu_1,2-3\nu_1-3\nu_2}(1,m)}{\xi(3\nu_1)\xi(3\nu_2)\xi(3\nu_1+3\nu_2-1)}\left[e\left(\frac{m\bar{h}}{q}\right)+(-1)^k e\left(\frac{-m\bar{h}}{q}\right)\right]\phi(m)$$
$$= V_1+V_2+V_3+V_4.$$

Taking partial derivatives $\frac{\partial^3}{\partial\nu_1\partial\nu_2^2}$ on both sides of the above formula and set $\nu_1=\nu_2=\frac{1}{3}$, we have the pre-Voronoi formula for the triple divisor function $\tau_3(m)$. All the notations are the same as in Theorem 1.1 and in the last proposition.

Proposition 2.2. *Let $k = 0,1$ and $\phi(x) \in C_c^\infty(0,\infty)$. Let $h,\bar{h},q$ be positive integers with $h\bar{h} \equiv 1 (\mathrm{mod}\ q)$. Then we have*

$$\sum_{m>0}\tau_3(m)\left[e\left(\frac{m\bar{h}}{q}\right)+(-1)^k e\left(\frac{-m\bar{h}}{q}\right)\right]\phi(m)$$
$$= (-i)^k \sum_{n|q}\sum_{m>0}\frac{q^{1+3k}\pi^{-3k-\frac{3}{2}}}{n^{2k+1}m^{k+1}}\sum_{n_1|n}\sum_{n_2|\frac{n}{n_1}}\sigma_{0,0}\left(\frac{n}{n_1n_2},m\right)S(m,h;qn^{-1})\Phi_k^*\left(\frac{mn^2}{q^3}\right)$$
$$+i^k \sum_{n|q}\sum_{m>0}\frac{q^{1+3k}\pi^{-3k-\frac{3}{2}}}{n^{2k+1}m^{k+1}}\sum_{n_1|n}\sum_{n_2|\frac{n}{n_1}}\sigma_{0,0}\left(\frac{n}{n_1n_2},m\right)S(-m,h;qn^{-1})\Phi_k^*\left(\frac{mn^2}{q^3}\right)$$
$$+\delta_{k,0}\sum_{n|q}S(0,h;qn^{-1})nq^{-2}\tau(n)\Bigg\{(\log n)^2-5\log q\log n+\frac{9}{2}(\log q)^2+3\gamma^2$$
$$-3\gamma_1+\tau(n)^{-1}\left[\sum_{d|n}\log d(\log n+\log q-5\gamma)-\frac{3}{2}\sum_{d|n}(\log d)^2\right]$$

$$+7\gamma\log n - 9\gamma\log q\Big\}\tilde{\phi}(1) + \delta_{k,0}\sum_{n|q} S(0,h;qn^{-1})nq^{-2}\tau(n)$$

$$\cdot\left[\frac{5}{3}\log n - 3\log q + 3\gamma - \frac{1}{3}\tau(n)^{-1}\sum_{d|n}\log d\right]\tilde{\phi}'(1)$$

$$+\frac{\delta_{k,0}}{2}\sum_{n|q} S(0,h;qn^{-1})nq^{-2}\tau(n)\tilde{\phi}''(1).$$

By combining the two cases $k = 0$ and $k = 1$ in the above proposition, we obtain Theorem 1.1 easily.

Acknowledgments

The author would like to thank Qing Lu for comments. These notes were written in 2006.

References

[1] D. Bump, Automorphic Forms on $GL(3,R)$, Lecture Notes in Mathematics, Vol. 1083, Springer-Verlag, Berlin, 1984.

[2] D. Bump, Barnes' second lemma and its application to Rankin-Selberg convolutions, Amer. J. Math. **110** (1988), No. 1, 179–185.

[3] J. W. Cogdell, Voronoi summation, available online at: `http://www.math.ohio-state.edu/~cogdell/`.

[4] W. Duke, J. B. Friedlander and H. Iwaniec, A quadratic divisor problem, Invent. Math. **115** (1994), No. 2, 209–217.

[5] D. Goldfeld, Automorphic Forms and L-Functions for the Group $GL(n,\mathbb{R})$, Cambridge Studies in Advanced Mathematics, Vol. 99, 2006.

[6] D. Goldfeld and X. Li, Voronoi formulas on $GL(n)$, Int. Math. Res. Not. (2006), Art. ID 86295.

[7] I. S. Gradshteyn and I. M. Ryzhik, Table of Integrals, Series, and Products, translated from the Russian, sixth edition, translation edited and with a preface by A. Jeffrey and D. Zwillinger, Academic Press, Inc., San Diego, CA, 2000.

[8] H. Iwaniec and E. Kowalski, Analytic Number Theory, Amer. Math. Soc. Colloq. Publ., Vol. 53, Amer. Math. Soc., Providence, RI, 2004.

[9] H. Iwaniec and P. Sarnak, Perspectives on the analytic theory of L-functions, Geom. Funct. Anal. 2000, Special Volume, Part II, 705–741.

[10] A. Ivić, On the ternary additive divisor problem and the sixth moment of the zeta-function, in: Sieve Methods, Exponential Sums, and Their Applications in Number Theory, London Math. Soc. Lecture Note Ser., Vol. 237, Cambridge University Press, Cambridge, 1997, 205–243.

[11] H. Jacquet, I. I. Piatetski-Shapiro and J. Shalika, Automorphic forms on $GL(3)$, I & II, Ann. Math. **109** (1979), 169–258.

[12] T. Meurman, On exponential sums involving the Fourier coefficients of Maass wave forms, J. Reine Angew. Math. **384** (1988), 192–207.

[13] S. D. Miller and W. Schmid, Summation formulas, from Poisson and Voronoi to the present, in: Noncommutative Harmonic Analysis, Progr. Math., Vol. 220, Birkäuser Boston, Boston, MA, 2004, 419–440.

[14] S. D. Miller and W. Schmid, Automorphic distributions, L-functions, and Voronoi summation formulas for $GL(3)$, Ann. Math. **164** (2006), 423–488.

[15] P. Sarnak, Estimates for Rankin-Selberg L-functions and quantum unique ergodicity, J. Funct. Anal. **184** (2001), No. 2, 419–453.

[16] E. Stade, On explicit integral formulas for $GL(n, R)$-Whittaker functions, with an appendix by D. Bump, S. Friedberg and J. Hoffstein, Duke Math. J. **60** (1990), No. 2, 313–362.

[17] A. I. Vinogradov and L. A. Takhtadzhyan, Theory of the Eisenstein series for the group $SL(3, R)$ and its application to a binary problem. I. Fourier expansion of the highest Eisenstein series. (Russian) Analytic number theory and the theory of functions, Zap. Nauchn. Sem. Leningrad. Otdel. Mat. Inst. Steklov. (LOMI) **76** (1978), 5–52, 216.

[18] G. N. Watson, A Treatise on the Theory of Bessel Functions, Cambridge University Press, Cambridge, England; The Macmillan Company, New York, 1944.

Automorphic Forms and L-functions
ALM 30, pp. 91–130

Linnik's Ergodic Method and the Hasse Principle for Ternary Quadratic Forms

Philippe Michel*

Abstract

In this lecture series, we explain Linnik's approach to the study of representation of integers by ternary quadratic forms. More precisely we explain how Linnik's ergodic method allows to prove that given a positive definite integral ternary quadratic form Q, any integer, which everywhere locally representable by Q, sufficiently large AND satisfying an additional congruence condition modulo some fixed prime is indeed representable over $\mathbb{Z}$.

2010 Mathematics Subject Classification: 11E12, 11E20, 11F70.

Keywords and Phrases: Quadratic forms, Automorphic forms, Adeles, Ergodic theory.

1 Foreword

This text is an expanded written version of a series of lectures given during the CIMPA summer school on analytic number theory held in Weihai in 2010. The objective of these methods was to present the so-called "ergodic method" on the problem of representing integers by ternary quadratic forms. The method was developped in the 50's by Linnik, Skubenko and their student collaborators and the basic reference is Linnik's book [11]. Recently, investigations by the author together with E. Lindenstrauss, M. Einsiedler and A. Venkatesh have lead us to revisit Linnik's method in the modern language of dynamics on arithmetic homogeneous space; the outcome has been new presentations of the method in related but slightly different contexts ([3] and [6] the later joint also with J. Ellenberg). The present notes are yet another survey of the method in the adelic language to obtain the integral Hasse principle for representation of large integers

*EPFL/SB/IMB/TAN, Station 8, Ecole Polytechnique Fédérale de Lausanne, CH-1015, Lausanne, Suisse, philippe.michel@epfl.ch.

(satisfying additional conditions) by ternary definite quadratic forms. These notes have therefore significant overleap with [3, 6] yet we hope that the repetition of the arguments with some variantions and in a slightly different context will be useful for students. By now, in the context of studying representation of integers by ternary forms, the ergodic method has been superseeded, thanks to breakthrough by Iwaniec [9] and Duke [1], by another approach which builds on harmonic analysis and analytic number theory (see the forthcoming [13] for a general presentation of this approach), however it seem that the Linnik ergodic method have further applications which at the moment seem inaccessible to other techniques (see [6, §1.6 and §3] and [5]).

To conclude this foreword it remains to thank E. Lindenstrauss, M. Einsiedler and A. Venkatesh, without which it would not have been possible for me to understand correctly Linnik's method and to translate it into a modern language, for our past and future collaboration. It would also like to thank the audience of the CIMPA 2010 summer on analytic number theory in Weihai for their attention and Jianya Liu for his invitation to lecture there and for his perfect organization of the event.

2 Integral Quadratic Forms

Let

$$q(x_1,\dots,x_n)=\sum_{i=1}^{n} m_{i,i}x_i^2+\sum_{i<j}2m_{i,j}x_ix_j$$

be an homogeneous polynomial of degree 2 in n variable with rational coefficients (in the sequel we may and will assume that $m_{i,j}=m_{j,i}$). This polynomial defines a quadratic form whose polarization is the symetric bilinear form

$$\langle \mathbf{x},\mathbf{x}'\rangle := q(\mathbf{x}+\mathbf{x}')-q(\mathbf{x})-q(\mathbf{x}').$$

Let $\mathcal{B}_0=\{\mathbf{e}_{0,1},\dots,\mathbf{e}_{0,n}\}$ denote the canonical basis of $\mathbb{Q}^n$; we have

$$\langle \mathbf{e}_{0,i},\mathbf{e}_{0,j}\rangle=\langle \mathbf{e}_{0,j},\mathbf{e}_{0,i}\rangle=2m_{i,j}.$$

We assume that q is *non-degenerate* meaning that the map

$$\begin{array}{ccc}\mathbb{Q}^n & \to & (\mathbb{Q}^n)^*,\\ \mathbf{x} & \mapsto & \langle \mathbf{x},\cdot\rangle\end{array}$$

is an isomorphism or equivalently that

$$\{\mathbf{x}\in\mathbb{Q}^n|\forall \mathbf{x}',\langle \mathbf{x},\mathbf{x}'\rangle=0\}=\{\mathbf{0}\}.$$

We also assume that q is *integral*, that is

$$q(\mathbb{Z}^n)\subset\mathbb{Z}$$

or equivalently

$$m_{i,i},\ 2m_{i,j}\in\mathbb{Z}.$$

We are interested in studying the fibers of the map $q : \mathbb{Z}^n \to \mathbb{Z}$ or in simpler terms, given $d \in \mathbb{Z}$ we are interested in the integral solutions of the polynomial equation

$$q(x_1, \dots, x_n) = d, \ (x_1, \dots, x_n) \in \mathbb{Z}^n.$$

Definition 2.1. An integer d is representable by q (in $\mathbb{Z}^n$) if there is $(x_1, \dots, x_n) \in \mathbb{Z}^n \backslash \{(0, \dots, 0)\}$ such that

$$d = q(x_1, \dots, x_n).$$

Such a vector $\mathbf{x} = (x_1, \dots, x_n)$ is a representation of d by q and the set of all such representations is denoted by

$$\mathcal{R}_q(d) = \{(x_1, \dots, x_n) \in \mathbb{Z}^n \backslash \{(0, \dots, 0)\} |\, q(x_1, \dots, x_n) = d\}.$$

Remark 2.1. The null vector $(0, \dots, 0)$ is always a representation of $d = 0$ by q this is the reason why we exclude it a priori from $\mathcal{R}_q(0)$.

The three basic questions we will consider are

(1) Under which conditions is $\mathcal{R}_q(d)$ non-empty?

(2) If is $\mathcal{R}_q(d)$ is non-empty how large is it?

(3) What can be said about its "shape"?

In these lectures we will be mainly concerned with the first question but let us just point out that the three questions are deeply intertwined and answering the first one often give substantial partial answers to the two others. For the discussion to come it will be useful to recall the notion of discriminant.

2.1 Discriminant

The discriminant of an integral quadratic form (relative to the lattice $\mathbb{Z}^n$) is an invariant measuring the arithmetic complexity of the form. It is defined as

$$\mathrm{disc}_{\mathbb{Z}^n}(q) := \begin{cases} (-1)^{n/2} \det((2m_{i,j})_{i,j \in [1,n]}), & \text{if } n \text{ is even,} \\ (-1)^{n(n-1)/2} \det((2m_{i,j})_{i,j \in [1,n]})/2, & \text{if } n \text{ is odd.} \end{cases}$$

Since q is non-degenerate, the discriminant is nonzero and is integral since q is: this is obvious when n is even since the matrix $(2m_{i,j})_{i,j \in [1,n]}$ is integral but also true when n is odd because $(2m_{i,j})_{i,j \in [1,n]}$ has even diagonal entries so its determinant is even.

3 The Hasse Principle

We discuss here the first question

Question. *Under which conditions is the set $\mathcal{R}_q(d)$ non-empty?*

A partial, yet significant answer is provided by the *Hasse-Minkowski* local-global principle which gives a necessary and sufficient condition for the existence of rational representations:

$$q(x_1, \ldots, x_n) = d, \ (x_1, \ldots, x_n) \in \mathbb{Q}^n.$$

For this we need to extend a bit the notion of representation: for any ring R with unit 1_R, because of the canonical ring homomorphism $\mathbb{Z} \to R$, $m \mapsto m.1_R$, the form q defines a quadratic form over $R^n = \mathbb{Z}^n \otimes_{\mathbb{Z}} R$ which we still denote by q.

Definition 3.1. An integer $d \in \mathbb{Z}$ is representable by q in R^n if there is $\mathbf{x} \in R^n$ such that

$$d.1_R = q(\mathbf{x}).$$

Such a vector $\mathbf{x}$ is then called a representation of d by q in R^n.

We recall now the

Theorem 3.1 (Hasse-Minkowski local-global principle). *An integer $d \in \mathbb{Z}$ is representable by q in $\mathbb{Q}^n$ if and only if d is representable by q in $\mathbb{R}^n$ and in $\mathbb{Q}_p^n$ for any prime p.*

Proof. We refer to [15] for a proof. □

Regarding now the existence of *integral* representations of d, an obvious necessary condition is that d it is *everywhere locally representable by q in $\mathbb{Z}^n$*:

d is representable by q in $\mathbb{R}^n$ and in $\mathbb{Z}_p^n$ for any prime p.

Remark 3.1. Observe that this condition is equivalent to being representable over $\mathbb{R}$ and in $(\mathbb{Z}/q\mathbb{Z})^n$ for any prime power $q = p^\alpha > 1$. Given q, and d the local representability can be checked by a finite amount of computations that can be performed in logarithmic time (w.r.t. $|d|$). For $n = 2$, the representability of d by q amounts to the computation of some quadratic residue symbol and therefore reduces to compute the congruence class of d modulo primes depending on q. For $n \geqslant 3$, representability in $\mathbb{R}^n$ is trivial to check and for the p-adic integers it is sufficient to check representability in $\mathbb{Z}_p^n$ for the primes dividing $\det((2m_{i,j})_{i,j\in[1,n]})$. By Newton's method, for any such prime p this reduces to check whether the equation

$$q(\mathbf{x}) = d$$

has a solution modulo p^α where α depends only on the p-adic valuation of the determinant.

It is natural to expect that this condition is sufficient.

Integral Hasse Principle. *If d is representable in $\mathbb{R}^n$ and in $\mathbb{Z}_p^n$ for any prime p – in which case we say that d is* everywhere locally representable *– then d should be representable by q in $\mathbb{Z}^n$.*

Unfortunately, the Integral Hasse Principle does not hold in this generality:

– When $n = 2$ for "most" quadratic forms q, there is a positive proportion of the integers d to which the Integral Hasse Principle fails: the quadratic forms in question are the ones obtained by considering the norm form restricted to proper ideals of orders having class number > 1 in quadratic fields. In any event the $n = 2$ case has been understood long time ago.

– When $n = 3, 4$, there are explicit examples of (definite) quadratic forms for which infinitely many integers do not satisfy the integral Hasse principle: for instance (cf. [7]) the form

$$q(X, Y, Z) = 4X^2 + 48Y^2 + 49Z^2 + 4XZ + 48YZ,$$

does not represents any odd prime $p \equiv 2 \pmod 3$ while these are everywhere locally representable.

On the positive side, one has the following

Theorem 3.2. *Given q an integral non-degenerate quadratic form and $d \neq 0$ an integer, the integral Hasse principle holds (for the pair (q, d)) whenever*

- *$n \geqslant 5$ and d is sufficiently large ($d \geqslant d(q)$ for some constant $d(q)$ depending on q).*
- *$n = 4$, d is sufficiently large and with bounded valuation at any prime p for which q is $\mathbb{Q}_p$-anisotropic.*
- *$n = 3$, d is sufficiently large, with bounded valuation at any prime p for which q is $\mathbb{Q}_p$-anisotropic and which does not belong to a finite, explicitable set of square classes $\{d_j\mathbb{Z}^2 \mid j \in J_q\}$ which depend on q only.*

Of all the cases above the hardest is arguably the ternary definite case ($n = 3$) which waited until the late 80's/early 90's to be solved by a combination of deep works of Duke [1], Iwaniec [9], Duke and Schulze-Pillot [2].

On the other hand back in the 50's, 60's, Y. V. Linnik invented his so-called *ergodic method* enabling him to obtain the following partial result which required an additional congruence condition (condition (3.1) below) for d modulo a fixed prime.

Theorem 3.3 (Linnik). *Let q be an integral non-degenerate positive quadratic form and $p \nmid \mathrm{disc}(q)$ be a prime. Let d be a square-free integer coprime with $\mathrm{disc}(q)$, such that*

$$p \text{ splits in the field } \mathbb{Q}(-\mathrm{disc}(q)d) \tag{3.1}$$

and d is locally representable by q in $\mathbb{Z}^n$, then if d is large enough (depending on p, q), d is representable by q (in $\mathbb{Z}^n$).

3.1 Equidistribution on Ellipsoids

Besides this existence result Linnik also proved a refinment regarding the distribution of these representations: let

$$V_{q,1}(\mathbb{R}) = \{\mathbf{x} \in \mathbb{R}^n \mid q(\mathbf{x}) = 1\}.$$

This is an ellipsoid (the unit sphere for the norm induced by q) and

$$d^{-1/2}\mathcal{R}_q(d) \subset V_{q,1}(\mathbb{R})$$

form a discrete finite set of points on it. One has the following refinment

Theorem 3.4 (Linnik). *Notations being as above; for d a square-free integer coprime with* disc(q), *locally representable by q in $\mathbb{Z}^n$ and satisfying* (3.1) *above, the set $d^{-1/2}\mathcal{R}_q(d)$ becomes equidistributed, as $d \to \infty$, w.r.t $\mu_{q,1}$, the unique $SO_q(\mathbb{Z})$-invariant probability measure on $V_{q,1}(\mathbb{R})$: for any continuous function $f \in \mathcal{C}(V_{q,1}(\mathbb{R}))$,*

$$\frac{1}{|\mathcal{R}_q(d)|}\sum_{\mathbf{x}\in\mathcal{R}_q(d)} f\left(\frac{\mathbf{x}}{d^{1/2}}\right) \to \int_{V_{q,1}(\mathbb{R})} f(x)d\mu_q(x).$$

Remark 3.2. As above, Duke [1], Iwaniec [9], Duke and Schulze-Pillot [2] obtained that result without the condition (3.1).

Remark 3.3. The statement which has been really proven is the following: let $SO_q(\mathbb{Z}) \subset SO_q(\mathbb{R})$ denote the stabilizer of $\mathbb{Z}^n$ in $SO_q(\mathbb{R})$, then as $d \to \infty$ the quotient set

$$SO_q(\mathbb{Z})\backslash d^{-1/2}\mathcal{R}_q(d) \subset SO_q(\mathbb{Z})\backslash V_{q,1}(\mathbb{R})$$

because equidistributed w.r.t. the quotient of the measure μ_q by the counting probability measure on $SO_q(\mathbb{Z})$. Since $SO_q(\mathbb{Z})$ is finite both statement are equivalent; the advantage of the former is that it generalize to the case where q is indefinite.

The objective of these lectures is to present the principles of the proof of Linnik's theorem.

4 Quadratic Forms over Lattices

This problem is a special case of a slightly more general one: $\mathbb{Z}^n$ is the lattice

$$\mathbb{Z}^n = \bigoplus_{i=1,\dots,n} \mathbb{Z}\mathbf{e}_{0,i} = L_0$$

generated by the canonical basis of $\mathbb{Q}^n$. More generally, given

$$L = \bigoplus_{i=1,\dots,n} \mathbb{Z}\mathbf{e}_i \subset \mathbb{Q}^n$$

some *lattice* (i.e. a finite generated $\mathbb{Z}$-module generating the $\mathbb{Q}$-vector space $\mathbb{Q}^n$) on which q takes integral values ($q(L) \subset \mathbb{Z}$), we consider the problem of determining the set

$$\mathcal{R}_q(d;L) = \{\mathbf{x} \in L\backslash\{0\} \mid q(\mathbf{x}) = d\}$$

of representations of d by q in L.

Remark 4.1. Decomposing a vector $\mathbf{x}$ in a basis $\mathcal{B}_L = \{\mathbf{e}_1, \dots, \mathbf{e}_n\}$ of L we have

$$q(\mathbf{x}) = q(\sum_i x_i \mathbf{e}_i) = q_{\mathcal{B}}(x_1, \dots, x_n)$$

is another homogeneous polynomial of degree 2 therefore studying the representations of d by q in L amounts to studying the representation of d by $q_{\mathcal{B}}$ in $\mathbb{Z}^n$ therefore we may possibly assume that $L = \mathbb{Z}^n$.

An obvious necessary condition for the existence of representations in L is that d is everywhere locally representable by q in lattice L:

There exists

$$\mathbf{x}_{\mathbb{R}} \in L \otimes_{\mathbb{Z}} \mathbb{R} = \mathbb{R}^n \textit{ such that } q(\mathbf{x}_{\mathbb{R}}) = d$$

and for any prime p, there exists

$$\mathbf{x}_p \in L \otimes_{\mathbb{Z}} \mathbb{Z}_p \textit{ such that } q(\mathbf{x}_p) = d.$$

The lattice (i.e. the finitely generated $\mathbb{Z}_p$-module generating $\mathbb{Q}_p^n$ as a $\mathbb{Q}_p$-vector space)

$$L_p = L \otimes_{\mathbb{Z}} \mathbb{Z}_p \subset \mathbb{Q}_p^n$$

is called the localizations of L at the place p; alternatively L_p is the closure of $L \subset \mathbb{Q}_p^n$ for the p-adic topology.

4.1 Lattices and Their Localization

Let $\mathcal{L}_n(\mathbb{Q})$ denote the set of lattices in $\mathbb{Q}^n$ (i.e. the finitely generated $\mathbb{Z}$-module generating $\mathbb{Q}^n$ as a $\mathbb{Q}$-vector space) and let $\mathcal{L}_n(\mathbb{Q}_p)$ denote the set of lattices in $\mathbb{Q}_p^n$ (i.e. the finitely generated $\mathbb{Z}_p$-module generating $\mathbb{Q}_p^n$ as a $\mathbb{Q}_p$-vector space).

To each lattice $L \in \mathcal{L}_n(\mathbb{Q})$, we associate its sequence of localizations $(L_p)_{p \in \mathcal{P}}$. In fact a lattice can be entirely reconstructed from these local data:

Theorem 4.1. *The map*

$$L \mapsto \widehat{L} := (L_p)_p$$

is a bijection between $\mathcal{L}_n(\mathbb{Q})$ and the restricted product

$$\prod_p{}' \mathcal{L}_n(\mathbb{Q}_p) = \{(L_p)_p \,|\, L_p \in \mathcal{L}_n(\mathbb{Q}_p),\ L_p = \mathbb{Z}_p^n \textit{ for almost every (a.e.) prime } p\}.$$

the inverse map is given by

$$\widehat{L} = (L_p)_p \mapsto \bigcap_p \mathbb{Q}^n \cap L_p \in \mathcal{L}_n(\mathbb{Q}).$$

4.2 The q-genus of a Lattice

Let us analyse the condition of being everywhere locally representable by a lattice. From now on, let us assume that d is representable by q over $\mathbb{R}$ (i.e. q is either indefinite or q is definite and d of the right sign).

Given any prime p, d is representable by q in $L_p \in \mathcal{L}_n(\mathbb{Q}_p)$ if and only if it is representable by any lattice $L'_p \in \mathcal{L}_n(\mathbb{Q}_p)$ which is *isometric* to L_p: is of the shape $L'_p = g_p \cdot L_p$ for $g_p \in SO_q(\mathbb{Q}_p)$. Indeed, if $\mathbf{x}_p \in L_p$ is such that $q(\mathbf{x}_p) = d$ then $g_p \cdot \mathbf{x}_p \in g_p \cdot L_p = L'_p$ and

$$q(g_p \cdot \mathbf{x}_p) = q(\mathbf{x}_p) = d.$$

This prompt us to introduce the following

Definition 4.1. Two lattices $L, L' \subset \mathbb{Q}^n$ are everywhere locally isometric if for every prime p there exists $g_p \in SO_q(\mathbb{Q}_p)$ such that

$$L'_p = g_p \cdot L_p.$$

We denote this relation by

$$L \sim_{q,loc} L'.$$

This is an equivalence relation in the space of rational lattices and the equivalence class of a lattice L is called the q-genus of L and is noted $\mathrm{gen}_q(L)$.

If q is integral on L, then this is the same for every $L' \in \mathrm{gen}_q(L)$: for every p,

$$q(L') \subset q(L'_p) = q(g_p \cdot L_p) = q(L_p) \subset \mathbb{Z}_p$$

and if d is everywhere representable by q in L then it is everywhere locally representable in every lattice in $\mathrm{gen}_q(L)$. The Hasse-Minkowski theorem implies a partial answer to the question of global representability:

Theorem 4.2. *If d is locally representable by q in the lattice L, then there exists $L' \in \mathrm{gen}_q(L)$ such that d is globally representable by q in L'.*

Proof. By the Hasse-Minkowski theorem, there exists $\mathbf{x}_\mathbb{Q} \in \mathbb{Q}^n$ such that $q(\mathbf{x}_\mathbb{Q}) = d$ and for any p, there exists $\mathbf{x}_p \in L_p$ such that $q(\mathbf{x}_p) = d$. By Witt's theorem there exists

$$g_p \in SO_q(\mathbb{Q}_p),\ \mathbf{x}_\mathbb{Q} = g_p \cdot \mathbf{x}_p.$$

In fact for a.e. p, $\mathbf{x}_\mathbb{Q} \in L_p = \mathbb{Z}_p^n$ and then, we can and will choose $g_p = \mathrm{Id}$ (the identity element). By construction, the sequence of lattices $(g_p \cdot L_p)_p = \widehat{L}'$ belong to $\prod_p{}' \mathcal{L}_n(\mathbb{Q}_p)$ and therefore correspond to some rational lattice

$$L' := \bigcap_p \mathbb{Q}^n \cap g_p L_p,$$

and for any p, $\mathbf{x}_\mathbb{Q} = g_p \cdot \mathbf{x}_p \in g_p \cdot L_p$, therefore $\mathbf{x}_\mathbb{Q} \in L'$. □

The genus of a given lattice is infinite but it can be decomposed in a smaller number of pieces according to a finer equivalence relation, *global isometry* which is

Definition 4.2. Two lattices $L, L' \subset \mathbb{Q}^n$ are globally isometric if there exists $g_{\mathbb{Q}} \in SO_q(\mathbb{Q})$ such that

$$L' = g_{\mathbb{Q}} \cdot L,$$

or in other terms if L and L' are in the same $SO_q(\mathbb{Q})$-orbit. We denote this relation by

$$L \sim_q L'.$$

Obviously globally isometric lattices are locally isometric (by taking $g_p = g_{\mathbb{Q}}$) therefore the q-genus of a lattice decomposes into global isometry classes: the quotient

$$[\mathrm{gen}_q(L)] := SO_q(\mathbb{Q})\backslash \mathrm{gen}_q(L)$$

is called the set of *genus classes of* q.

We have the following finiteness theorem which generalize Gauss theorem on the finiteness of the class number of binary quadratic forms.

Theorem 4.3 (Hermite-Minkowski). *The set* $[\mathrm{gen}_q(L)]$ *is finite. Its cardinality is called the* q*-class number of the lattice* L *and is noted*

$$h_q(L) = |[\mathrm{gen}_q(L)]|.$$

Obviously is d is (globally) representable by q in some lattice L', then it is representable in any lattice globally isometric to it. Therefore, choosing

$$\{L_1 = L, \dots, L_h\},\ h = h_q(L)$$

a set of representatives of the various genus classes of L containing the original lattice L we obtain the following approximation to the integral Hasse principle:

Theorem 4.4. *Given* L *a lattice, there exists a finite set of lattices*

$$\{L_1 = L, \dots, L_h\},\ h = h_q(L)$$

containing L *such that if* d *is locally representable by* q *in* L, d *is representable by* q *in some lattice* L_i, $i = 1, \dots, h$. *In particular, if* $h = 1$, *the integral Hasse principle holds.*

Unfortunately for any $n \geqslant 2$ there are only finitely many integral quadratic lattices (L, q) with class number 1. Important (historic) examples of quadratic lattices with class number 1 are the euclidean quadratic forms in 2, 3 and 4 variables

$$(\mathbb{Z}^2, x_1^2 + x_2^2),\ (\mathbb{Z}^3, x_1^2 + x_2^2 + x_3^2),\ (\mathbb{Z}^4, x_1^2 + x_2^2 + x_3^2 + x_4^2).$$

4.3 Representations by the Genus

In order to give more quantitative results about the representations of d, let us introduce

$$\mathcal{R}(d; \mathrm{gen}_q(L)) = \{(L', \mathbf{x}') \mid L' \in \mathrm{gen}_q(L),\ \mathbf{x}' \in \mathcal{R}_q(d; L')\}$$

the set of all the representations of d by all the genus classes of L. By the Hasse-Minkowski theorem that set is non-empty if and only if d is locally representable by L.

Observe that if non-empty $\mathcal{R}(d;\mathrm{gen}_q(L))$ is even infinite the reason being that the group $SO_q(\mathbb{Q})$ acts diagonally on the set of such pairs:

$$g_{\mathbb{Q}}\cdot(L',\mathbf{x}')=(g_{\mathbb{Q}}\cdot L',g_{\mathbb{Q}}\cdot\mathbf{x}').$$

As for the q-genus we can decompose that set into global isometry classes and consider

$$[\mathcal{R}(d;\mathrm{gen}_q(L))]=SO_q(\mathbb{Q})\backslash\mathcal{R}(d;\mathrm{gen}_q(L))$$

the set of its $SO_q(\mathbb{Q})$-orbits.

Consider the projection map

$$\begin{array}{rcl}\pi_q:\mathcal{R}(d;\mathrm{gen}_q(L)) & \mapsto & \mathrm{gen}_q(L),\\ (L',\mathbf{x}') & \mapsto & L';\end{array}\tag{4.1}$$

that map induces a map on $SO_q(\mathbb{Q})$-orbits (which we note the same)

$$\pi_q:[\mathcal{R}(d;\mathrm{gen}_q(L))]\to[\mathrm{gen}_q(L)]$$

and we would like to study the fibers of these maps and in particular show that these are non-empty, i.e. π_q is surjective.

To be concrete let us fix

$$\{L_1=L,\dots,L_h\},\ h=h_q(L)$$

be a set of representatives of genus classes as above. For L a lattice, let

$$SO_q(L)=GL(L)\cap SO_q(\mathbb{Q})$$

denote the stabilizer of the lattice L in $SO_q(\mathbb{Q})$. We have

Proposition 4.1. *The inclusion*

$$\{(L_i,\mathbf{x}_i)\,|\,i=1,\dots,h,\ \mathbf{x}_i\in\mathcal{R}_q(d;L_i)\}\hookrightarrow\mathcal{R}(d;\mathrm{gen}_q(L))$$

induces an identification

$$\bigsqcup_{i=1,\dots,h}SO_q(L_i)\backslash\mathcal{R}_q(d;L_i)\simeq SO_q(\mathbb{Q})\backslash\mathcal{R}(d;\mathrm{gen}_q(L)).$$

Proof. Given any $(L',\mathbf{x}')$ with $\mathbf{x}'\in\mathcal{R}_q(L')$, there is $g_{\mathbb{Q}}\in SO_q(\mathbb{Q})$ and some i such that $L'=g_{\mathbb{Q}}\cdot L_i$ and $(L',\mathbf{x}')=g_{\mathbb{Q}}(L_i,g_{\mathbb{Q}}^{-1}\mathbf{x}')$ so the map is surjective. Suppose now that $(L_i,\mathbf{x}_i),(L_j,\mathbf{x}_j)$ are in the same $SO_q(\mathbb{Q})$-orbit: for some $g_{\mathbb{Q}}\in SO_q(\mathbb{Q})$ we have

$$L_i=g_{\mathbb{Q}}L_j,\ \mathbf{x}_i=g_{\mathbb{Q}}\mathbf{x}_j$$

so $L_i=L_j$ and $g_{\mathbb{Q}}\in SO_q(L_i)$ and $\mathbf{x}_i$ and $\mathbf{x}_j$ are in the same $SO_q(L_i)$-orbit so the map is injective. □

4.4 Siegel's Mass Formula

A generalization of Gauss's theorem (on the finiteness of the class number) states that the sets $SO_q(L_i)\backslash\mathcal{R}_q(d; L_i)$, $i = 1, \ldots, h$ are finite and therefore, by the previous proposition, the set

$$[\mathcal{R}_q(d; \mathrm{gen}_q(L))]$$

is finite; the mass formula of Siegel gives an estimate for its size:

Theorem 4.5 (Siegel). *For q definite of rank $n \geqslant 2$, d square-free and everywhere locally representable by q we have*

$$|[\mathcal{R}_q(d; \mathrm{gen}_q(L))]| = \sum_{i=1}^{h} |SO_q(L_i)\backslash\mathcal{R}_q(d; L_i)| = c_q|d|^{\frac{n-2}{2}+o(1)}, \ c_q > 0.$$

Actually the Mass formula is much more precise and expresses $|[\mathcal{R}_q(d; \mathrm{gen}_q(L))]|$ as a product of local densities. This fact will be appear more natural when we discuss the adelic interpretation of this problem. In particular, for $n = 3$ we have

$$|[\mathcal{R}_q(d; \mathrm{gen}_q(L))]| \gg_q |d|^{1/2+o(1)} \to \infty. \tag{4.2}$$

We will prove the following theorem

Theorem 4.6 (Linnik). *Let q be an integral non-degenerate definite quadratic form in $n = 3$ variables and let $p \nmid \mathrm{disc}(q)$ be a prime. As $d \to \infty$, amongst the integers which are locally representable by q, coprime with $\mathrm{disc}(q)$ and such that*

$$p \text{ splits in the field } \mathbb{Q}(-\mathrm{disc}(q)d), \tag{4.3}$$

we have for any L_i, $i = 1, \ldots, h$

$$\frac{|[\mathcal{R}_q(d; L_i)]|}{|[\mathcal{R}_q(d; \mathrm{gen}_q(L))]|} \to \frac{1/|SO_q(L_i)|}{\sum_j 1/|SO_q(L_j)|}.$$

In particular, if d is large enough as above, $|\mathcal{R}_q(d; L_i)| \neq 0$ and d is representable by any lattice in $\mathrm{gen}_q(L)$ (in particular by L itself).

This more precise statement may be interpreted by saying that as $d \to \infty$ the set of classes of representations of d by the genus get equidistributed over the various genus classes with respect to the probability measure $\mu_{\mathrm{gen}_q(L)}$ which assign the mass

$$\mu_{[\mathrm{gen}_q(L)]}([L_i]) := \frac{1/|SO_q(L_i)|}{\sum_j 1/|SO_q(L_j)|}$$

to the genus class $[L_i]$.

5 Equidistribution on Adelic Quotient

In this section we introduce the ring of adèles, building on the natural action of the linear group on the space of rational lattices $\mathcal{L}_n(\mathbb{Q})$ and of the local-global description of that space given in Theorem 4.1.

5.1 p-adic Linear Groups Acting on Lattices

We recall first that the group $GL_n(\mathbb{Q})$ act transitively on $\mathcal{L}_n(\mathbb{Q})$ via

$$(\gamma, L) \mapsto \gamma(L) \in \mathcal{L}_n(\mathbb{Q});$$

the stabilizer of $\mathbb{Z}^n$ is $GL_n(\mathbb{Z})$ and therefore we have the identification

$$\mathcal{L}_n(\mathbb{Q}) = GL_n(\mathbb{Q}) \cdot \mathbb{Z}^n \simeq GL_n(\mathbb{Q})/GL_n(\mathbb{Z}).$$

Similarly, for any prime p, the space $\mathcal{L}_n(\mathbb{Q}_p)$ of p-adic lattices is acted on transitively by the p-adic linear group $GL_n(\mathbb{Q}_p)$ and the stabilizer of $\mathbb{Z}_p^n$ is $GL_n(\mathbb{Z}_p)$:

$$\mathcal{L}_n(\mathbb{Q}_p) = GL_n(\mathbb{Q}_p) \cdot \mathbb{Z}_p^n \simeq GL_n(\mathbb{Q}_p)/GL_n(\mathbb{Z}_p).$$

As we have seen in Theorem 4.1, we have a natural identification

$$\mathcal{L}_n(\mathbb{Q}) \simeq \prod_p{}' \mathcal{L}_n(\mathbb{Q}_p) = \{(L_p)_p \,|\, L_p \in \mathcal{L}_n(\mathbb{Q}_p),\ L_p = \mathbb{Z}_p^n \text{ a.e. } p\} \tag{5.1}$$

induced by the sequence of maps

$$L \mapsto L_p = L \otimes_{\mathbb{Z}} \mathbb{Z}_p \subset \mathbb{Q}_p^n.$$

the p-adic group $GL_n(\mathbb{Q}_p)$ then acts on the restricted product $\prod_p{}' \mathcal{L}_n(\mathbb{Q}_p)$ via its action on the p-th factor and therefore on $\mathcal{L}_n(\mathbb{Q})$: given $g_p \in GL_n(\mathbb{Q}_p)$ and $L \in \mathcal{L}_n(\mathbb{Q})$, a lattice represented by its sequence of localizations $(L_{p'})_{p'}$, $g_p \cdot L$ is defined to be the lattice L' such that

$$L'_p = g_p \cdot L_p,\ L'_{p} = L_p,\ p' \neq p.$$

For different primes p, p' the actions of $GL_n(\mathbb{Q}_p)$ and $GL_n(\mathbb{Q}_{p'})$ commute, and therefore for any finite subset of the primes $S \subset \mathcal{P}$, the product $\prod_{p \in S} GL_n(\mathbb{Q}_p)$ acts on $\mathcal{L}_n(\mathbb{Q})$.

5.2 The Adelic Linear Group Acting on the Space of Lattices

The full product $\prod_p GL_n(\mathbb{Q}_p)$ however does *not* act on $\mathcal{L}_n(\mathbb{Q})$: if $(g_p)_p$ is an arbitrary sequence, the terms of the sequence of lattices $(g_p \cdot \mathbb{Z}_p^n)_p$ might not be equal to $\mathbb{Z}_p^n$ for almost every p. On the other hand the following restricted product subgroup

$$\prod_p{}' GL_n(\mathbb{Q}_p) = \left\{ (g_p)_p \in \prod_p GL_n(\mathbb{Q}_p) \,\middle|\, g_p \in GL_n(\mathbb{Z}_p) \text{ a.e. } p \right\}$$

does act: if $(L_p)_p$ and $(g_p)_p$ are such that $L_p = \mathbb{Z}_p^n$ and $g_p \in GL_n(\mathbb{Z}_p)$ for a.e. p, then $g_p \cdot L_p = \mathbb{Z}_p^n$ for a.e. p. Given $g_f = (g_p)_p$ and $L \simeq (L_p)_p$ this action is given by

$$g_f \cdot L \simeq (g_p)_p \cdot (L_p)_p = (g_p \cdot L_p)_p \simeq L' = \bigcap_p \mathbb{Q}^n \cap g_p \cdot L_p.$$

This restricted product is the group of *finite adelic points* of the linear group and is noted

$$GL_n(\mathbb{A}_f) = \prod_p{}' GL_n(\mathbb{Q}_p).$$

We observe that given $g_{\mathbb{Q}} \in GL_n(\mathbb{Q})$, $g_{\mathbb{Q}} \in GL_n(\mathbb{Z}_p)$ for almost every prime p: given any lattice L, $g_{\mathbb{Q}} \cdot L$ is a lattice and for any prime p,

$$(g_{\mathbb{Q}} \cdot L)_p = g_{\mathbb{Q}} \cdot L_p$$

and for almost every p,

$$L_p = \mathbb{Z}_p^n = g_{\mathbb{Q}} \cdot L_p.$$

In other terms the group $GL_n(\mathbb{Q})$ embeds into $GL_n(\mathbb{A}_f)$ via the diagonal map

$$\Delta : g_{\mathbb{Q}} \in GL_n(\mathbb{Q}) \hookrightarrow (g_{\mathbb{Q}}, g_{\mathbb{Q}}, \dots)_{p \in \mathcal{P}} \in GL_n(\mathbb{A}_f)$$

and this embedding intertwines the usual action of $GL_n(\mathbb{Q})$ on $\mathcal{L}_n(\mathbb{Q})$ with the adelic action: for any lattice L one has

$$\gamma \cdot L = \Delta(\gamma) \cdot L.$$

In particular, we verify easily the following

Proposition 5.1. *The group $GL_n(\mathbb{A}_f)$ acts transitively on $\mathcal{L}_n(\mathbb{Q})$ and the stabilizer of $\mathbb{Z}^n$ is the subgroup*

$$GL_n(\widehat{\mathbb{Z}}) = \prod_p GL_n(\mathbb{Z}_p) \text{ and } \mathcal{L}_n(\mathbb{Q}) \simeq GL_n(\mathbb{A}_f)/GL_n(\widehat{\mathbb{Z}}).$$

In particular since $GL_n(\mathbb{Q})$ acts transitively on $\mathcal{L}_n(\mathbb{Q})$ we have the decomposition

$$GL_n(\mathbb{A}_f) = GL_n(\mathbb{Q}) \cdot GL_n(\widehat{\mathbb{Z}}) \text{ and } GL_n(\mathbb{Q}) \cap GL_n(\widehat{\mathbb{Z}}) = GL_n(\mathbb{Z}). \tag{5.2}$$

More generally for $L \simeq \widehat{L} = (L_p)_p$ a lattice, we denote by

$$GL_n(\widehat{L}) = \mathrm{Stab}_{GL_n(\mathbb{A}_f)}(L) = \prod_p GL_n(L_p),\ GL_n(L_p) = \mathrm{Stab}_{GL_n(\mathbb{Q}_p)}(L_p),$$

the stabilizer of $\widehat{L}$ in $GL_n(\mathbb{A}_f)$. This subgroup is $GL_n(\mathbb{Q})$-conjugate to $GL_n(\widehat{\mathbb{Z}})$ and

$$GL_n(\mathbb{Q}) \cap GL_n(\widehat{L}) = \mathrm{Stab}_{GL_n(\mathbb{Q})}(L) = GL_n(L)$$

(say). Let

$$SO_q(\mathbb{A}_f) = \prod_p{}' SO_q(\mathbb{Q}_p) = \{g_f = (g_p)_p \in GL_n(\mathbb{A}_f) \,|\, g_p \in SO_q(\mathbb{Q}_p)\}$$

be the subgroup of transformations preserving the quadratic form q in $GL_n(\mathbb{A}_f)$; we note the stabilizer of L in that group by

$$SO_q(\widehat{L}) = \mathrm{Stab}_{SO_q(\mathbb{A}_f)}(L) = SO_q(\mathbb{A}_f) \cap GL_n(\widehat{L}).$$

We have the following interpretation of the q-genus of a lattice:

Proposition 5.2. *The q-genus (or SO_q-genus) of L is the following orbit in $\mathcal{L}_n(\mathbb{Q})$*

$$\mathrm{gen}_q(L) = SO_q(\mathbb{A}_f) \cdot L \subset \mathcal{L}_n(\mathbb{Q}) = GL_n(\mathbb{A}_f) \cdot L.$$

In particular, the set of genus classes of L (the set of lattices locally isometric to L modulo global isometries) is identified with the quotient

$$[\mathrm{gen}_q(L)] \simeq SO_q(\mathbb{Q})\backslash SO_q(\mathbb{A}_f) \cdot L \simeq SO_q(\mathbb{Q})\backslash SO_q(\mathbb{A}_f)/SO_q(\widehat{L}). \tag{5.3}$$

5.3 The Affine Adelic Space

We observe first that the group of adelic points $GL_n(\mathbb{A}_f)$ acts on the n-dimensional *adelic space*

$$\mathbb{A}_f^n = \prod_p{}' \mathbb{Q}_p^n = \{\mathbf{x}_f = (\mathbf{x}_p)_p \,|\, \mathbf{x}_p \in \mathbb{Q}_p^n,\ \mathbf{x}_p \in \mathbb{Z}_p^n,\ \text{a.e. } p\}$$

by

$$g_f \cdot \mathbf{x}_f = (g_p \mathbf{x}_p)_p.$$

Observe that for any $x \in \mathbb{Q}$, $x \in \mathbb{Z}_p$ for a.e. p (the primes not dividing the denominator of x), it follows that the rational space $\mathbb{Q}^n$ embeds diagonally into that adelic space:

$$\begin{array}{rcl} \Delta_f : \mathbb{Q}^n & \hookrightarrow & \mathbb{A}_f^n, \\ \mathbf{x} & \hookrightarrow & \Delta_f(\mathbf{x}) = (\mathbf{x})_p \end{array}$$

so, for $\mathbf{x} \in \mathbb{Q}^n$, $g_f \cdot \mathbf{x}$ is a well defined element of $\mathbb{A}_f^n$ (but not a rational one in general).

For $g_{\mathbb{Q}} \in GL_n(\mathbb{Q})$ and $\mathbf{x} \in \mathbb{Q}^n$ we have

$$\Delta_f(g_{\mathbb{Q}}) \cdot \Delta_f(\mathbf{x}) = \Delta_f(g_{\mathbb{Q}} \cdot \mathbf{x}),$$

therefore we will usually omit to write the embedding Δ_f. We also observe that

$$\mathrm{Stab}_{GL_n(\mathbb{A}_f)}(\mathbb{Q}^n) = GL_n(\mathbb{Q}).$$

5.4 Representations by the Genus as Adelic Orbits I

In this section we explain how the set of representations of d by $\mathrm{gen}_q(L)$ can be realized as an orbit for this adelic action.

To ease notation we write G for SO_q and for $\mathbf{x} \in \mathcal{R}_q(d; L)$ some representation[1], let

$$\mathrm{H}_{\mathbf{x}} = SO_{q,\mathbf{x}} = \mathrm{Stab}_{SO_q}(\mathbf{x}) = \{g \in SO_q,\ g\mathbf{x} = \mathbf{x}\}$$

be the stabilizer of $\mathbf{x}$ in SO_q; if $\mathbf{x}$ is understood we denote that group simply by H.

The group H is a $\mathbb{Q}$-algebraic subgroup of G isomorphic to an orthogonal group in $n-1$ variables, namely

$$SO_{q,\mathbf{x}} \simeq SO_{q_{|\mathbf{x}^\perp}},$$

where $\mathbf{x}^\perp$ denote the subspace of $\mathbb{Q}^n$ orthogonal to $\mathbf{x}$ and $q_{|\mathbf{x}^\perp}$ denote the restriction of q to this subspace.

We denote by

$$\mathrm{H}(\mathbb{A}_f) = \prod_p{}' \mathrm{H}(\mathbb{Q}_p) = \{(h_p)_p \in SO_q(\mathbb{A}_f) \,|\, h_p \cdot \mathbf{x} = \mathbf{x} \text{ for every } p\} \subset SO_q(\mathbb{A}_f)$$

the corresponding subgroup of finite adelic points.

Let $(L, \mathbf{x})$ be a representation of d by q in L. For any $h_f \in \mathrm{H}(\mathbb{A}_f)$, the lattice $L' = h_f \cdot L$ is in $\mathrm{gen}_q(L)$ and $\mathbf{x} \in L'$:

$$\mathbf{x} \in L \Rightarrow \mathbf{x} \in \widehat{L} \Rightarrow h_f \cdot \mathbf{x} = \mathbf{x} \in h_f\widehat{L} = \widehat{L}'.$$

Therefore $(L', \mathbf{x})$ is another representation of d in $\mathrm{gen}_q(L)$. We denote by

$$\mathrm{gen}_{q,\mathbf{x}}(L) := \mathrm{H}_{\mathbf{x}}(\mathbb{A}_f) \cdot L, \text{ resp. } [\mathrm{gen}_{q,\mathbf{x}}(L)] = \mathrm{H}_{\mathbf{x}}(\mathbb{Q})\backslash \mathrm{H}_{\mathbf{x}}(\mathbb{A}_f) \cdot L,$$

the $\mathrm{H}_{\mathbf{x}}$-genus of L, resp. the set of $\mathrm{H}_{\mathbf{x}}$-genus classes of L:

Proposition 5.3. *The map*

$$L' \in \mathrm{gen}_{q,\mathbf{x}}(L) \mapsto (L', \mathbf{x}) \in \mathcal{R}_q(d; \mathrm{gen}_q(L))$$

induces an injective map

$$[\mathrm{gen}_{q,\mathbf{x}}(L)] \hookrightarrow [\mathcal{R}_q(d; \mathrm{gen}_q(L))]$$

whose image depends only on the $SO_q(\mathbb{Q})$-orbit of $(L, \mathbf{x})$. The set $[\mathcal{R}_q(d; \mathrm{gen}_q(L))]$ is a disjoint union of $SO_q(\mathbb{Q})$-orbits of such maps (associated to a possibly to different pairs $(L', \mathbf{x}') \in \mathcal{R}_q(d; \mathrm{gen}_q(L))$ and subgroup $\mathrm{H}_{\mathbf{x}'}$).

---

[1] a prioiri we know only that there exist a representation of d in some $L' \in \mathrm{gen}_q(L)$ but since $\mathrm{gen}_q(L) = \mathrm{gen}_q(L')$ we may assume for notational simplicity that $L' = L$.

Proof. Clearly if h_f is replaced by $h_{\mathbb{Q}}h_f$ for some $h_{\mathbb{Q}} \in \mathrm{H}(\mathbb{Q})$ then the pair $(L', \mathbf{x})$ is replaced by $(h_{\mathbb{Q}}L', \mathbf{x}) = h_{\mathbb{Q}}(L', \mathbf{x})$ so we have a well defined map from $\mathrm{H}(\mathbb{Q})\backslash\mathrm{H}(\mathbb{A}_f)\cdot L$ to $SO_q(\mathbb{Q})\backslash\mathcal{R}_q(d;\mathrm{gen}_q(L))$. Now, suppose that for $L', L'' \in \mathrm{H}(\mathbb{A}_f)\cdot L$ and some $s_{\mathbb{Q}} \in SO_q(\mathbb{Q})$, we have

$$(L'', \mathbf{x}) = s_{\mathbb{Q}}(L', \mathbf{x});$$

then $\mathbf{x} = s_{\mathbb{Q}}\mathbf{x}$ so that $s_{\mathbb{Q}} \in \mathrm{H}(\mathbb{Q})$ and $L'' \in \mathrm{H}(\mathbb{Q})\cdot L'$.

For the second part, let us assume that $(L', \mathbf{x}') \in SO_q(\mathbb{Q})(\mathrm{H}(\mathbb{A}_f)L, \mathbf{x})$. This means that for some $s_{\mathbb{Q}} \in SO_q(\mathbb{Q})$,

$$\mathbf{x}' = s_{\mathbb{Q}}\mathbf{x},\ L' = s_{\mathbb{Q}}h_f\cdot L.$$

In particular $\mathrm{H}_{\mathbf{x}'} = SO_{q,\mathbf{x}'} = s_{\mathbb{Q}}\mathrm{H}s_{\mathbb{Q}}^{-1}$ and therefore

$$\mathrm{H}_{\mathbf{x}'}(\mathbb{A}_f)\cdot L' = s_{\mathbb{Q}}\mathrm{H}(\mathbb{A}_f)s_{\mathbb{Q}}^{-1}s_{\mathbb{Q}}h_f\cdot L = s_{\mathbb{Q}}\mathrm{H}(\mathbb{A}_f)\cdot L$$

and the images of the sets $[\mathrm{gen}_{q,\mathbf{x}}(L)]$ and $[\mathrm{gen}_{q,\mathbf{x}'}(L')]$ in $[\mathcal{R}_q(d;\mathrm{gen}_q(L))]$ are the same. □

The image of $[\mathrm{gen}_{q,\mathbf{x}}(L)]$ by the projection π_q of (4.1) is the set

$$\mathrm{G}(\mathbb{Q})\backslash\mathrm{G}(\mathbb{Q})\mathrm{H}(\mathbb{A}_f)\cdot L \subset \mathrm{G}(\mathbb{Q})\backslash\mathrm{G}(\mathbb{A}_f)L = [\mathrm{gen}_q(L)].$$

We will show that for d large enough and under the assumptions of Linnik's theorem, this inclusion is an equality: this will prove that d is representable by every $L' \in \mathrm{gen}_q(L)$.

6 Properties of the Adeles

In this section we review general facts about adeles and adelic points of algebraic groups which were introduced in the previous section.

We denote by $V = \mathcal{P}\cup\{\infty\}$ the set of places of the field $\mathbb{Q}$ (the set of equivalence classes of absolute values on $\mathbb{Q}$).

The ring of *finite adeles* of $\mathbb{Q}$ is the restricted product

$$\mathbb{A}_f = \prod_p{}' \mathbb{Q}_p = \{x_f = (x_p)_p \,|\, x_p \in \mathbb{Q}_p, x_p \in \mathbb{Z}_p \text{ for a.e. } p\}.$$

The ring of *adeles* of $\mathbb{Q}$ is the product

$$\mathbb{A} = \mathbb{R}\times\mathbb{A}_f = \{(x_\infty, x_f) = (x_v)_{v\in V} \,|\, x_v \in \mathbb{Q}_v,\ x_p \in \mathbb{Z}_p \text{ for a.e. } p\}.$$

Since any rational number $x \in \mathbb{Q}$ belongs to $\mathbb{Z}_p$ for a.e. p the field of rationals $\mathbb{Q}$ embeds diagonally into $\mathbb{A}_f$ and $\mathbb{A}$:

$$\Delta_f : x \in \mathbb{Q} \to (x)_p \in \mathbb{A}_f,\ \Delta : x \in \mathbb{Q} \to (x, (x)_p) \in \mathbb{A}$$

so these two rings have the structure of a $\mathbb{Q}$-algebras. The groups of units of these algebras are

$$\mathbb{A}_f^\times = \prod_p{}' \mathbb{Q}_p = \{x_f = (x_p)_p \,|\, x_p \in \mathbb{Q}_p^\times, x_p \in \mathbb{Z}_p^\times \text{ for a.e. } p\}$$

and

$$\mathbb{A}^\times = \mathbb{R}^\times \times \mathbb{A}_f^\times$$

are called the groups of (*finite*) *idèles* of $\mathbb{Q}$. We denote by $\widehat{\mathbb{Z}}$ the subring of $\mathbb{A}_f$,

$$\widehat{\mathbb{Z}} = \prod_p \mathbb{Z}_p.$$

6.1 Adeles over a Number Field

If K be a finite dimensional $\mathbb{Q}$-vector space, we can form the free $\mathbb{A}$-module $K(\mathbb{A}) = \mathbb{A} \otimes_{\mathbb{Q}} K$; if $\{\mathbf{e}_1, \ldots, \mathbf{e}_n\}$ is any $\mathbb{Q}$-basis of K, and $\mathscr{O} = \mathbb{Z}\mathbf{e}_1 + \cdots + \mathbb{Z}\mathbf{e}_n$ the lattice generated by this basis, one has

$$\begin{aligned} K(\mathbb{A}) &= \prod_v{}' K_v = \{(\mathbf{v}_v)_v \,|\, \mathbf{v}_v \in K_v = K \otimes_{\mathbb{Q}} \mathbb{Q}_v,\ \mathbf{v}_p \in \mathscr{O}_p \text{ for a.e. } p\} \\ &= K_\infty \times K(\mathbb{A}_f). \end{aligned}$$

where $\mathscr{O}_p = \mathscr{O} \otimes_{\mathbb{Z}} \mathbb{Z}_p$. We set

$$\widehat{\mathscr{O}} := \mathscr{O} \otimes_{\mathbb{Z}} \widehat{\mathbb{Z}} = \sum_i \widehat{\mathbb{Z}}\mathbf{e}_i = \prod_p \mathscr{O}_p.$$

This applies in particular to K a number field; in this case $K(\mathbb{A})$ is a K-algebra called the ring of adeles of K and is also noted $\mathbb{A}_K$; the ring $K(\mathbb{A}_f)$ is called the ring of finite adeles of K and is noted $\mathbb{A}_{K,f}$. The groups of invertible elements of these rings $\mathbb{A}_K^\times$ and $\mathbb{A}_K^\times$ are the rings of finite ideles and of ideles of K. If the lattice $\mathscr{O}$ is an order, then $\widehat{\mathscr{O}}$ is a subring.

We have the following

Proposition 6.1. *Let K be a number field and $\mathscr{O} \subset K$ an order, one has a bijection*

$$K^\times \backslash \mathbb{A}_{K,f}^\times / \widehat{\mathscr{O}}^\times \simeq \mathrm{Pic}(\mathscr{O}),$$

where $\mathrm{Pic}(\mathscr{O})$ *denote the Picard group of $\mathscr{O}$: the group of $K^\times$-homothety classes of locally principal ideals of $\mathscr{O}$. In particular, if $\mathscr{O} = \mathscr{O}_K$ (the ring of integers of K) or is monomial (of the shape $\mathbb{Z}[x]$, $x \in \mathscr{O}_K$), then* $\mathrm{Pic}(\mathscr{O})$ *is just the ideal class group of $\mathscr{O}$.*

Proof. The bijection is induced by the map

$$\begin{aligned} \alpha_f = (\alpha_p)_p \in \mathbb{A}_{K,f}^\times &= (\alpha_p \mathscr{O}_p)_p = (I_p)_p \\ &\to I = \bigcap_p K \cap I_p \subset \{\text{locally principal } \mathscr{O}\text{-ideals}\}. \end{aligned}$$

If $\mathscr{O} = \mathscr{O}_K$ or is monomial, it is known that any finitely generated $\mathscr{O}$-module $I \subset K$ is locally principal (i.e. for any p, $I_p = \alpha_p \mathscr{O}_p$ for some $\alpha_p \in K_p^\times$). □

6.2 Adelic Points of an Algebraic Group

Let $G_{\mathbb{Q}}$ be an algebraic variety defined over $\mathbb{Q}$, since $\mathbb{A}_f$ and $\mathbb{A}$ are $\mathbb{Q}$-algebras one may speak of $G(\mathbb{A}_f)$, $G(\mathbb{A})$ the set of $\mathbb{A}_f$ and $\mathbb{A}$-points of G. More concretely, if

$$G \hookrightarrow \mathbf{A}^m_{\mathbb{Q}}$$

is an affine variety, defined by some ideal $\mathcal{I}_G \subset \mathbb{Q}[X_1, \dots, X_m]$ one has

$$G(\mathbb{A}) = \{\mathbf{x} \in \mathbb{A}^m \,|\, P(\mathbf{x}) = 0,\ \forall\ P \in \mathcal{I}_G\}.$$

Given such an affine embedding, we define similarly

$$G(\mathbb{A}_f), G(\widehat{\mathbb{Z}}) = \prod_p G(\mathbb{Z}_p) = \prod_p G(\mathbb{Q}_p) \cap \mathbb{Z}_p^m.$$

Observe that the later set $G(\widehat{\mathbb{Z}})$ is not intrinsic but depend on the affine embedding of G.

In particular, if $G = \mathrm{G}$ a linear algebraic group, G is affine and $\mathrm{G}(\mathbb{A})$ and $\mathrm{G}(\mathbb{A}_f)$ are defined as above: for instance, let us recall why GL_n is affine: this is a subvariety of the (n^2+1)-dimensional affine space of $M_{n,\mathbb{Q}} \times \mathbf{A}^1_{\mathbb{Q}}$ ($M_{n,\mathbb{Q}}$ the space of $(n \times n)$-matrices) defined by the equation

$$\det(g)t = 1.$$

If $\mathrm{G} \subset GL_n$ is a linear algebraic group defined over $\mathbb{Q}$, one has

$$\mathrm{G}(\mathbb{A}) = \prod_v{}' \mathrm{G}(\mathbb{Q}_v) = \{(g_p)_p \,|\, g_p \in \mathrm{G}(\mathbb{Q}_p),\ g_p \in \mathrm{G}(\mathbb{Z}_p) \text{ a.e. } p\},$$

where

$$\mathrm{G}(\mathbb{Z}_p) = \mathrm{G}(\mathbb{Q}_p) \cap GL_n(\mathbb{Z}_p),$$

we also set

$$\mathrm{G}(\widehat{\mathbb{Z}}) := \mathrm{G}(\mathbb{A}_f) \cap GL_n(\widehat{\mathbb{Z}}).$$

More generally if $L \in \mathcal{L}_n(\mathbb{Q})$ is a lattice L corresponding to $\widehat{L} = (L_p)_p \subset \mathbb{A}_f^n$, $L_p \in \mathcal{L}_n(\mathbb{Q}_p)$ we set

$$\mathrm{G}(\widehat{L}) = \prod_p \mathrm{G}(L_p),\ \mathrm{G}(L_p) = \mathrm{Stab}_{\mathrm{G}(\mathbb{Q}_p)}(L_p),$$

thus $\mathrm{G}(\widehat{\mathbb{Z}}) = \mathrm{G}(\widehat{\mathbb{Z}^n})$.

6.2.1 The Multiplicative Group

Let $(\mathbb{G}_m, \times)$ denote the multiplicative group: it maybe realized as GL_1 and we have

$$\mathbb{G}_m(\mathbb{A}_f) = GL_1(\mathbb{A}_f) = \mathbb{A}_f^\times,$$

$$\mathbb{G}_m(\mathbb{A}) = GL_1(\mathbb{A}) = \mathbb{A}^\times = \mathbb{R}^\times \times \mathbb{A}_f^\times$$

is the group of units of $\mathbb{A}$.

6.2.2 The Additive Group

Let $(\mathbb{G}_a, +)$ denote the additive group. $\mathbb{G}_a$ maybe realized as the subgroup of GL_2 of triangular unipotent matrices

$$\left\{\begin{pmatrix} a & b \\ c & d \end{pmatrix} \in GL_2 \,\middle|\, c = a - 1 = d - 1 = 0\right\}.$$

We have

$$\mathbb{G}_a(\mathbb{A}) = \mathbb{A},\ \mathbb{G}_a(\mathbb{A}_f) = \mathbb{A}_f.$$

6.2.3 The Torus Associated to a Number Field

Let K be a number field of degree n; the choice of a $\mathbb{Q}$-basis of K, $K = \mathbb{Q}\mathbf{e}_1 + \cdots + \mathbb{Q}\mathbf{e}_n$ yields an identification

$$\iota : K \simeq \mathbb{Q}^n$$

and the action on K by multiplication induces a $\mathbb{Q}$-algebra embedding

$$\theta : K \hookrightarrow \mathrm{M}_n(\mathbb{Q}),$$

whose image $\theta(K)$ is precisely the set of elements of $\mathrm{M}_n(\mathbb{Q})$ commuting with $\theta(\mathbf{e}_1), \ldots, \theta(\mathbf{e}_n)$. These conditions of commutation are algebraic condition and restricting to the linear group GL_n, we see that $\theta(K^\times) = T_K(\mathbb{Q})$ is the set of rational points of an algebraic group

$$T_K \simeq \mathrm{res}_{K/\mathbb{Q}}\mathbb{G}_{mK}.$$

Namely the restriction of scalars to $\mathbb{Q}$ of the multiplicative group over K: for any $\mathbb{Q}$-algebra A,

$$T_K(A) = (K \otimes_{\mathbb{Q}} A)^\times.$$

In particular $T_K(\mathbb{C}) \simeq (\mathbb{C}^\times)^n$ (so T_K is a torus) and

$$T_K(\mathbb{Q}) \simeq K^\times,\ T_K(\mathbb{A}) \simeq \mathbb{A}_K^\times,\ T_K(\widehat{\mathbb{Z}}) \simeq \widehat{\mathscr{O}}^\times,$$

where $\mathscr{O}$ is the order of the $\mathbb{Z}$-module $I = \mathbb{Z}\mathbf{e}_1 + \cdots + \mathbb{Z}\mathbf{e}_n$: the set of $\lambda \in K$ such that $\lambda I \subset I$.

6.3 The Adelic Topology

The rings of adeles and finite adeles are equipped with the restricted product topology (a basis of open sets is given by products of the shape

$$\Omega = \prod_v \Omega_v,\ \Omega_v \subset \mathbb{Q}_v \text{ open and } \Omega_p = \mathbb{Z}_p \text{ for a.e. } p)$$

and one can verify that these are locally compact algebras.

One verifies the following facts

- For every $v \in \mathcal{V}$, the local field $\mathbb{Q}_v$ embeds as a closed subset of $\mathbb{A}$ via
$$x_v \hookrightarrow (0, \dots, 0, x_v, 0, \dots) \in \mathbb{A}.$$
- $\mathbb{Q} \hookrightarrow \mathbb{A}_f$ (diagonally embedded) is dense in $\mathbb{A}_f$ (a reformulation of the Chinese Reminder Theorem).
- The closure of $\mathbb{Z}$ in $\mathbb{A}_f$ is the ring $\widehat{\mathbb{Z}} = \prod_p \mathbb{Z}_p$. $\widehat{\mathbb{Z}}$ is a open-compact subring of $\mathbb{A}_f$ maximal for this property.
- On the other hand $\mathbb{Q} \hookrightarrow \mathbb{A}$ is discrete ($\mathbb{Q} \cap (]-1/2, 1/2[\times\widehat{\mathbb{Z}}) = \{0\} \in \mathbb{A}$) but cocompact (i.e. $\mathbb{Q}\backslash\mathbb{A}$ is compact: $\mathbb{Q}\backslash\mathbb{A}/\widehat{\mathbb{Z}} \simeq \mathbb{R}/\mathbb{Z}$).

One immediate derive similar properties for the product spaces $\mathbb{A}^n$ equipped with the product topology. In particular

- $\mathbb{Q}^n$ is dense in $\mathbb{A}_f^n$ and discrete and cocompact in $\mathbb{A}^n$.
- For every lattice $L \in \mathcal{L}_n(\mathbb{Q})$, the closure of L in $\mathbb{A}_f^n$ is $\widehat{L} = \prod_p L_p$. $\widehat{\mathbb{Z}}$ is a open-compact subring of $\mathbb{A}_f$ maximal for this property.

The set of adelic points of an affine algebraic variety inherit the topology as a closed subset of the affine adelic space it is embedded in (the topology is independent of the embedding). For G a linear algebraic group, $\mathrm{G}(\mathbb{A})$ and $\mathrm{G}(\mathbb{A}_f)$ become then locally compact topological groups.

We have

- For every $v \in \mathcal{V}$, the local group $\mathrm{G}(\mathbb{Q}_v)$ embeds as a closed subgroup of $\mathrm{G}(\mathbb{A})$ via
$$g_v \hookrightarrow (e_{\mathrm{G}}, \dots, e_{\mathrm{G}}, g_v, e_{\mathrm{G}}, \dots) \in \mathrm{G}(\mathbb{A}).$$
- $\mathrm{G}(\mathbb{Q})$ is discrete in $\mathrm{G}(\mathbb{A})$. However, $\mathrm{G}(\mathbb{Q})\backslash\mathrm{G}(\mathbb{A})$ is in general *not* compact: for instance $\mathbb{Q}^\times\backslash\mathbb{A}^\times/\widehat{\mathbb{Z}}^\times \simeq \mathbb{R}^\times/\{\pm 1\}$. See below for several important compactness statements.
- The subgroup $\mathrm{G}(\widehat{\mathbb{Z}})$ and more generally $\mathrm{G}(\widehat{L})$ for L a lattice is an open compact subgroup of $\mathrm{G}(\mathbb{A}_f)$. The subgroups $GL_n(\widehat{L})$ are the maximal open-compact subgroups of $GL_n(\mathbb{A}_f)$; there are all conjugate.
- If $K_f, K_f' \subset \mathrm{G}(\mathbb{A}_f)$ are open compact subgroups, they are commensurable: $K_f \cap K_f'$ is open compact and of finite index in K_f and K_f'.

One main problem in number thoery is to understand the structure of the topological quotient $[\mathrm{G}] = \mathrm{G}(\mathbb{Q})\backslash\mathrm{G}(\mathbb{A})$ as a $\mathrm{G}(\mathbb{A})$-homogeneous space – for the action of $\mathrm{G}(\mathbb{A})$ by right multiplication – and the space of continuous functions on it (as a representation of $\mathrm{G}(\mathbb{A})$).

For instance basic question are whether $[\mathrm{G}]$ is compact: it is the case when $\mathrm{G} = \mathbb{G}_a$, the additive group or when G is nilpotent but, for instance
$$[\mathbb{G}_m] = \mathbb{Q}^\times\backslash\mathbb{A}^\times \simeq \mathbb{R}^\times\widehat{\mathbb{Z}}^\times/\{\pm 1\}$$
is not compact. We begin with the following general finiteness result:

6.4 Borel Finiteness Theorem

Let us recall the decomposition

$$GL_n(\mathbb{Q}) \cdot GL_n(\widehat{\mathbb{Z}}) = GL_n(\mathbb{A}_f)$$

or in other terms the double quotient

$$GL_n(\mathbb{Q})\backslash GL_n(\mathbb{A}_f)/GL_n(\widehat{\mathbb{Z}})$$

has only one element. More generally the double quotient

$$GL_n(\mathbb{Q})\backslash GL_n(\mathbb{A}_f)/K_f$$

is finite for $K_f \subset GL_n(\mathbb{A}_f)$ any open compact subgroup (since it is commensurable with $GL_n(\widehat{\mathbb{Z}})$). This is a special case of a much more general result:

Theorem 6.1 (Borel). *If* G *is reductive, then for any open compact subgroup* $K_f \subset \mathrm{G}(\mathbb{A}_f)$ *(for instance* $\mathrm{G}(\widehat{\mathbb{Z}})$*), the double quotient*

$$\mathrm{G}(\mathbb{Q})\backslash \mathrm{G}(\mathbb{A}_f)/K_f \simeq \mathrm{G}(\mathbb{Q})\backslash \mathrm{G}(\mathbb{A})/\mathrm{G}(\mathbb{R})K_f$$

is finite. In particular $[\mathrm{G}]$ *is compact if* $\mathrm{G}(\mathbb{R})$ *is compact.*

In particular taking $\mathrm{G} = SO_q$ one see via Proposition 5.2 that Borel's theorem is a wide generalization of the Hermite-Minkowski Theorem 4.3 on the finiteness of the q-genus of a lattice.

6.5 Measures on Adelic Groups

As a way to discuss the "size" of $[\mathrm{G}]$ we would like to put a measure on that space which is invariant under the $\mathrm{G}(\mathbb{A})$-action. such measure is obtained as a quotient of *Haar measures* on the appropriate groups. For this we need to discuss first such Haar measures (which exist since $\mathrm{G}(\mathbb{A})$ is locally compact) and their modulus.

6.5.1 The Adelic Modulus

Let

$$|x_\infty|_\infty = \max\{x_\infty, -x_\infty\}$$

denote the usual absolute value on $\mathbb{Q}_\infty = \mathbb{R}$ and

$$|x_p| = p^{-v_p(x_p)},\ v_p(x_p) = \max_{\alpha\in\mathbb{Z}}\{p^{-\alpha}x_p \in \mathbb{Z}_p\}$$

is the usual non-archimedean absolute value on $\mathbb{Q}_p$. The *adelic modulus* function is defined on $\mathbb{A}^\times$ by

$$|x|_{\mathbb{A}} = |(x_v)_v|_{\mathbb{A}} = \prod_v |x_v|_v$$

(observe that $|x_p|_p = 1$ for a.e. p). The modulus $|\cdot|_{\mathbb{A}} \to \mathbb{R}^\times$ is a continuous group homomorphism whose kernel is denoted $\mathbb{A}^{(1)}$ (the ideles of modulus 1). For $x_{\mathbb{Q}} \in \mathbb{Q}^\times \subset \mathbb{A}^\times$, one has (product formula)

$$|x_{\mathbb{Q}}|_{\mathbb{A}} = |x_{\mathbb{Q}}| \prod_p p^{-v_p(x_{\mathbb{Q}})} = 1$$

so

$$\mathbb{Q}^\times \subset \mathbb{A}^{(1)}.$$

In particular, since $|x_\infty|_{\mathbb{A}} = |x_\infty|$ we see that $\mathbb{Q}^\times \backslash \mathbb{A}^\times$ is not compact, on the other hand

Lemma 6.1. *The quotient $\mathbb{Q}^\times \backslash \mathbb{A}^{(1)}$ is compact.*

6.5.2 The Haar Measure

Given G a linear algebraic group, the groups $G(\mathbb{A})$ and $G(\mathbb{A}_f)$ are locally compact topological groups hence admit (left-invariant) Haar measures. Such measure can be realized as a product of measures on the factors $G(\mathbb{Q}_v)$: for any place v let μ_{G_v} denote some Haar measure on $G(\mathbb{Q}_v)$; it is defined up to a scalar and we choose it so that

$$\mu_{G_p}(G(\mathbb{Z}_p)) = 1 \text{ for a.e. } p$$

then the product measure

$$\mu_G = \mu_{G_\infty} \otimes \mu_{G_f} = \mu_{G,\infty} \otimes \bigotimes_p \mu_{G_p}$$

defines a Haar measure on $G(\mathbb{A})$. Of course other choices are possible and there is a canonical one: the *Tamagawa* measure.

The closed (discrete) subgroup $G(\mathbb{Q})$ will be equipped with its Haar measure which is the *counting measure.*

Example 6.1. The Haar measure on $\mathbb{Q}_\infty = \mathbb{R}$, is the usual Lebesgue measure dx (say normalized to give mass 1 to the interval $[0,1]$); on $\mathbb{Q}_p$, we note dx_p the one which gives mass 1 to $\mathbb{Z}_p$; a Haar measure on $\mathbb{A}$ is $\prod_v dx_v$.

The Haar measure on $\mathbb{R}^\times$ is $d^\times x = dx/|x|_\infty$ and on $\mathbb{Q}_p^\times$ is $d^\times x_p = dx_p/|x_p|_p$; the Haar measure on $\mathbb{A}^\times$ is $\prod_v d^\times x_v$.

6.5.3 The Modulus

In general $G(\mathbb{A})$ is non-abelian and the Haar measure has a modulus $\Delta_{G(\mathbb{A})}$: it is given as the product of moduli $\Delta_{G(\mathbb{Q}_v)}$ and can be expressed in terms of the adelic modulus as

$$\Delta_{G(\mathbb{A})}(g) = |\det(\mathrm{Ad}(g)|\mathfrak{N}_{\mathbb{Q}} \otimes \mathbb{A})|_{\mathbb{A}} = \prod_v |\det(\mathrm{Ad}(g_v)|\mathfrak{N}_{\mathbb{Q}} \otimes \mathbb{Q}_v)|,$$

where $\mathrm{Ad}(g)$ denote the adjoint action of g on the adelic points of the Lie algebra, $\mathfrak{N}_{\mathbb{Q}}$ of the unipotent radical of G. In particular,

$$\Delta_{\mathrm{G}(\mathbb{A})}(\mathrm{G}(\mathbb{Q})) = \Delta_{\mathrm{G}(\mathbb{A})}(\mathrm{G}(\widehat{\mathbb{Z}})) = \{1\}$$

($\mathrm{G}(\widehat{\mathbb{Z}})$ being compact) and if G is *reductive*, $\mathrm{G}(\mathbb{A})$ (and $\mathrm{G}(\mathbb{A}_f)$ as well as all $\mathrm{G}(\mathbb{Q}_v)$) is unimodular.

In the sequel we will equip the quotient $\mathrm{G}(\mathbb{Q})\backslash\mathrm{G}(\mathbb{A})$ with the quotient of the Haar measure on $\mathrm{G}(\mathbb{A})$ by the counting measure on $\mathrm{G}(\mathbb{Q})$: since $\Delta_{\mathrm{G}(\mathbb{A})}(\mathrm{G}(\mathbb{Q})) = 1$ the quotient measure exists : we denote it by $\mu_{[\mathrm{G}]}$ (abusing notations slightly we will also denote by $\mu_{[\mathrm{G}]}$ the quotient (or the restriction) of that measure by subgroups such as the center).

6.5.4 The Borel-Harish-Chandra Theorem

We have

$$\mathbb{Q}^\times\backslash\mathbb{A}^\times \simeq \mathbb{R}_{>0} \times \widehat{\mathbb{Z}}^\times$$

and its Haar measure is $d^\times x_\infty \times d^\times x_f$, and therefore

$$\mu_{[\mathrm{G}]}(\mathbb{Q}^\times\backslash\mathbb{A}^\times) = \infty.$$

The Borel-Harish-Chandra theorem states that for a general reductive group, this is essentially the only obstruction to finiteness of the Haar measure:

Theorem 6.2 (Borel-Harish-Chandra). *Let* G *be reductive, and* $\mathrm{Z}(\mathbb{A})$ *be its center then*

$$\mu_{[\mathrm{G}]}(\mathrm{Z}(\mathbb{A})\mathrm{G}(\mathbb{Q})\backslash\mathrm{G}(\mathbb{A})) < \infty.$$

An equivalent formulation is the following: let $\mathrm{G}(\mathbb{A})^{(1)}$ denote the closed subgroup

$$\mathrm{G}(\mathbb{A})^{(1)} = \bigcap_{\chi\in X^*(\mathrm{G})_{\mathbb{Q}}} \ker(|\chi(\cdot)|_{\mathbb{A}}) \subset \mathrm{G}(\mathbb{A}),$$

where χ ranges over the group of $\mathbb{Q}$-characters of G; $\mathrm{G}(\mathbb{A})^{(1)}$ is the analog of the group of ideles of modulus 1 and in particular $\mathrm{G}(\mathbb{A})^{(1)}$ contains $\mathrm{G}(\mathbb{Q})$ as a discrete subgroup.

Theorem 6.3 (Borel-Harish-Chandra). *Let* G *be reductive, then*

$$\mu_{[\mathrm{G}]}(\mathrm{G}(\mathbb{Q})\backslash\mathrm{G}(\mathbb{A})^{(1)}) < \infty.$$

if, in addition, $\mathrm{G}(\mathbb{Q})$ *has no nontrivial unipotent element, then* $\mathrm{G}(\mathbb{Q})\backslash\mathrm{G}(\mathbb{A})^{(1)}$ *is compact.*

Remark 6.1. If $\mathrm{G}(\mathbb{Q})$ has no nontrivial unipotent element, then one can show using Mahler's compactness criterion that $\mathrm{G}(\mathbb{Q})\backslash\mathrm{G}(\mathbb{A})^{(1)}$ and $\mathrm{Z}(\mathbb{A})\mathrm{G}(\mathbb{Q})\backslash\mathrm{G}(\mathbb{A})$ are compact.

7 The Hasse Integral Principle and Equidistribution of Adelic Orbits

We consider now specifically the case of $\mathrm{G} = SO_q$ is the special orthogonal group of a non-degenerate quadratic form of rank $n \geqslant 3$: from the previous section the quotient $\mathrm{G}(\mathbb{Q})\backslash\mathrm{G}(\mathbb{A})$ is compact if (and only if) q is anisotropic (does not represent 0 in $\mathbb{Q}^n$) and in any case it always has finite $\mu_{[\mathrm{G}]}$-measure. We will always normalize $\mu_{[\mathrm{G}]}$ to be a probability measure:

$$\mu_{[\mathrm{G}]}(\mathrm{G}(\mathbb{Q})\backslash\mathrm{G}(\mathbb{A})) = 1.$$

Let us recall that for L a lattice, one has the identification

$$[\mathrm{gen}_q(L)] \simeq \mathrm{G}(\mathbb{Q})\backslash\mathrm{G}(\mathbb{A})/\mathrm{G}(\mathbb{R})\mathrm{G}(\widehat{L})$$

which is finite (by the Hermite-Minkowski theorem or Borel finiteness theorem). This identification extend the following homeomorphism of topological spaces: for $\{L_1, L_2, \dots, L_h\}$ a set of representatives of $[\mathrm{gen}_q(L)]$ we have

$$\mathrm{G}(\mathbb{Q})\backslash\mathrm{G}(\mathbb{A})/\mathrm{G}(\widehat{L}) \simeq \bigsqcup_{i=1,\dots,h} \mathrm{G}(L_i)\backslash\mathrm{G}(\mathbb{R}).$$

The above space is equipped with a natural $\mathrm{G}(\mathbb{R})$-invariant probability measure $\mu_{[\mathrm{G}],L}$: that measure is described either as the quotient of $\mu_{[\mathrm{G}]}$ by the probability Haar measure of $\mathrm{G}(\widehat{L})$ (which is open-compact) or simply as the restriction of $\mu_{[\mathrm{G}]}$ to the closed subspace of continuous functions on $\mathrm{G}(\mathbb{Q})\backslash\mathrm{G}(\mathbb{A})$ which are right-$\mathrm{G}(\widehat{L})$ invariant.

Proposition 7.1. *Under the above homeomorphism, the measure $\mu_{[\mathrm{G}],L}$ correspond to the sum of the measures*

$$\frac{1}{\mathrm{vol}}\sum_i \mu_i,$$

where $\mu_i = \mu_{\mathrm{G}(L_i)\backslash\mathrm{G}(\mathbb{R})}$ is the quotient of the counting measure on the (discrete) subgroup $\mathrm{G}(L_i) \subset \mathrm{G}(\mathbb{R})$ by μ_{G_∞} with

$$\mathrm{vol} = \sum_i \mu_i(\mathrm{G}(L_i)\backslash\mathrm{G}(\mathbb{R})).$$

Remark 7.1. When q is positive definite, $\mathrm{G}(\mathbb{R})$ is compact, and the (finite dimensional) space of function on $[\mathrm{gen}_q(L)]$ is identified with the subspace of functions on $\mathrm{G}(\mathbb{Q})\backslash\mathrm{G}(\mathbb{A})/\mathrm{G}(\widehat{L})$ which are right $\mathrm{G}(\mathbb{R})$-invariant. The restriction of $\mu_{[\mathrm{G}],L}$ to that subspace, we obtain precisely the probability measure $\mu_{[\mathrm{gen}_q(L)]}$.

7.1 Representations by the Genus as Adelic Orbits II

Let $\mathbf{x} \in \mathcal{R}_q(L)$ be a representation of d in L and

$$\mathrm{H} = \mathrm{Stab}_{\mathrm{G}}(\mathbf{x}) = SO_{q|\mathbf{x}^\perp}.$$

We will assume that the rank-$(n-1)$ quadratic form, $q_{|\mathbf{x}^\perp}$, is anisotropic (which for $n = 3$ is the case if and only if $-\mathrm{disc}_L(q)d$ is not a square in $\mathbb{Q}$) so that $[\mathrm{H}] = \mathrm{H}(\mathbb{Q})\backslash\mathrm{H}(\mathbb{A})$ is compact and has an $\mathrm{H}(\mathbb{A})$-invariant probability measure noted $\mu_{[\mathrm{H}]}$ or $\mu_{\mathbf{x}}$ in the sequel.

As we have seen (Proposition 5.3), the quotient

$$[\mathrm{gen}_{q,\mathbf{x}}(L)] = \mathrm{H}(\mathbb{Q})\backslash\mathrm{H}(\mathbb{A}_f)\cdot L \simeq \mathrm{H}(\mathbb{Q})\backslash\mathrm{H}(\mathbb{A})/\mathrm{H}(\mathbb{R})\mathrm{H}(\widehat{L})$$

is identified with a subset of $[\mathcal{R}_q(d, \mathrm{gen}_q(L))]$; moreover the image of $\mathrm{H}(\mathbb{A})$ inside $\mathrm{G}(\mathbb{Q})\mathrm{G}(\mathbb{A})/\mathrm{G}(\mathbb{R})\mathrm{G}(\widehat{L})$ correspond to a set of genus classes of L admitting a representation of d by q.

Since $\mathrm{H}(\mathbb{Q})\backslash\mathrm{H}(\mathbb{A})$ injects into $\mathrm{G}(\mathbb{Q})\backslash\mathrm{G}(\mathbb{A})$ as a closed subspace, the measure $\mu_{\mathbf{x}}$ defines a probability measure on $\mathrm{G}(\mathbb{Q})\backslash\mathrm{G}(\mathbb{A})$ and on $\mathrm{G}(\mathbb{Q})\backslash\mathrm{G}(\mathbb{A})/\mathrm{G}(\widehat{L})$ upon restriction to the subspace of $\mathrm{G}(\widehat{L})$-invariant functions. Linnik's theorem is a consequence of the following equidistribution result:

Theorem 7.1. *Let q be a non-degenerate ternary quadratic form which is integral on some lattice L and $p \nmid \mathrm{disc}(q)$ be a prime. Let $d \to \infty$, ranging amongst the square-free integers, coprime with $\mathrm{disc}(q)$, which admit a representation by q, $\mathbf{x} \in L$, and such that*

$$p \text{ splits in the field } K = \mathbb{Q}(-\mathrm{disc}(q)d), \tag{7.1}$$

the measure $\mu_{\mathbf{x}}$ weak-$*$ converge to the measure $\mu_{[\mathrm{G}],L}$ on $\mathrm{G}(\mathbb{Q})\backslash\mathrm{G}(\mathbb{A})/\mathrm{G}(\widehat{L})$. That is, for any continuous function $\phi : \mathrm{G}(\mathbb{Q})\backslash\mathrm{G}(\mathbb{A}) \to \mathbb{C}$ which is right $\mathrm{G}(\widehat{L})$-invariant,

$$\int_{\mathrm{H}(\mathbb{Q})\backslash\mathrm{H}(\mathbb{A})} \phi(t)dt \to \int_{\mathrm{G}(\mathbb{Q})\backslash\mathrm{G}(\mathbb{A})} \phi(g)dg.$$

In particular, if q is definite positive, $\mathrm{G}(\mathbb{R})$ is compact and Linnik's theorem 4.6, follows by restricting the above statement to the functions which are right $\mathrm{G}(\mathbb{R})\mathrm{G}(\widehat{\mathbb{Z}})$-invariant.

Remark 7.2. This theorem hold in a much greater generality: it is possible to replace the group $\mathrm{G}(\widehat{L})$ with any open-compact subgroup $K_f \subset \mathrm{G}(\mathbb{A}_f)$ and the condition d *square-free* with the condition $d \notin \{d_i(\mathbb{Q}^\times)^2 \,|\, i = 1, \ldots, s_{K_f}\}$ *some finite set of square classes depending only on K_f*. In addition, due to the work of Iwaniec [9], Duke [1], Duke and Schulze-Pillot [2], this theorem, these relaxed conditions, is valid without Linniks' condition (7.1). See [4, Theorem 4.6] for a general statement implying the above claim.

7.2 Interpretation in Terms of Quaternions

Binary ternary and quaternary quadratic spaces can be interpreted in terms of quadratic and quaternion algebras. We review here this interpretation which will be useful in the sequel. Recall the following (see [10] for this as well as [16] for an introduction to quaternions):

Proposition 7.2. *Given (V, q) a non-degenerate binary quadratic space over a field k of characteristic $\neq 2$, there exists a quadratic algebra K/k (with involution noted σ), a scalar $\lambda \in k^\times$ and an isometry*

$$(V, q) \simeq (K, \lambda\,\mathrm{Nr}_K).$$

Here Nr_K denote the K/k-norm form on K (the determinant for the action of K on itself by multiplication). The form q is anisotropic (admits no nonzero isotropic vector) if and only if K is a field (otherwise $K = k \times k$).

Moreover one has an isomorphism of k-algebraic groups

$$SO(V, q) \simeq SO(K, \mathrm{Nr}_K) \simeq K^{(1)} = \{t \in K^\times \mid \mathrm{Nr}(t) = 1\} = \left\{ \frac{t}{\sigma(t)} \,\middle|\, t \in K^\times \right\},$$

the group $K^{(1)}$ acting on K by multiplication.

Proposition 7.3. *Given (V, q) a non-degenerate ternary quadratic space over a field of characteristic $\neq 2$, there exists a quaternion algebra B, a scalar $\lambda \in k^\times$ and an isometry*

$$(V, q) \simeq (\mathrm{B}^0, \lambda\,\mathrm{Nr}).$$

Here B^0 denote the subspace of trace 0 quaternions and Nr denote the reduced norm on B. Moreover one has an isomorphism of k-algebraic groups

$$SO(V, q) \simeq SO(\mathrm{B}^0, \mathrm{Nr}) \simeq \mathrm{PB}^\times,$$

where $\mathrm{PB}^\times$ denote the projective group of units of B acting on B^0 by conjugation: for any k-algebra F,

$$\mathrm{PB}^\times(F) = \mathrm{B}^\times(F)/F^\times.$$

Moreover, q is anisotropic if and only if B is a skew field (otherwise B is the algebra of 2×2 matrices $M_{2,k}$, $\mathrm{PB}^\times = PGL_2$).

In particular we have

$$\mathrm{disc}_V(q) \equiv \lambda\,\mathrm{disc}_{rd}(\mathrm{B})^2 \equiv \lambda \mod (\mathbb{Q}^\times)^2,$$

where $\mathrm{disc}_{rd}(\mathrm{B})$ denote the reduced discriminant of B.

We will make free use of this identification for our given quadratic form $(\mathbb{Q}^3, q)$; for notational simplicity, we will assume that $\lambda = 1$. We will identify the lattice L with some lattice in $\mathrm{B}^0(\mathbb{Q})$ (also noted L) such that $\mathrm{Nr}_\mathrm{B}(L) \subset \mathbb{Z}$. Notice that for q definite, $\mathrm{B}(\mathbb{R})$ is isomorphic to the algebra of Hamilton quaternions and that for any $p \nmid \mathrm{disc}_L(q)$ $(\mathbb{Q}_p^3, q)$ is isotropic and therefore $\mathrm{B}(\mathbb{Q}_p) = M_2(\mathbb{Q}_p)$ is the split quaterion algebra (equipped with the determinant quadratic form).

It will be also useful to identify certain subgroups of SO_q in terms of quaternions. Given $\mathbf{x} \in \mathbb{Q}^3$ be some rational representation of d,

$$q(\mathbf{x}) = d.$$

$\mathbf{x}$ correspond to some trace zero quaternion still noted $\mathbf{x}$ which satifies

$$\mathrm{Nr}_{\mathrm{B}}(\mathbf{x}) = d = -\mathbf{x}^2.$$

Therefore the subfield $\mathbb{Q}(\mathbf{x})$ generated by $\mathbf{x}$ in $\mathrm{B}(\mathbb{Q})$ is isomorphic to the quadratic field $K = \mathbb{Q}(\sqrt{-d})$.

Let H be the stabilizer of $\mathbf{x}$ in SO_q; H is a $\mathbb{Q}$-algebraic group, in fact, the special orthogonal group of the two dimensional quadratic space $\mathbf{x}^{\perp}$ (the subspace of elements of $\mathbb{Q}^3$ perpendicular to $\mathbf{x}$); in terms of quaternions, the set of its $\mathbb{Q}$-point is identified with the image in $\mathrm{PB}^{\times}(\mathbb{Q})$ of the invertible quaternions commuting with $\mathbf{x}$, i.e. the $\mathbf{y} \in \mathrm{B}^{\times}(\mathbb{Q})$ satisfying

$$\mathbf{yx} = \mathbf{xy}.$$

This is precisely $\mathbb{Q}(\mathbf{x})^{\times}$. More generally, for any $\mathbb{Q}$-algebra k, one has

$$\mathrm{H}(k) \simeq k^{\times} \backslash (K \otimes k)^{\times},$$

or said otherwise, one has an isomorphism of algebraic groups

$$\mathrm{H} \simeq \mathrm{res}_{K/\mathbb{Q}} \mathbb{G}_{m,K} / \mathbb{G}_{m,\mathbb{Q}}.$$

7.3 Counting the Number of Representations

In this section, we assume that $d > 0$ (so that K is imaginary quadratic) and estimate the size of

$$[\mathrm{gen}_{q,\mathbf{x}}(L)] = \mathrm{H}(\mathbb{Q}) \backslash \mathrm{H}(\mathbb{A}_f) \mathrm{H}(\widehat{L})$$

for $\mathrm{H}(\widehat{L})$ the stabilizer of $\widehat{L} \subset \mathrm{B}^0(\mathbb{A}_f)$ in $\mathrm{H}(\mathbb{A}_f)$. This subgroup contain the image in $\mathrm{PB}^{\times}(\mathbb{A}_f)$ of the $\lambda \in (\mathbb{Q}(x) \otimes_{\mathbb{Q}} \mathbb{A}_f)^{\times}$ such that

$$\lambda \cdot \widehat{L} = \widehat{L} \cdot \lambda$$

and in particular the image of the group of units $\widehat{\mathbb{Z}}[\mathbf{x}]^{\times}$. From this description, we see that the group (H is abelian since $n - 1 = 2$)

$$\mathrm{H}(\mathbb{Q}) \backslash \mathrm{H}(\mathbb{A}_f) / \mathrm{H}(\widehat{L})$$

injects into

$$\mathbb{A}_f^{\times} K^{\times} \backslash \mathbb{A}_{f,K}^{\times} / \widehat{\mathscr{O}_K}^{\times} \simeq \mathrm{Pic}(\mathscr{O}_K)$$

and it is also possible to see that the former has index bounded by $d^{o(1)}$. Therefore for $d > 0$, the estimate (4.2)

$$|[\mathrm{gen}_{q,\mathbf{x}}(L)]| = d^{1/2+o(1)}$$

follows from Dirichlet class number formula and Siegel's theorem

$$L(\chi_K, 1) = d^{o(1)}.$$

Observe that in this case, since H is abelian the measure $\mu_{\mathbf{x}}$ on

$$\mathrm{H}(\mathbb{Q})\backslash\mathrm{H}(\mathbb{A})/\mathrm{H}(\mathbb{R})\mathrm{H}(\widehat{L}) = \mathrm{H}(\mathbb{Q})\mathrm{H}(\mathbb{R})\mathrm{H}(\widehat{L})\backslash\mathrm{H}(\mathbb{A})$$

is the uniform probability measure.

8 The Ergodic Method

We assume that q is definite positive. We will not prove Theorem 7.1 in its strongest form here but rather the equidistribution statement restricted to function ϕ on $\mathrm{G}(\mathbb{Q})\backslash\mathrm{G}(\mathbb{A})$ which are $\mathrm{G}(\mathbb{R})\mathrm{G}(\widehat{L})$-invariant which is equivalent to proving the integral Hasse principle.

Let us introduce the following notations: for $L' \in \mathrm{gen}_q(L)$ we denote by

$$[L'] \in [\mathrm{gen}_q(L)] = SO_q(\mathbb{Q})\backslash SO_q(\mathbb{A}_f) \cdot L$$

its image under the canonical projection. For $L' \in \mathrm{H}(\mathbb{A}_f) \cdot L$ we denote by

$$[L']_{\mathbf{x}} \in \mathrm{H}(\mathbb{Q})\backslash\mathrm{H}(\mathbb{A}_f) \cdot L$$

its image under the canonical projection. We denote by π_q the natural projection map

$$\begin{array}{rclcl} \pi_q : \mathrm{H}(\mathbb{Q})\backslash\mathrm{H}(\mathbb{A}_f) \cdot L & \to & [\mathrm{gen}_q(L)] & = SO_q(\mathbb{Q})\backslash SO_q(\mathbb{A}_f) \cdot L, \\ [L']_{\mathbf{x}} & \to & [L']. & \end{array}$$

This map is the restriction to $\mathrm{H}(\mathbb{Q})\backslash\mathrm{H}(\mathbb{A}_f) \cdot L$ of the map

$$\pi_q : [\mathcal{R}_q(d; \mathrm{gen}_q(L))] \to [\mathrm{gen}_q(L)]$$

discussed in Section 4.3. Our main objective will be to show that for d large enough (and under Linnik's condition) the map π_q is surjective. We prove this by proving a stronger statement, namely that the measure $\mu_{\mathbf{x}}$ (viewed as a measure on $\mathrm{H}(\mathbb{Q})\mathrm{H}(\mathbb{A})$ then as a measure on $\mathrm{G}(\mathbb{Q})\backslash\mathrm{G}(\mathbb{A}) \hookleftarrow \mathrm{H}(\mathbb{Q})\mathrm{H}(\mathbb{A})$, then as a measure on $[\mathrm{gen}_q(L)]$ by restriction to the $\mathrm{G}(\mathbb{R})\mathrm{G}(\widehat{L})$-invariant functions) weak-$*$ converge to $\mu_{[\mathrm{gen}_q(L)]}$.

8.1 The Meaning of Linnik's Condition

Let us recall that we need to investigate the distribution properties of (the projection of) the adelic orbit $\mathrm{H}(\mathbb{A}_f) \cdot L$ on $[\mathrm{gen}_q(L)]$ and wish to exploit the action of $\mathrm{H}(\mathbb{A}_f)$ on the space of lattice. The principle of Linnik's method is to exploit only a very small part of this group action namely the action of the (much smaller) subgroup $\mathrm{H}(\mathbb{Q}_p)$.

By Linnik's splitting assumption (7.1), we have

$$K_p = K \otimes_{\mathbb{Q}} \mathbb{Q}_p \simeq \mathbb{Q}_p^2$$

and an isometry of quadratic spaces

$$(K_p, \mathrm{Nr}_{K/\mathbb{Q}}) \simeq (\mathbb{Q}_p \times \mathbb{Q}_p, XY).$$

This quadratic form is isotropic, and since (K, Nr_K) is embedded into $(\mathrm{B}^0, \mathrm{Nr}_{\mathrm{B}})$ the space $(\mathrm{B}^0(\mathbb{Q}_p), \mathrm{Nr}_{\mathrm{B}})$ is isotropic: in particular we may make the following identifications

$$\mathrm{B}^0(\mathbb{Q}_p) \simeq M_2^0(\mathbb{Q}_p),\ L_p \simeq M_2^0(\mathbb{Z}_p),$$
$$\mathrm{PB}^\times(\mathbb{Q}_p) \simeq PGL_2(\mathbb{Q}_p),$$
$$\mathrm{PB}^\times(L_p) \simeq PGL_d(\mathbb{Z}_p)$$

(indeed p is coprime with the discriminant of q); since $-d = \alpha_p^2$ for some $\alpha_p \in \mathbb{Z}_p^\times$, one may even choose the above identifications so that

$$\mathrm{H}(\mathbb{Q}_p) \simeq \mathrm{Diag}_2(\mathbb{Q}_p)/\mathbb{Q}_p^\times \simeq \mathbb{G}_m(\mathbb{Q}_p).$$

We denote by $t_p \subset \mathrm{H}(\mathbb{Q}_p)$ the preimage of the matrix $\begin{pmatrix} p & 0 \\ 0 & 1 \end{pmatrix} \in PGL_2(\mathbb{Q}_p)$ under the above identification; therefore we have

$$\mathrm{H}(\mathbb{Q}_p) = t_p^{\mathbb{Z}} \mathbb{H}(L_p).$$

Therefore, what Linnik's condition provide us with, is the action by a rank-one (split) p-adic torus $\mathrm{H}(\mathbb{Q}_p)$ at the specific prime p. We call this action the *Linnik flow.*

The principle of the ergodic method is to study the distribution properties of the images in $[\mathrm{gen}_q(L)]$ of the suborbits $t_p^{\mathbb{Z}} \cdot L' \subset \mathrm{H}(\mathbb{A}_f) \cdot L' \subset \mathrm{gen}_q(L)$ for $L' \in \mathrm{H}(\mathbb{A}_f) \cdot L$; we call this image $[t_p^{\mathbb{Z}} \cdot L']$ the *trajectory* associated to L'; in order to show that these trajectories fill the whole space as $d \to \infty$ we will see that they satisfy a certain "repulsion property" with respect to each other.

8.2 The Graph Structure on $[\mathrm{gen}_q(L)]$

To explain this we need to remark that $[\mathrm{gen}_q(L)]$ is endowed with an additional graph structure: recall that $[\mathrm{gen}_q(L)]$ is identified with the finite quotient of $\mathrm{G}(\mathbb{A}_f)/\mathrm{G}(\widehat{L})$ by $\mathrm{G}(\mathbb{Q})$ and

$$\mathrm{G}(\mathbb{A})/\mathrm{G}(\widehat{L}) = \prod_{p' \neq p}{}' \mathrm{G}(\mathbb{Q}_{p'}) / \prod_{p' \neq p} \mathrm{G}(L_{p'}) \times \mathrm{G}(\mathbb{Q}_p)/\mathrm{G}(\mathbb{Z}_p).$$

The quotient

$$\mathcal{T}_p = \mathrm{G}(\mathbb{Q}_p) \cdot L_p \simeq SO_q(\mathbb{Q}_p)/SO_q(L_p) \simeq PGL_2(\mathbb{Q}_p)/PGL_d(\mathbb{Z}_p)$$

has the structure of a $(p+1)$-regular tree wrt the following equivalent definitions of the adjacency relation (this is an instance of a Bruhat-Tits tree):

- Let $\mathbf{v} \in L'_p - pL'_p$ be such that $q(\mathbf{v}) \equiv 0(p)$ ($\mathbf{v}$ defines an isotropic line in the $\mathbb{F}_p$-quadratic space $(L'_p/pL'_p, q \pmod p)$); one can find $\mathbf{z}_0 \in L'_p$ such that $q(\mathbf{v} + p\mathbf{z}_0) \equiv 0 \pmod{p^2}$. Then

$$L'_{\overline{\mathbf{v}}} = \frac{1}{p}\mathbb{Z}_p(\mathbf{v} + p\mathbf{z}_0) + \{\mathbf{z} \in L'_p \,|\, \langle \mathbf{v}, \mathbf{z}\rangle \equiv 0 \pmod p\}$$

  is a lattice adjacent to L'_p which depends only on the isotropic line generated by $\overline{\mathbf{v}} = \mathbf{v} \pmod p$ in L'_p/pL'_p.
- Two lattices $L'_p, L''_p \in \mathcal{T}_p$ are adjacent if and only if $L'_p \cap L''_p$ is of index p in L'_p (or equivalently in L''_p); More generally, given two lattices $L'_p, L''_p \in \mathrm{G}(\mathbb{Q}_p) \cdot L_p$, the tree distance in $\mathcal{T}_p$ is given by

$$d(L'_p, L''_p) = \log([L'_p : L'_p \cap L''_p]) / \log p.$$

  the group $SO_q(\mathbb{Q}_p)$ acts by isometries relative to this distance.
- Under the identification $(L_p, q) \simeq (M_2^0(\mathbb{Z}_p), \det)$, L'_p is adjacent to L''_p if and only if there exists $g_p \in GL_2(\mathbb{Q}_p)$ with

$$|\det(g_p)|_p = p^{\pm 1}$$

  so that

$$L''_p = g_p L'_p g_p^{-1}.$$

- Under the above identification, if L'_p and L''_p are represented by the classes $g'PGL_d(\mathbb{Z}_p)$ and $g''PGL_d(\mathbb{Z}_p)$, there exists g_p as above such that

$$g'' = g_p g' = g' g''_p, \; g''_p = (g'^{-1} g_p g').$$

Since $\mathrm{G}(\mathbb{Q})$ acts on the tree $\mathrm{G}(\mathbb{Q}_p) \cdot L_{0,p}$ by tree automorphisms it follows that the quotient

$$[\mathrm{gen}_q(L)] \simeq \mathrm{G}(\mathbb{Q})\backslash \mathrm{G}(\mathbb{A})/\mathrm{G}(\mathbb{R})\mathrm{G}(\widehat{L})$$

has a structure of a $(p+1)$-regular graph (not necessarily connected): two lattice classes $[L']$, $[L''] \in [\mathrm{gen}_q(L)]$ are adjacent if and only if they are represented by two lattices $L', L'' \in \mathrm{gen}_q(L)$ whose localizations at p, L'_p and L''_p are adjacent in $\mathcal{T}_p$. In the sequel we denote this graph $\mathcal{G}_p$. We denote its set of vertices by $V(\mathcal{G}_p)$ and its set of edges by $E(\mathcal{G}_p)$. Recall that $V(\mathcal{G}_p)$ is identified with $[\mathrm{gen}_q(L)]$ and is endowed with the probability measure $\mu_{[\mathrm{gen}_q]}$ (which is not apriori the uniform measure).

8.2.1 The Structure of the Spectrum of $\mathcal{G}_p$ and Relation to Automorphic Forms

Let us recall the definition of the adjacency operator on $L^2(V(\mathcal{G}_p), \mu_{\mathrm{gen}_q})$: given $\phi : e \in V(\mathcal{G}_p) \mapsto \mathbb{C}$ a function

$$T_p\phi : e \mapsto \sum_{e \frown e'} f(e'),$$

where e' runs over the $p+1$ neighbors of the vertice e (counted with multiplicity).

This adjacency operator admits an adelic description: first, let us recall that the group $\mathrm{G}(\mathbb{A})$ acts on the L^2-space $L^2(\mathrm{G}(\mathbb{Q})\backslash\mathrm{G}(\mathbb{A}))$ (relative to the probability measure $\mu_{\mathrm{G}(\mathbb{Q})\backslash\mathrm{G}(\mathbb{A})}$) by right multiplications: given $g\in\mathrm{G}(\mathbb{A})$ and $\phi\in L^2(\mathrm{G}(\mathbb{Q})\backslash\mathrm{G}(\mathbb{A}))$

$$g\cdot\phi\in L^2(\mathrm{G}(\mathbb{Q})\backslash\mathrm{G}(\mathbb{A})):\ g'\mapsto\phi(g'g).$$

This is a unitary representation and as pointed out above, one has an isometry

$$L^2([\mathrm{gen}_q(L)],\mu_{[\mathrm{gen}_q]})\simeq L^2(\mathrm{G}(\mathbb{Q})\backslash\mathrm{G}(\mathbb{A}))^{\mathrm{G}(\mathbb{R})\mathrm{G}(\widehat{L})}$$

the subspace of $\mathrm{G}(\mathbb{R})\mathrm{G}(\widehat{L})$-invariant vectors in this representation.

By the definition of the adjacency relation and the above identification we see that T_p is given by the following convolution operator

$$\begin{aligned}T_p\phi(g)&=\frac{1}{\mu_{\mathrm{G}(\mathbb{Q}_p)}(\mathrm{G}(L_p))}1_{\mathrm{G}(L_p)g_p\mathrm{G}(L_p)}\star\phi(g)\\&=\frac{1}{\mu_{\mathrm{G}_p}(\mathrm{G}(L_p))}\int_{\mathrm{G}(L_p)g_p\mathrm{G}(L_p)}\phi(gg')d\mu_{\mathrm{G}_p}g',\end{aligned}$$

where g_p is the image in $\mathrm{G}(\mathbb{Q}_p)\simeq PGL_2(\mathbb{Q}_p)$ of any matrix such that $|\det(g_p)|_p=p^{\pm1}$; to see that the above convolution is a sum one write

$$\mathrm{G}(L_p)g_p\mathrm{G}(L_p)=\bigsqcup_{i=1}^{p+1}g_{p,i}\mathrm{G}(L_p)$$

and noting that ϕ is right-$\mathrm{G}(L_p)$ invariant we get

$$\frac{1}{\mu_{\mathrm{G}(\mathbb{Q}_p)}(\mathrm{G}(L_p))}\int_{\mathrm{G}(L_p)g_p\mathrm{G}(L_p)}\phi(gg')d\mu_{\mathrm{G}_p}g'=\sum_{i=1}^{p+1}\phi(gg_{p,i}).$$

In other term T_p is the p-th Hecke operator; from this description and the invariance of the Haar measure, it is elementary to see that this operator is selfadjoint.

The theory of automorphic representations of $\mathrm{G}(\mathbb{A})$ then gives important informations one the structure of the spectrum of T_p: one has the following orthogonal decomposition into T_p-eigenspaces

$$\begin{aligned}&L^2(\mathrm{G}(\mathbb{Q})\backslash\mathrm{G}(\mathbb{A}))^{\mathrm{G}(\mathbb{R})\mathrm{G}(\widehat{\mathbb{Z}})}\\&=L^2(\mathrm{G}(\mathbb{Q})\backslash\mathrm{G}(\mathbb{A}))_{\mathrm{fin}}^{\mathrm{G}(\mathbb{R})\mathrm{G}(\widehat{\mathbb{Z}})}\bigoplus L^2(\mathrm{G}(\mathbb{Q})\backslash\mathrm{G}(\mathbb{A}))_{\mathrm{cusp}}^{\mathrm{G}(\mathbb{R})\mathrm{G}(\widehat{\mathbb{Z}})}.\end{aligned}\tag{8.1}$$

Here

$$L^2(\mathrm{G}(\mathbb{Q})\backslash\mathrm{G}(\mathbb{A}))_{\mathrm{fin}}^{\mathrm{G}(\mathbb{R})\mathrm{G}(\widehat{\mathbb{Z}})}=\bigoplus_{\chi^2=1}\mathbb{C}\chi\circ\mathrm{Nr},\tag{8.2}$$

where the sum runs over suitable Hecke characters of order at most 2, $\chi:\mathbb{Q}^\times\backslash\mathbb{A}^\times\to\mathbb{C}^\times$ and $\chi\circ\mathrm{Nr}$ is the function on $\mathrm{B}^\times(\mathbb{Q})\backslash\mathrm{B}^\times(\mathbb{A})$ given by[2]

$$\chi\circ\mathrm{Nr}(g)=\chi(\mathrm{Nr}(g)),$$

[2]since $\chi^2=1$, this function descent to $\mathrm{PB}^\times(\mathbb{A})$.

and

$$L^2(\mathrm{G}(\mathbb{Q})\backslash\mathrm{G}(\mathbb{A}))^{\mathrm{G}(\mathbb{R})\mathrm{G}(\widehat{\mathbb{Z}})}_{\mathrm{cusp}} = \bigoplus_{\phi} \mathbb{C}\phi. \tag{8.3}$$

for ϕ varying over an orthonormal set of T_p-eigenforms such that, for each ϕ, the set of its translates $\mathrm{G}(\mathbb{A}) \cdot \phi = \{g \cdot \phi \,|\, g \in \mathrm{G}(\mathbb{A})\}$ generate an irreducible infinite dimensional sub-representation of $L^2(\mathrm{G}(\mathbb{Q})\backslash\mathrm{G}(\mathbb{A}))$.

Since $\mathrm{Nr} : \mathrm{B}^\times(\mathbb{Z}_p) = GL_2(\mathbb{Z}_p) \mapsto \mathbb{Z}_p^\times$ is surjective any χ occurring in (8.2) above has to be $\mathbb{Z}_p^\times$-invariant, hence we see that for such χ

$$T_p\chi \circ \mathrm{Nr}(\cdot) = (p+1)\chi(p) = \pm(p+1).$$

On the other hand, for ϕ occurring in (8.3) write

$$T_p\phi = \lambda_\phi(p);$$

it follows from the work of Jacquet-Langlands that $\lambda_\phi(p)$ is the eigenvalue of the Hecke operator of a standard holomorphic modular form of weight 2 and level coprime with p; it then follows from Deligne's work that

$$|\lambda_\phi(p)| \leqslant 2\sqrt{p} < p+1.$$

In particular we obtain

Proposition 8.1. *The number of connected components of the graph $\mathcal{G}_p$ is precisely the number of χ occurring in* (8.2) *such that* $\chi(p) = 1$.

Proof. Indeed the number of connected components of $\mathcal{G}_p$ is precisely the multiplicity of the $(p+1)$-eigenvalue of the adjacency operator. □

We will use this proposition to prove the following partial result towards Theorem 7.1.

Theorem 8.1. *Under the assumption of Theorem* 7.1, *one has*

$$\int_{\mathrm{H}(\mathbb{Q})\backslash\mathrm{H}(\mathbb{A})} \phi(t)dt \to \int_{\mathrm{G}(\mathbb{Q})\backslash\mathrm{G}(\mathbb{A})} \phi(g)dg,\ d \to \infty$$

for ϕ the characteristic function of any connected component of $\mathcal{G}_p$. In particular, for d large enough, each connected component of $\mathcal{G}_p$ intersect the image of $\mathrm{H}(\mathbb{Q})\backslash\mathrm{H}(\mathbb{A})$ *in $\mathcal{G}_p$ in a set of positive measure.*

Proof. The proof of this result is elementary does not require Linnik's ergodic method: the subspace generated by of characteristic functions of the connected components is precisely the subspace generated by the functions $\chi \circ \mathrm{Nr}(\cdot)$ from Proposition 8.1. It is therefore sufficient to show that for any such χ

$$\int_{\mathrm{H}(\mathbb{Q})\backslash\mathrm{H}(\mathbb{A})} \chi(\mathrm{Nr}(t))dt \to \int_{\mathrm{G}(\mathbb{Q})\backslash\mathrm{G}(\mathbb{A})} \chi(\mathrm{Nr}((g))dg \text{ as } d \to \infty.$$

If χ is trivial the result is obvious. If χ is not trivial, the integral on the right hand side is zero. Indeed, if χ is nontrivial there is a prime q such that $\chi(q) = -1$ (here

q is viewed as the element $(x_v)_v$ with $x_v = 1$ for $v \neq q$ and $x_q = q$) and picking $g_q \in \mathrm{B}^\times(\mathbb{Q}_q)$ such that $\mathrm{Nr}(g_q) = q$ we see that

$$\begin{aligned}\int_{\mathrm{G}(\mathbb{Q})\backslash\mathrm{G}(\mathbb{A})} \chi(\mathrm{Nr}(g))dg &= \int_{\mathrm{G}(\mathbb{Q})\backslash\mathrm{G}(\mathbb{A})} \chi(\mathrm{Nr}(gg_q))dg \\ &= \chi(\mathrm{Nr}(g_q)) \int_{\mathrm{G}(\mathbb{Q})\backslash\mathrm{G}(\mathbb{A})} \chi(\mathrm{Nr}(g))dg \\ &= -\int_{\mathrm{G}(\mathbb{Q})\backslash\mathrm{G}(\mathbb{A})} \chi(\mathrm{Nr}(g))dg.\end{aligned}$$

Therefore, it suffice to show that the integral on the left hand side is zero for d large enough; again is sufficient to see that there exists $t_0 \in \mathrm{H}(\mathbb{A})$ such that

$$\chi(\mathrm{Nr}(t_0)) = -1$$

for then

$$\begin{aligned}\int_{\mathrm{H}(\mathbb{Q})\backslash\mathrm{H}(\mathbb{A})} \chi(\mathrm{Nr}(t))dt &= \int_{\mathrm{H}(\mathbb{Q})\backslash\mathrm{H}(\mathbb{A})} \chi(\mathrm{Nr}(tt_0))dt \\ &= -\int_{\mathrm{H}(\mathbb{Q})\backslash\mathrm{H}(\mathbb{A})} \chi(\mathrm{Nr}(t))dt.\end{aligned}$$

Let K_χ be the quadratic field corresponding to χ: for any prime q does not dividing the discriminant of K_χ, we have $\chi(q) = \pm 1$ depending on whether q splits in K_χ or not. Now, if d is square-free and large enough the quadratic fuel $\mathbb{Q}(\mathbf{x})$ is different from K_χ and there exists a prime q which splits in $\mathbb{Q}(\mathbf{x})$ but which is inert in K_χ. Since q splits in $\mathbb{Q}(\mathbf{x})$ there exists $t_q \in \mathrm{H}(\mathbb{Q})$ with $\mathrm{Nr}(t_q) = q$ and

$$\chi(\mathrm{Nr}(t_q)) = -1. \qquad \square$$

For the facts about automorphic forms enunciated in this section we refer to the book [12] and in particular to its appendix by J. Rogawsky.

8.3 Trajectories of the Linnik Flow

From the above description, given

$$L_1 \in \mathrm{H}(\mathbb{A}_f) \cdot L \subset \mathrm{gen}_q(L),$$

we associate to it is trajectory in $\mathrm{gen}_q(L)$

$$t_p^{\mathbb{Z}} \cdot L_1 = (\dots, t_p^{-1} L_1, L_1, t_p \cdot L_1, t_p^2 \cdot L_1, \dots)$$

and for $l \geqslant 1$ an integer we associate its truncated trajectory of length $2l+1$

$$t_p^{[-l,l]} \cdot L_1 = (t_p^{-l} L_1, \dots, t_p^{-1} L_1, L_1, t_p \cdot L_1, \dots, t_p^l L_1).$$

The classes $[t_p^i L_1]$ and $[t_p^{i+1} \cdot L_1]$ are adjacent vertices in the graph $\mathcal{G}_p$ and we obtains non-backtracking[3] walks $[t_p^{\mathbb{Z}} \cdot L_1]$ and $[t_p^{[-l,l]} \cdot L_1]$ on $\mathcal{G}_p$ (each such walk is contained in some connected component of $\mathcal{G}_p$).

We would like to study the map

$$\begin{aligned} \mathbf{w}_{2l+1} : \mathrm{H}(\mathbb{A}_f) \cdot L &\to \mathrm{NbtW}_{2l+1}(\mathcal{G}_p), \\ L_1 &\to \mathbf{w}_{2l+1}(L_1) = [t_p^{[-l,l]} \cdot L_1], \end{aligned} \tag{8.4}$$

where $\mathrm{NbtW}_{2l+1}(\mathcal{G}_p)$ is the set of non-backtracking walks of length $2l+1$ on $\mathcal{G}_p$ in particular the size of its image and the size of its fibers.

Remark 8.1. Let us insist on the fact that a *walk* $\mathbf{w} = \cdots e_i \curvearrowright e_{i+1} \curvearrowright e_{i+2} \cdots$ on $\mathcal{G}_p$ is more than the sequence of connected edges $(e_i)_i$: this is the sequence of the consecutive edges $(e_i)_i$ AND of the sequence of the vertices $(v_i)_i$ connecting the consecutive edges $e_i \curvearrowright e_{i+1}$.

Remark 8.2. Observe that if we consider $s_{\mathbb{Q}}$ in $SO_q(\mathbb{Q})$ and set $\mathbf{x}' = s_{\mathbb{Q}}\mathbf{x}$, $L' = s_{\mathbb{Q}} \cdot L$, $\mathrm{H}' = s_{\mathbb{Q}} \mathrm{H} s_{\mathbb{Q}}^{-1}$ and $t'_p = s_{\mathbb{Q}} t_p s_{\mathbb{Q}}^{-1}$, the walk in $\mathcal{G}_p$ generated by $t_p'^{\mathbb{Z}} \cdot s_{\mathbb{Q}} \cdot L_1$ is the same as the one generated by $t_p^{\mathbb{Z}} \cdot L_1$. Therefore the image of the map (8.4) depends only on $SO_q(\mathbb{Q})(L, \mathbf{x})$. In addition since $\mathbb{H}$ is commutative, if $s_{\mathbb{Q}} = h_{\mathbb{Q}} \in \mathrm{H}(\mathbb{Q})$ we see that the walk generated by $t_p^{\mathbb{Z}} \cdot h_{\mathbb{Q}} L_1$ is the same as the walk generated by $t_p^{\mathbb{Z}} \cdot L_1$, therefore the map (8.4) factors through the quotient $\mathrm{H}(\mathbb{Q}) \backslash \mathrm{H}(\mathbb{A}_f) \cdot L$.

8.3.1 Size of the Fibers

In this section we evaluate the size of the fibers of the map (8.4). We consider $L_1, L_2 \in \mathrm{H}(\mathbb{A}_f) \cdot L$, $l \geqslant 1$ an integer and their two truncated walks

$$\mathbf{w}_{2l+1}(L_1) = [t_p^{[-l,l]} L_1], \ \mathbf{w}_{2l+1}(L_2) = [t_p^{[-l,l]} L_2],$$

and we determine under which conditions these two truncated trajectories can coincide over an interval of time $[-l, l]$; by this we mean not only that

$$[L_1^i] = [L_2^i], \ \forall i \in [-l, l]$$

but also that the edge connecting $[L_1^i]$ to $[L_1^{i+1}]$ is the same as the one connecting $[L_2^i]$ to $[L_2^{i+1}]$ for all $i \in [-l, l-1]$.

More precisely, since $[L_1] = [L_2]$ there is $s_{\mathbb{Q}} \in SO_q(\mathbb{Q})$ such that $L_2 = s_{\mathbb{Q}} L_1$. Saying that the two trajectories coincide from time $-l$ to time l means that, for $i = -l, \ldots, l$, one has

$$t_p^i L_2 = s_{\mathbb{Q}} t_p^i L_1$$

or equivalently for $i = -l, \ldots, l$, one has

$$t_p'^i L_1 = t_p^i L_1, \tag{8.5}$$

[3]A walk $\mathbf{w} = \cdots e_i \curvearrowright e_{i+1} \curvearrowright e_{i+2} \cdots$ on a non-directed graph is non-backtracking if for any i the edge connecting e_{i+1} to e_{i+2} is not the edge connecting e_i to e_{i+1} in the reverse direction.

for

$$t'_p = s_{\mathbb{Q}}^{-1} t_p s_{\mathbb{Q}} \in \mathrm{H}'(\mathbb{Q}_p), \ \ \mathrm{H}' = s_{\mathbb{Q}}^{-1} \mathrm{H} s_{\mathbb{Q}}.$$

Let

$$\mathbf{x}' = s_{\mathbb{Q}}^{-1}\mathbf{x} \in L' = s_{\mathbb{Q}}^{-1} \cdot L;$$

we have $q(\mathbf{x}') = d$ and H' is the stabilizer of $\mathbf{x}'$ in SO_q. Moreover, since $\mathbf{x} \in L_1 \cap L_2$ we have

$$\text{for any } i, j \in \mathbb{Z}, \ \mathbf{x} \in t_p^i L_1 \cap t_p^j L_2$$

and by (8.5) we have for $|j| \leqslant l$,

$$\mathbf{x}' = s_{\mathbb{Q}}^{-1}\mathbf{x} \in s_{\mathbb{Q}}^{-1} t_p^j L_2 = t_p^j L_1$$

therefore

$$\mathbf{x}, \mathbf{x}' \in t_p^l L_1 \cap t_p^{-l} L_1. \tag{8.6}$$

We have the following

Lemma 8.1. *For any $l \geqslant 0$,*

$$t_p^l L_1 \cap t_p^{-l} L_1 = \{\mathbf{x}' \in L_1 \,|\, \mathbf{x}' \in \mathbb{Z}_p \mathbf{x} + p^l L_{1,p}\}$$

is the preimage of a line in $L_{1,p}/p^l L_{1,p}$ under the obvious projection map.

Proof. Since $(2d, p) = 1$, under Linnik's condition d is a square in $\mathbb{Z}_p^\times$. We will prove the lemma in the special case $L_p = M_2^0(\mathbb{Z}_p)$, $\mathbf{x} = \begin{pmatrix} \sqrt{d} & 0 \\ 0 & -\sqrt{d} \end{pmatrix}$ for $\sqrt{d} \in \mathbb{Z}_p^\times$ a choice of a square-root. Then $\mathrm{H} = \mathrm{diag}_2(\mathbb{Q}_p) \subset PGL_2(\mathbb{Q}_p)$ and we may take $t_p = \begin{pmatrix} p & 0 \\ 0 & 1 \end{pmatrix}$. It follows that

$$t_p^l L_p \cap t_p^{-l} L_p = \left\{ \begin{pmatrix} a & bp^l \\ cp^l & -a \end{pmatrix} \,\middle|\, a \in \mathbb{Z}_p, \ b, c \in \mathbb{Z}_p \right\}.$$

As for the general case, one can show that, under our assumptions, it is always possible to reduce to that case. □

This lemma, (8.6) and $p > 2$ imply that

$$\mathbf{x}' \equiv \pm\mathbf{x} \,(\mathrm{mod}\, p^l) \text{ and } \langle \mathbf{x}, \mathbf{x}' \rangle \equiv \pm 2d \,(\mathrm{mod}\, p^{2l}). \tag{8.7}$$

Indeed there is $\lambda \in \mathbb{Z}_p$ such that $\mathbf{x}' - \lambda\mathbf{x} \in p^l L_{1,p}$, so that

$$q(\mathbf{x}' - \lambda\mathbf{x}) = q(\mathbf{x}') - \lambda\langle \mathbf{x}, \mathbf{x}' \rangle + \lambda^2 q(\mathbf{x}) = d(1 + \lambda^2) - \lambda \langle \mathbf{x}, \mathbf{x}' \rangle \equiv 0 \,(\mathrm{mod}\, p^{2l}).$$

Therefore we have the following

Lemma 8.2. *The number of pairs $(L_1, L_2) \in (\mathrm{H}(\mathbb{Q})\backslash\mathrm{H}(\mathbb{A}_f))^2$ such that $\mathbf{w}_{2l+1}(L_1) = \mathbf{w}_{2l+1}(L_2)$ is bounded by the number of $SO_q(\mathbb{Q})$-orbits of triples*

$$SO_q(\mathbb{Q})\backslash\{(L', \mathbf{x}_1, \mathbf{x}_2) \,|\, L' \in \mathrm{gen}_q(L), \ \mathbf{x}_1, \mathbf{x}_2 \in \mathcal{R}_q(d, L'), \ \langle \mathbf{x}_1, \mathbf{x}_2 \rangle \equiv \pm 2d \,(\mathrm{mod}\, p^{2l})\}.$$

This later set decompose as the disjoint union

$$\bigsqcup_{a\equiv\pm 2d\,(\mathrm{mod}\,p^{2l})} SO_q(\mathbb{Q})\backslash\mathcal{R}_q(d,a;\mathrm{gen}_q(L)),$$

where

$$\mathcal{R}_q(d,a;\mathrm{gen}_q(L))=\{(L',\mathbf{x}_1,\mathbf{x}_2)\,|\,L'\in\mathrm{gen}_q(L),\ q(\mathbf{x}_1)=q(\mathbf{x}_2)=d,\ \langle\mathbf{x}_1,\mathbf{x}_2\rangle=a\}$$

which is the set of isometric embeddings

$$(\mathbb{Z}^2,dX^2+aXY+dY^2)\hookrightarrow(L',q)$$

of the binary quadratic lattice $(\mathbb{Z}^2,dX^2+aXY+dY^2)$ into some lattice L' in the genus of L (the embedding being the $\mathbb{Z}$-linear map $(1,0)\mapsto\mathbf{x}_1,\ (0,1)\mapsto\mathbf{x}_2$).

Let us evaluate the size of these sets:

– observe that since q is positive definite, $|a|\leqslant 2d$ by the Cauchy-Schwarz inequality.

– If $a=\pm 2d$, $dX^2+aXY+dY^2=d(X\pm Y)^2$ is degenerate but we are in the equality case of the Cauchy-Schwarz inequality $\mathbf{x}'=\pm\mathbf{x}$ and therefore

$$|SO_q(\mathbb{Q})\backslash\mathcal{R}_q(d,\pm 2d;\mathrm{gen}_q(L))|=|SO_q(\mathbb{Q})\backslash\mathcal{R}_q(d;\mathrm{gen}_q(L))|=d^{1/2+o(1)}.$$

– On the other hand, if $a\neq\pm 2d$, $dX^2+aXY+dY^2$ is non-degenerate and in [3, Proposition 3.4] the following result has been proven:

Proposition 8.2. *If $a\neq\pm 2d$, we have for $d\to\infty$,*

$$|SO_q(\mathbb{Q})\backslash\mathcal{R}_q(d,a;\mathrm{gen}_q(L))|=O_q(d^{o(1)}).$$

Corollary 8.1. *The number of pairs $(L_1,L_2)\in(\mathrm{H}(\mathbb{Q})\backslash\mathrm{H}(\mathbb{A}_f))^2$ such that $\mathbf{w}_{2l+1}(L_1)=\mathbf{w}_{2l+1}(L_2)$ is bounded by*

$$\ll d^{o(1)}\left(d^{1/2}+\frac{d}{p^{2l}}\right)$$

and the size of the image of the map (8.4) is

$$|\mathbf{w}_{2l+1}(\mathrm{H}(\mathbb{Q})\backslash\mathrm{H}(\mathbb{A}_f)\cdot L)|\gg\frac{d^{1/2+o(1)}}{1+d^{1/2}/p^{2l}}.$$

Proof. The first part is an immediate consequence of the previous theorem and of the previous lemma. As for the second we have, counting fiber by fiber

$$\begin{aligned}d^{1+o(1)}&=|\mathrm{H}(\mathbb{Q})\backslash\mathrm{H}(\mathbb{A}_f)\cdot L|^2\\&\leqslant|\mathbf{w}_{2l+1}(\mathrm{H}(\mathbb{Q})\backslash\mathrm{H}(\mathbb{A}_f)\cdot L)|\sum_{\mathbf{w}\in\mathrm{NbtW}_{2l+1}}|\mathbf{w}_{2l+1}^{-1}(\mathbf{w})|^2\qquad(8.8)\\&\ll|\mathbf{w}_{2l+1}(\mathrm{H}(\mathbb{Q})\backslash\mathrm{H}(\mathbb{A}_f)\cdot L)|d^{o(1)}\left(d^{1/2}+\frac{d}{p^{2l}}\right).\end{aligned}$$

□

In particular, when

$$p^{2l+1} \asymp d^{1/2}, \tag{8.9}$$

we have

$$d^{1/2-o(1)} \ll_q |\mathbf{w}_{2l+1}(\mathrm{H}(\mathbb{Q})\backslash\mathrm{H}(\mathbb{A}_f)\cdot L)| \leqslant |\mathrm{NbtW}_{2l+1}(\mathcal{G}_p)| = (p+1)p^{2l-1} \ll d^{1/2+o(1)}$$

therefore the fibers of the map $\mathbf{w}_{2l+1}$ are of size $d^{o(1)}$ and the image of that map is of size $|\mathrm{NbtW}_{2l+1}(\mathcal{G}_p)|d^{-o(1)}$ therefore that map is almost bijective.

8.4 Large Deviation Inequalities

Let $\mathcal{G}_p^\circ \subset \mathcal{G}_p$ be some connected component of the graph $\mathcal{G}_p$. By Theorem 8.1, for d sufficiently large a positive proportion of the (image of) orbit $\mathrm{H}(\mathbb{Q})\backslash\mathrm{H}(\mathbb{A}_f)\cdot L$ is contained in $\mathcal{G}_p^\circ$. In particular, choosing l as in (8.9), the corresponding lattices generate at least

$$d^{1/2-o(1)} = |\mathrm{NbtW}_{2l+1}(\mathcal{G}_p^\circ)|^{1-o(1)} \tag{8.10}$$

distinct non-backtracking walks in that connected component.

For $\mathcal{G}$ a connected d-regular finite graph and $\gamma = \cdots e_i \curvearrowright e_{i+1} \curvearrowright e_{i+2} \cdots$ a walk of length $l \geqslant 1$ on $\mathcal{G}$ define

$$t(\gamma;\Omega) = |\{i \,|\, e_i \in \Omega\}| \in [0,l];$$

this is the portion of the time spend by γ in Ω; if l is large and γ is choosen at random, one expects that $t(\gamma;\Omega)$ is close to $\mu_{\mathcal{G}}(\Omega)l$. The large deviation inequalities confirm this expactation in a strong form; in [6, Proposition 12.4] one can find a proof of the following result:

Proposition 8.3 (Large deviation inequalities). *Let $\mathcal{G}$ be a connected d-regular graph and $\Omega \subset V(\mathcal{G})$ a subset of vertices of measure $\mu_{\mathcal{G}}(\Omega)$. For any $\varepsilon > 0$, there exists a constant $\eta = \eta(\mathcal{G},\varepsilon) > 0$ such that the number of non-backtracking walks γ of length l satisfying*

$$|t(\gamma;\Omega) - \mu_{\mathcal{G}}(\Omega)l| \geqslant \varepsilon l$$

is bounded by

$$O_{\mathcal{G},\varepsilon}(|\mathrm{NbtW}_l(\mathcal{G})|^{1-\eta}).$$

8.5 Conclusion

Applying Proposition 8.3 to any non-empty subset $\Omega \subset \mathcal{G}_p^\circ$ and for $\varepsilon = \frac{1}{2}\mu_{\mathcal{G}_p^\circ}(\Omega)$ we see from (8.10) that the number of walks $\mathbf{w}_{2l+1}(L_1)$ contained in that component and such that

$$t(\mathbf{w}_{2l+1}(L_1);\Omega) \geqslant \frac{1}{2}\mu_{\mathcal{G}_p}(\Omega)l$$

is at least

$$|\mathrm{NbtW}_{2l+1}(\mathcal{G}_p^\circ)|^{1-o(1)} - O_{\mathcal{G},\varepsilon}(|\mathrm{NbtW}_l(\mathcal{G})|^{1-\eta}) \sim |\mathrm{NbtW}_{2l+1}(\mathcal{G}_p^\circ)|^{1-o(1)}.$$

Doing this for all the components of $\mathcal{G}_p = [\mathrm{gen}_q(L)]$ we obtain that, for any $[L'] \in [\mathrm{gen}_q(L)]$ and for d large enough satisfying Linnik's conditions, there exist $L'' \in \mathrm{H}(\mathbb{A}_f) \cdot L \cap [L']$ and therefore there exists a representation ($\mathbf{x} \in L''$) of d by q in the genus class $[L']$.

The proof of equidistribution follows the argument given in [6, §2.10]; for notational simplicity we will assume that $\mathcal{G}_p$ is connected; in the general case it suffice to repeat the argument below using (8.10).

For $\delta > 0$ let Ω_δ be the subset of $[\mathrm{gen}_q(L)]$ such that for $[L'] \in \Omega_\delta$

$$\mu_{\mathbf{x}}(\pi_q^{-1}([L'])) - \mu_{[\mathrm{gen}_q(L)]}([L']) \geqslant \delta\mu_{[\mathrm{gen}_q(L)]}([L'])$$

(the "bad" elements whose preimage under the projection π_q have measure which exceed the expected measure $\mu_{[\mathrm{gen}_q(L)]}([L'])$ by a factor δ). We assume that

$$\mu_{[\mathrm{gen}_q(L)]}(\Omega_\delta) \geqslant \varepsilon$$

for some $\varepsilon > 0$. For any $\Omega \in [\mathrm{gen}_q(L)]$ we have

$$\begin{aligned}\mu_{\mathbf{x}}(\pi_q^{-1}(\Omega)) &= \sum_{[L']_{\mathbf{x}}} \mu_{\mathbf{x}}([L']_{\mathbf{x}}) 1_{\pi_q(L')\in\Omega} \\ &= \frac{1}{2l+1} \sum_{[L']_{\mathbf{x}}} \mu_{\mathbf{x}}([L']_{\mathbf{x}}) \sum_{i=-l+1}^{l} 1_{\pi_q(t_p^i L')\in\Omega} \\ &= \sum_{[L']_{\mathbf{x}}} \mu_{\mathbf{x}}([L']_{\mathbf{x}}) \frac{t(\mathbf{w}_{2l+1}(L');\Omega)}{2l+1}.\end{aligned}$$

We applying this to Ω_δ: by definition we have

$$\mu_{\mathbf{x}}(\pi_q^{-1}(\Omega)) \geqslant \mu_{[\mathrm{gen}_q(L)]}(\Omega_\delta)(1+\delta) \geqslant \mu_{[\mathrm{gen}_q(L)]}(\Omega_\delta) + \delta\varepsilon$$

and since

$$\frac{t(\mathbf{w}_{2l+1}(L');\Omega_\delta)}{2l+1} \in [0,1]$$

this implies that the inequality

$$\frac{t(\mathbf{w}_{2l+1}(L');\Omega_\delta)}{2l+1} \geqslant \mu_{[\mathrm{gen}_q(L)]}(\Omega_\delta) + \frac{\delta\varepsilon}{2}$$

holds for a set of $[L']_{\mathbf{x}}$ of positive $\mu_{\mathbf{x}}$-measure (depending on δ and ε). Taking l as in (8.9), we see that the number of such walk $\mathbf{w}_{2l+1}([L']_{\mathbf{x}})$ is

$$\geqslant |\mathrm{NbtW}_{2l+1}(\mathcal{G}_p^\circ)|^{1-o(1)}.$$

On the other hand, by Proposition 8.3, the number of such walks is

$$\ll |\mathrm{NbtW}_{2l+1}(\mathcal{G}_p^\circ)|^{1-\eta}$$

for some $\eta = \eta(\varepsilon, \delta, \mathcal{G}_p) > 0$, contradiction. □

References

[1] W. Duke, Hyperbolic distribution problems and half-integral weight Maass forms, Invent. Math. **92** (1988), No. 1, 73–90.

[2] W. Duke and R. Schulze-Pillot, Representation of integers by positive ternary quadratic forms and equidistribution of lattice points on ellipsoids, Invent. Math. **99** (1990), No. 1, 49–57.

[3] M. Einsiedler, E. Lindenstrauss, P. Michel and A. Venkatesh, The distribution of closed geodesics on the modular surface and Duke's theorem, L'enseignement Mathématique **58** (2012), No. 2, 249–313.

[4] M. Einsiedler, E. Lindenstrauss, P. Michel and A. Venkatesh, Distribution of periodic torus orbits on homogeneous spaces III: Duke's theorem for cubic fields, Ann. Math. **173** (2011), No. 2, 815–885.

[5] M. Einsiedler, E. Lindenstrauss, P. Michel and A. Venkatesh, On representations of binary quadratic forms by quaternary quadratic forms, in preparation.

[6] J. Ellenberg, P. Michel and A. Venkatesh, Equidistribution of integral points on spheres, to appear in Proceedings of the 2012 International Math. Colloquium (TIFR); preprint available at `http://arxiv.org/abs/1001.0897`.

[7] J. Hanke, Local densities and explicit bounds for representability by a quadratic form, Duke Math. J. **124** (2004), No. 2, 351–388.

[8] J. Hanke, Some recent results about (ternary) quadratic forms, in: CRM Proc. Lecture Notes, Vol. 36, Amer. Math. Soc., Providence, RI, 2004, 147–164.

[9] H. Iwaniec, Fourier coefficients of modular forms of half-integral weight, Invent. Math. **87** (1987), No. 2, 385–401.

[10] Y. Kitaoka, Arithmetic of Quadratic Forms, Cambridge Tracts in Mathematics, Vol. 106, Cambridge University Press, Cambridge, 1993.

[11] Yu. V. Linnik, Ergodic Properties of Algebraic Fields, translated from the Russian by M. S. Keane, Ergebnisse der Mathematik und ihrer Grenzgebiete, Band 45, Springer-Verlag, New York Inc., New York, 1968.

[12] A. Lubotzky, Discrete Groups, Expanding Graphs and Invariant Measures, Progress in Mathematics, Vol. 125, with an appendix by J. D. Rogawski, Birkhäuser Verlag, Basel, 1994.

[13] P. Michel and A. Venkatesh, Subconvexity, Periods and Equidistribution, in preparation.

[14] V. Platonov and A. Rapinchuk, Algebraic Groups and Number Theory, Pure and Applied Mathematics, Vol. 139, translated from the 1991 Russian original by R. Rowen, Academic Press Inc., Boston, MA, 1994.

[15] J. -P. Serre, A Course in Arithmetic, translated from the French, Graduate Texts in Mathematics, No. 7, Springer-Verlag, New York, 1973.

[16] M. -F. Vignéras, Arithmétique des Algèbres de Quaternions, Lecture Notes in Mathematics, Vol. 800, Springer-Verlag, Berlin, 1980.

Automorphic Forms and
L-functions
ALM 30, pp. 131–148

Automorphic Periods and Representation Theory

(from the analytic perspective)

Andre Reznikov*

Abstract

In these notes, we discuss periods of automorphic functions from the analytic perspective. We consider an effective version of the Frobenius reciprocity of Gelfand and Fomin. It turns out that it is related to the classical Sobolev restriction theorem, but gives results which go beyond results known in PDE theory. We also discuss the problem of restricting Maass forms to orbits of one-dimensional subgroups. Our emphasis is on the use of methods form representation theory.

2000 Mathematics Subject Classification: Primary 11F67, 22E45; Secondary 11F70, 11M26.

Keywords and Phrases: Automorphic functions, Periods, Restrictions, Representation theory.

1 Automorphic Representations and Frobenius Reciprocity

We will discuss Frobenius reciprocity in the framework of automorphic representations. Our aim is to emphasize the (functional-)analytic side of this notion. The corresponding Frobenius reciprocity was formulated in works of Gelfand, Graev, and Piatetski-Shapiro in the 60's (see [6]) and later elaborated by G. Ol'shanski (see [10]). We will follow mostly [4], and consider the simplest setup of automorphic representations of $PGL_2(\mathbb{R})$.

*Bar-Ilan University, Ramat Gan, Israel, reznikov@math.biu.ac.il.

1.1 Automorphic Representations

We quickly recall the notion of automorphic representations. We refer to [6, 8, 16] for standard facts which we mention without attributions.

Let $G = PGL_2(\mathbb{R})$ and $\Gamma \subset G$ be a lattice. We denote by $X = \Gamma \setminus G$ the automorphic space, and for simplicity (unless stated otherwise) assume that X is compact (i.e., Γ is co-compact). We fix a G-invariant measure μ_X on X, and consider the vector space $L^2(X) = L^2(X, \mu_X)$. The measure could be normalized by the requirement $\mathrm{vol}(X) = 1$.

Definition 1.1. An irreducible (unitary) representation π of G in a Hilbert space L_π is called *automorphic* if it is contained in $L^2(X)$, i.e., $\exists$ G-morhpism $\nu : L_\pi \to L^2(X)$.

We always assume that we fix a G-invariant Hermitian norm $\|\cdot\|_\pi$ on L_π, and that ν *is isometric* with respect to the L^2-norm $\|\cdot\|_X$ on X. In fact, we call a pair (π, ν) an automorphic representation. Hence $\nu(v) \in L^2(X)$ for any vector $v \in L_\pi$.

It is well-known that for the compact space X, the space $L^2(X)$ decomposes into a direct sum of irreducible (unitary) representations of G, all, except the trivial one, are infinite-dimensional.

Classical automorphic functions (or forms) are obtained as special vectors in the space L_π (e.g., Maass forms correspond to K-fixed vectors; see below).

For a given *abstract* representation π, there might be different morphisms ν into $L^2(X)$. First of all, there might be morphisms with different image. This is called multiplicity, determination of which is a very important and difficult problem in automorphic functions. Second, given ν we can multiply it by a complex number of norm one. This also creates an ambiguity.

It is not always convenient to work with L^2 functions. For example, in order to take restrictions of functions to subsets (the main subject of present notes), one has to consider functions satisfying certain smoothness conditions. Hence we consider vectors in a (dense) subspace of L_π. There is a well-known notion of *smooth* vectors in a representation of a *real* Lie group (over p-adics it is somewhat different).

Definition 1.2. A vector $v \in L_\pi$ is called smooth if the function $\xi_v(g) = \pi(g)v$ is a smooth function from G to L_π.

It is a well-known fact that the space of all smooth vectors $V_\pi = L_\pi^\infty \subset L_\pi$ is a *dense* subspace on which G *acts* (in fact it is a continuous Frechét representation of G).

An important classical fact is :

Fact. *For an automorphic representation* (π, ν), *we have* $\nu : V_\pi \hookrightarrow C^\infty(X)$.

In fact, the passage to smooth vectors looses no automorphic representations, i.e., we have (assuming X is compact) $\mathrm{Mor}_G(L_\pi, L^2(X)) \simeq \mathrm{Mor}_G(V_\pi, C^\infty(X))$.

1.2 Frobenius Reciprocity

Since a smooth vector $v \in V_\pi$ is realized by a smooth function $\nu(v) \in C^\infty(X)$, we can take its *restrictions* to subsets in X. This leads to an important isomorphism:

Frobenius Reciprocity. Let (π, ν) be an automorphic representation on a compact space $X = \Gamma \setminus G$, then $\mathrm{Mor}_G(V_\pi, C^\infty(X)) \simeq \mathrm{Mor}_\Gamma(V_{\pi,\mathbb{C}})$.

The isomorphism is given by the restriction to a point on X. Namely, we have the association $\{\nu : V_\pi \to C^\infty(X)\} \rightsquigarrow \{I : V_\pi \to \mathbb{C},\ I(v) = \nu(v)(\bar{e})\}$, where $\bar{e} \in X = \Gamma e \in \Gamma \setminus G$ is the image of the identity in X.

In the opposite direction, we recover ν for a given I, by $\nu(v)(g) = I(\pi(g)v)$.

Hence we can "forget" about ν and work with I. The advantage of working with I is that it does not need the complicated space X.

Since I is a functional on a (Frechét) space V_π of smooth vectors, it is natural to study its properties from the point of view of Functional Analysis.

1.3 Functional Analysis

We have a functional on V_π, i.e., $I \in V_\pi^*$ is a vector in the dual space (it is not a smooth vector as we will see). In fact, it is natural to consider the functional I as a (distributional) vector in the contragredient representation $\tilde{\pi}$ to π (although for $PGL_2(\mathbb{R})$ we can identify π and $\tilde{\pi}$).

So from the point of view of Functional Analysis it would be a natural question to ask what is the norm of I. What kind of a norm? *Not* the invariant Hermitian norm $\|\cdot\|_\pi$ on V_π defining the unitary structure.

We recall that for a norm N on a space V, the norm of a functional $l \in V^*$ is defined by

$$\|l\|_N := \sup_{v \in V} \frac{|l(v)|}{N(v)} .$$

In particular, for any $v \in V_\pi$ and a norm N on V_π, $|\nu(v)(g)| \leqslant \|I\|_N \cdot N(\pi(g) \cdot v)$. Hence if $\|I\|_N$ is finite, this would give a bound for *automorphic* functions on X. Certainly there could be no *pointwise* bounds for L^2-functions. This is the main reason we switched from the space L_π to the dense subspace V_π. Hence we have to look for other norms on V_π.

To discuss norms on V_π, let us recall (one of) its realization(s): irreducible (not necessarily unitary) representations of $PGL_2(\mathbb{R})$ are parameterized by complex numbers $\lambda \in \mathbb{C}$: for $\lambda \in \mathbb{C}$ consider the space V_λ of (smooth) homogeneous of the homogeneous degree $\lambda - 1$ functions on $\mathbb{R}^2 \setminus 0$ (i.e., $f(tx, ty) = |t|^{\lambda-1} f(x, y)$) with the natural action of G. Namely, the action is given by

$$\pi_\lambda(g) f \begin{pmatrix} x \\ y \end{pmatrix} = f\left(g^{-1} \begin{pmatrix} x \\ y \end{pmatrix}\right) \cdot |\det g|^{\frac{\lambda-1}{2}}, \tag{1.1}$$

the twist makes it a representation of $PGL_2(\mathbb{R})$.

The space V_λ is irreducible for $\lambda \notin \mathbb{Z}$ (otherwise, one has to take a subspace/subquotient; in such a way one obtains discrete series and finite-dimensional representations). For $\lambda \in i\mathbb{R}$, V_λ is unitarizable, with the invariant form given by

$\int_{S^1\subset\mathbb{R}^2\setminus 0} f(t)\bar{g}(t)dt$ – these are called *principal* (*unitary*) *series*. For λ real and $-1<\lambda<1$, V_λ is also unitarizable – these are called *complementary* (*unitary*) *series*. An integer λ corresponds to discrete series representations.

Let us assume for *simplicity* that $\lambda \in i\mathbb{R}$. By restricting functions to $S^1 \subset \mathbb{R}^2\setminus 0$, we get the space $V_\lambda \simeq C^\infty_{\text{even}}(S^1)$ (we restrict to even functions since $\begin{pmatrix} -1 & 0 \\ 0 & -1 \end{pmatrix}$ acts trivially). This is called the compact model of V_λ, since it is realized on the compact subgroup $K \simeq S^1$ of G (i.e., $K = PSO(2)$ is the connected component of the maximal compact subgroup of G). In particular, in such a model the action of K is given by rotations.

Another useful model is obtained by restricting to the line $\mathbb{R} \simeq N = \{(x,1) \mid x \in \mathbb{R}\} \subset \mathbb{R}^2\setminus 0$. This is called the unipotent model of V_λ and provides the realization $V_\lambda \simeq C^\infty_\lambda(\mathbb{R})$ in the space of smooth functions on $\mathbb{R}$ decaying as $|x|^{\lambda-1}$ at infinity. In such a model, the action of the upper triangular matrices is easy to describe.

At this point we are ready to mention a few examples of the use of the functional I in order to obtain familiar objects from the theory of automorphic functions.

1.3.1 Examples-Exercises

1. I is a functional on $V_\pi \simeq C^\infty_\lambda(\mathbb{R})$. In particular, for $\Gamma = PSL_2(\mathbb{Z})$, the functional I is $\begin{pmatrix} 1 & 1 \\ 0 & 1 \end{pmatrix}$-invariant, i.e., periodic with respect to $x \mapsto x+1$. Hence we can decompose it into the Fourier series on $[0,1]$: $I\mid_{[0,1]} = \sum a_n \cdot e^{2\pi i x}$. The representation (π,ν) is cuspidal if $a_0 = 0$.

Exercise: Show that the coefficients a_n coincide (up to a normalization!) with the classical Fourier coefficients of the corresponding Maass form.

2. Let $\gamma \in PSL_2(\mathbb{Z})$ be a hyperbolic element. Then I is γ-periodic (γ acts by the multiplication in the unipotent model!) and hence we obtain another set of coefficients b_n such that coefficients $I\mid_{\mathbb{R}/\langle\gamma\rangle} = \sum b_n \cdot |x|^{-1/2-\lambda/2+inq}$.

Exercise: What is q?

Not an exercise: What are coefficients b_n? In fact, these are related to some L-functions! This is a highly nontrivial theorem of Waldspurger (see [15]). The same fact holds for an elliptic element in $PSL_2(\mathbb{Z})$ (e.g., $\begin{pmatrix} 0 & -1 \\ 1 & 0 \end{pmatrix}$).

3. Exercise: Write I for the Eisenstein series E_s (as a distribution on $\mathbb{R}$).

1.4 Sobolev Norms

We return to the study of I via methods of Functional Analysis. We need to choose a norm N on the space V_π such that $||I||_N$ will be finite. We assume that V_π is an irreducible unitary representation of the principal series, i.e., $V_\pi \simeq V_\lambda$ for $\lambda \in i\mathbb{R}$.

Whatever norm we want to consider, it should be related to the action of G on V_λ. This means that we want to consider a norm on $V_\lambda \simeq C^\infty_{\text{even}}(S^1)$ such that the action (1.1) of G (via π_λ) on V_λ is continuous in this norm. There are zillions of such norms, but not too many *Hermitian* ones. In fact, there is a natural family of norms coming from the action of the Lie algebra. These are (L^2)-Sobolev norms (or derived norms).

Norms on V_π is the same as norms on $C^\infty_{\text{even}}(S^1)$ (for principal unitary series). We also can assume that the norm is K-invariant. This is because we can integrate any norm over the action of the *compact* group K and obtain an equivalent norm, assuming the initial norm was G-continuous. A *Hermitian* K-invariant norm on $C^\infty_{\text{even}}(S^1)$ is easy to describe by its Fourier series.

For $n \geqslant 0$, we define the n-th L^2-Sobolev norm by $S_n^2(f) = \|f\|^2_{L^2(S^1)} + \|f^{(n)}\|^2_{L^2(S^1)}$ (classically on $\mathbb{R}^m$, it is given by $S_n^2(f) = \sum\limits_{|\alpha|\leqslant n} \|D^\alpha f\|^2_{L^2(\mathbb{R}^m)}$).

In fact, one also can use the Fourier transform $\hat{f}$ of f to define these norms (i.e., $S_\sigma^2(f) = \displaystyle\int |\hat{f}(\xi)|^2(1+|\xi|^{2\sigma})d\xi$). This makes sense for any *real* σ (these are fractional Sobolev norms; one also can define such norms via derivatives by interpolation of norms).

On S^1 this is given by Fourier series:

$$S_\sigma^2(f) = \sum |\hat{f}(n)|^2(1+|n|^\sigma)^2,$$

where $\hat{f}(n) = \displaystyle\int_{S^1} f(t)e^{-int}dt$. An equivalent norm which is more convenient is the one given by $\sum |\hat{f}(n)|^2(1+|n|)^{2\sigma}$.

Since we are interested in Sobolev norms of *functionals* and not functions, we note that the L^2-Sobolev norm of a functional $l \in C^{-\infty}(S^1)$ with respect to S_σ is given by $\|l\|^2_{S_\sigma} = S^2_{-\sigma}(l) = \sum\limits_n |\hat{l}(n)|^2(1+|n|)^{-2\sigma}$, where $\hat{l}(n) = l(e^{int})$ are Fourier coefficients of l.

Example. The Dirac δ-function satisfies $\hat{\delta}(n) = \delta(e^{int}) = 1$. Hence $S_\sigma^2(\delta) = \sum |1|^2(1+|n|)^{2\sigma} < \infty$ iff $\sigma < -\dfrac{1}{2}$.

The fundamental theorem about Sobolev norms is the celebrated

Sobolev Restriction Theorem. *Let M be a manifold (or a domain in $\mathbb{R}^m$), $\dim M = m$. For any $f \in C^\infty(M)$, the following bound holds:*

$$|f(x)| \leqslant C_M \cdot S_{\frac{m}{2}}(f) \quad \textit{for any} \quad x \in M.$$

We note that this bound is sharp (i.e., for a general function one can not improve the bound above). In our setup we have $\dim X = 3$, and hence the functional I (which is given as the evaluation at a point!) should be bounded in the 3/2-Sobolev norm. In fact, all functions in the space V_π satisfy the second order partial differential equation (the Casimir equation), and the theory of partial differential equations (which we will not discuss here) allows one to "reduce" dimensions by one, i.e., the functional I is in fact bounded by the first Sobolev norm on X. It turns out that from representation theory we have an even stronger bound:

Theorem 1.1 ([4]). *Let (π, ν) be an automorphic representation (of a principal series on a compact space $X = \Gamma \setminus G$), and I the corresponding Γ-invariant functional on $V_\pi \simeq C^\infty_{\text{even}}(S^1)$. The functional I is bounded with respect to the norm S_σ for any $\sigma < -\frac{1}{2}$.*

This means that the functional I is in the same class as the δ-function (with respect to L^2-Sobolev norms only!): if $I = \sum C_n e^{int}$ is the Fourier expansion of the functional I, then

$$S^2_\sigma(I) = \sum |C_n|^2(1+|n|)^{2\sigma} < \infty ,$$

if (and only if) $\sigma < -\frac{1}{2}$. Note that I is invariant with respect to the group Γ acting *ergodically* on S^1, while δ-function is supported at one point!

Remarks. (1) In fact, Theorem 1.1 is completely general: Let $\Gamma \subset G$ be a lattice in a real reductive group, and (π, ν) be a (unitary) irreducible automorphic representation appearing discretely in $L^2(X)$. Assume that the invariant Hermitian form P_π is isometric to the standard Hermitian form on $L^2(X)$. Let N be a G-continuous *Hermitian* norm on V_π. Then $\|I\|_N \leqslant C_\Gamma \cdot \text{tr}(P|N)$, where $\text{tr}(\cdot|\cdot)$ is the relative trace of Hermitian forms. For $P > 0$ and $0 \leqslant N \leqslant P$ two Hermitian forms, there is an operator A_N which defines N via P. In that case we have $\text{tr}(P|N) = \text{tr}\, A_N$ (see the Appendix in [4]).

For example, S_σ is defined by the operator $A(e_n) = (1+|n|)^{2\sigma} e_n$ with respect to the unitary form $\langle \cdot, \cdot \rangle_{L_\pi}$ on π. Hence we have $S_\sigma(e_n) = \langle A(e_n), e_n \rangle_{L_\pi} = (1+|n|)^{2\sigma}$.

(2) The bound in Theorem 1.1 implies a bound for the growth of automorphic functions. Namely we have

$$|\nu(\sigma)(g)| = |I(\pi(g)v)| \leqslant C_\Gamma \cdot \|I\|_N \cdot \underbrace{N(\pi(g)v)}_{\text{“computable”}} .$$

The quantity $N(\pi(g)v)$ could be estimated explicitly in the model of N on $C^\infty(S^1)$, while C_Γ and $\|I\|_N$ are more difficult to estimate effectively. See [12–14] for some examples.

(3) What about Eisenstein series? The corresponding Frobenius morphism $\nu_s : \pi_s \to L^2(X)$ is not “defined” (i.e., it is not into $L^2(X)$), and hence the standard Hermitian form on π_s (for unitary π_s) does not correspond to the Hermitian form on $L^2(X)$. This example has not been worked out yet.

1.5 A Curious Application: Meromorphic Continuation of a Double Dirichlet Series

We quickly consider a curious application of Frobenius reciprocity discussed above (here we follow [12]). The aim is to show meromorphic continuation of a Dirichlet series constructed out of periods of automorphic functions. Our automorphic representations and the corresponding functionals will come from Eisenstein series, although the same considerations could be applied to cusp forms.

Let $K/\mathbb{Q}$ be a CM-field. We consider the simplest example of $K = \mathbb{Q}(i)$. This case allows an "elementary" treatment in terms of the Lerch zeta function, but the approach we present below applies in general, including cuspidal representations.

Consider Hecke characters of K given by $\chi_n = \left(\frac{z}{|z|}\right)^{4n}$, $n \in \mathbb{Z}$. The set of such Hecke characters could be described as the set $\mathcal{X}_{un}$ of maximally unramified Hecke characters of $\mathbb{Q}(i)$. The corresponding Hecke L-function is given by

$$L(s,\chi) = \sum_{a \in I_{\mathcal{O}}^{x} = \mathbb{Z}[i]/\{\pm 1, \pm i\}} \chi(a) N(a)^{-s}$$

for $R(s) > 1$, and then analytically continued.

Consider a Double Dirichlet Series

$$D_E(s,w) = L(s,\chi_0) + \sum_{n \in \mathbb{Z}\setminus 0} L(s,\chi_n) \cdot |n|^{-w}, \quad \text{for} \quad \mathrm{Re}(w) \gg 1.$$

Claim. $D_E(s,w)$ defines a meromorphic function on $\mathbb{C}^2$.

1.5.1 Torus Periods of Eisenstein Series

We start with the classical
Observation : $E_s(i) = \xi_K(s)$.

In fact, a more general relation holds. We recall a classical result due to Hecke and Maass. We present it in (a more transparent) adelic language. Let $G = GL_2$. By specifying an isomorphism $K^\times \subset \mathrm{Aut}_{\mathbb{Q}}(K) \simeq \mathrm{Aut}_{\mathbb{Q}}(\mathbb{Q}^2)$, we obtain the corresponding tori $T_K \subset GL_2$ defined over $\mathbb{Q}$. For $s \in \mathbb{C}$, $s \neq 1$, let $\mathcal{E}_s \simeq \hat{\otimes}_{p \leqslant \infty} E_{s,p}$ be the automorphic representation of $G_{\mathbb{A}_\mathbb{Q}}$ corresponding to the classical *normalized* Eisenstein series (given by $E_s(z) = \sum_{c,d} y^s/|cz+d|^{2s}$ for $\mathrm{Re}(s) > 1$, where the summation is over $(c,d) \in \mathbb{Z}^2\setminus(0,0)$). In this normalization, the unitary Eisenstein series corresponds to $\mathrm{Re}(s) = \frac{1}{2}$. We consider vectors in $\mathcal{E}$ which are pure tensors of the form $v_\infty \otimes v_f \in \mathcal{E}_s$ where $v_f = \otimes_{p<\infty} v_p$ is the standard K_f-fixed vector for the maximal compact subgroup K_f of G over finite adels, and v_∞ is an arbitrary vector in the infinite component $E_{s,\infty}$ of $\mathcal{E}_s$. Recall that the theory of Eisenstein series provides the automorphic realization $E_s(v,g)$ (i.e., a function on $X_\mathbb{A}$) for a vector $v \in \mathcal{D}_s$ in the principal series representation $\mathcal{D}_s$ of $G_{\mathbb{A}_\mathbb{Q}}$. The space $\mathcal{D}_s$ is realized in the space of homogeneous functions with respect to $\mathbb{A}_\mathbb{Q}^\times$ action on the space $Z_{\mathbb{A}_\mathbb{Q}} = N_{\mathbb{A}_\mathbb{Q}} \setminus G_{\mathbb{A}_\mathbb{Q}} = \prod'_{p \leqslant \infty} N_{\mathbb{Q}_p} \setminus G_{\mathbb{Q}_p}$. The space $\mathcal{D}_s$ has the natural structure of the (restricted) tensor product $\mathcal{D}_s \simeq \hat{\otimes}_{p \leqslant \infty} D_{s,p}$ coming from the above product structure of $Z_{\mathbb{A}_\mathbb{Q}}$ (unlike the space $\mathcal{E}_s$ where the tensor product is not described in terms of the underlying space $X_\mathbb{A} = G_\mathbb{Q} \setminus G_\mathbb{A}$). Here local components are the spaces $D_{s,p}$ of homogeneous functions on $N_p \setminus G_p$. Hence when talking about models of the local representations $E_{s,p}$, we use the spaces $D_{s,p}$ (in fact, we only use $p = \infty$ since we will not discuss ramified Hecke characters).

Proposition 1.1 (Hecke-Maass-Siegel). *Let $v = v_\infty \otimes v_f \in \mathcal{E}_s$ be a vector with everywhere unramified standard finite components. We have then*

$$\int_{Z_G(\mathbb{A}_\mathbb{Q})T_K(\mathbb{Q})\backslash T_K(\mathbb{A}_\mathbb{Q})} E_s(v,t)\chi(t)dt = I_\infty(s,\chi_\infty,v_\infty)\cdot L(s,\chi) \tag{1.2}$$

for any character $\chi = \chi_\infty \otimes \chi_f : Z_G(\mathbb{A}_\mathbb{Q})T_K(\mathbb{Q}) \setminus T_K(\mathbb{A}_\mathbb{Q}) \to \mathbb{C}^\times$ (i.e., for a Hecke character of K trivial on $\mathbb{A}_\mathbb{Q}^\times \subset \mathbb{A}_K^\times$). Here the functional $I_\infty(s,\chi_\infty,\cdot) : E_{s,\infty} \to \mathbb{C}$ is given by

$$I_\infty(s,\chi_\infty,v_\infty) = \int_{Z_G(\mathbb{R})\backslash T_K(\mathbb{R})} v_\infty(t)\chi_\infty(t)dt \tag{1.3}$$

for a vector $v \in D_{s,\infty}$.

Proof. This follows from, the standard by now, unfolding (see [5]). For a general K and a Hecke character χ of K, one has to take into account the ramification of χ. In that case, the right hand side of (1.2) takes the form $I_\infty(s,\chi_\infty,v_\infty)\prod_{\nu\in S} I_\nu(s,\chi_\nu, v_\nu)\cdot L_S(s,\chi)$ where S is the set of ramified for χ primes, and $L_S(s,\chi)$ is the partial L-function. □

Note that for $K = \mathbb{Q}(i)$, the group $Z_G(\mathbb{R})T_K(\mathbb{R})$ could be naturally identified with $\mathbb{R}^\times \cdot O(2,\mathbb{R}) \subset GL_2(\mathbb{R})$. Hence for the Hecke character χ_n, the functional $I_\infty(s,\chi_{n,\infty},\cdot)$ could be identified with the projection to the standard n-th K_∞-type in the representation $E_{s,p}$ (here $K_\infty = O(2,\mathbb{R})$ is the maximal compact subgroup of $G_\infty = GL_2(\mathbb{R})$). Moreover, in the realization of $E_{s,\infty}$ as $D_{s,\infty}$, this functional is given by the integration against the character *itself* on the image of the compact subgroup $K_\infty \subset Z_\infty = N_\infty \setminus G_\infty$ coming from the archimedean part $(T_K)_\infty \subset G_\infty$ of torus T_K. For other CM-fields, we obtain a compact subgroup conjugated to $O(2,\mathbb{R})$, and hence simply consider types with respect to that subgroup.

By abuse of notations, we denote by $e \in G_\mathbb{Q} \setminus G_{\mathbb{A}_\mathbb{Q}}$ the image of the identity. From the Plancherel formula for $Z_G(\mathbb{A}_\mathbb{Q})T_K(\mathbb{Q}) \setminus T_K(\mathbb{A}_\mathbb{Q})$ (i.e., the Fourier expansion w.r.t. characters of $T_K(\mathbb{A}_\mathbb{Q})$ trivial on $Z_G(\mathbb{A}_\mathbb{Q})T_K(\mathbb{Q})$), and Proposition 1.1 above, we see that the following expansion holds for a vector $v = v_\infty \otimes v_f \in \mathcal{E}_s$

$$E_s(v,e) = \sum_{\chi_n \in \mathcal{X}_{un}} L(s,\chi_n)\cdot I_\infty(s,\chi_{n,\infty},v_\infty)\ . \tag{1.4}$$

1.5.2 Test Vectors

In order to realize the series $D_K(s,w)$ as the right hand side of formula (1.4), we need to construct a vector v_w in the principal series representation satisfying certain properties. We construct such a vector and make computations in the model $D_{s,\infty}$.

The Plane Realization. The basic affine space Z_∞ is isomorphic to the punctured plane $\mathbb{R}^2 \setminus 0$. This leads to the standard realization of the principal series representation in homogeneous functions on the plane.

For a complex parameter $\tau \in \mathbb{C}$, consider the (smooth part of the) representation π_τ of principal series whose realizations we discussed in Section 1.3. This means that we realize the space of the representation as the space of smooth even functions $C^\infty_{\text{even}}(S^1)$ on the circle S^1 (or on K_∞). Hence in such a model the functional $I_\infty(s, \chi_{n,\infty}, \cdot)$ is given by the integration against the exponent $e^{in\theta}$, i.e., the scalar product with the standard n-th K_∞-type.

An easy computation shows that in the above described normalization of the principal series, the infinity component $E_{s,\infty}$ is isomorphic to the representation of the principal series with the parameter $\tau = 1 - 2s$.

In order to realize the series $D_K(s, w)$, we need to construct the test vector v_w in the representation of the principal series π_{1-2s} of $G_\infty = GL_2(\mathbb{R})$ (with the trivial central character) with the K_∞-types satisfying $\hat{v}_w(n) = |n|^{-w}$ for even integer $n \neq 0$, and $\hat{v}_w(0) = 1$. Clearly such a vector exists in an appropriate completion of π_{1-2s}. For the unitary principal series (i.e., $\text{Re}(s) = \frac{1}{2}$), the vector v_w belongs to the L^2-Sobolev space $S_{\text{Re}(w)-1/2}(\pi_{1-2s})$ (see [4]).

Moreover, it is easy to see that such a vector has "local" singularities in a natural model of the representation π_{1-2s}. Namely, the (smooth part of the) representation π_{1-2s} of principal series has the above mentioned realization in the space of smooth even functions $C^\infty_{\text{even}}(S^1)$ on the circle S^1 (or on K_∞).

Let θ be the local parameter on S^1 (e.g., $\theta \in [0, 2\pi)$). Consider the following vector in the circle model

$$u_w(\theta) = |\theta|^{1+w} f(w) \ , \tag{1.5}$$

where $f \in C^\infty_{\text{even}}(S^1)$ is a smooth even function which is supported in a small neighborhood of points $\theta = 0,\ \pi$. We have then, for even integer $n \neq 0$,

$$\hat{u}_w(n) = \int_{S^1} u_w(\theta) e^{-in\theta} d\theta = |n|^{-w} + r(f, w, n) \ ,$$

where $r(f, w, n)$ is a holomorphic in w function rapidly decaying in $|n|$ for every fixed f and w. Moreover, the function r could be effectively bounded in terms of derivatives of f. Together with relation (1.4) and known properties of K-finite Eisenstein series (moderate growth in the type and analyticity in s), this implies the following relation

$$E_s(u_w, e) = D_K(s, w) + R(f, s, w) \ , \tag{1.6}$$

where the function R is entire for $(s, w) \in \mathbb{C}^2$. This relation holds as long as $E_s(u_w, e)$ is well-defined. The theory of smooth Eisenstein series ([2, 7]) implies that the value at a point for an Eisenstein series for a non-K-finite data is well-defined as long as the defining vector is smooth enough (e.g., belongs to a certain Sobolev space). In particular, $E_s(u_w, e)$ is well-defined for big enough $\text{Re}(w)$ *depending* on s.

Hence in order to meromorphically continue the series $D_K(s, w)$, it is enough to continue $E_s(u_w, e)$. We do it strip by strip in the variable w for each *fixed* s. Analyticity in s comes from the theory of smooth Eisenstein series.

1.5.3 Automorphic Functionals

It is easier now to switch to a more classical language of representations of $GL_2(\mathbb{R})$ (or in fact of $PGL_2(\mathbb{R})$ since we are interested in representations with the trivial center character). Let $X_{\mathbb{R}} = \Gamma \setminus G_{\mathbb{R}}$, for $\Gamma = GL_2(\mathbb{Z})$. We view automorphic representations through the Frobenius reciprocity (see [4]). Let $E_s(g)$ be the classical (normalized) non-holomorphic Eisenstein series for $GL_2(\mathbb{Z})$ as in Section 1.5.1. The theory of (smooth) Eisenstein series implies that the function $E_s(g)$ generates an irreducible (for $s \neq 1$) smooth representation $\mathrm{Eis}_s \subset C^\infty(X_{\mathbb{R}})$ which is isomorphic to the (generalized) principal series representation π_{1-2s}. Hence the evaluation at $e \in X_{\mathbb{R}}$ defines a Γ-invariant functional $\ell_{2s-1} \in \mathrm{Hom}_\Gamma(\pi_{1-2s}, \mathbb{C})$. This is Frobenius reciprocity. The automorphic function (i.e., the automorphic realization) φ_v corresponding to a vector $v \in V_{\pi_{1-2s}}$ is given by $\varphi_v(x) = \ell_{2s-1}(\pi_{1-2s}(g)v)$.

It is natural to view the functional ℓ_{2s-1} as a (generalized) vector in the dual representation π_{2s-1} (i.e., in the usual notations $\ell_{2s-1} \in V^{-\infty}_{\pi_{2s-1}}$). We have the canonical pairing $\langle \cdot, \cdot \rangle : \pi_{1-2s} \otimes \pi_{2s-1} \to \mathbb{C}$. We assume that this pairing coincides with the pairing on automorphic functions. Hence we have $\langle \ell_{2s-1}, v \rangle = E_s(v, g)|_{g=e}$ for a vector $v \in V_{\pi_{1-2s}}$.

1.5.4 Hecke Operators

We consider Hecke operators acting on automorphic representations of $G_{\mathbb{R}}$. The theory of Hecke operators provides for each integer prime p, a collection of elements $\gamma_i \in GL_2(\mathbb{Q})$, $0 \leqslant i \leqslant p$ such that the Hecke operator acting on the space $C^\infty(X_{\mathbb{R}})$ is given by

$$T_p(f)(x) = \frac{1}{\sqrt{p}} \sum_i f(\gamma_i x) \ . \tag{1.7}$$

The Eisenstein series $E_s(g)$ is an eigenvector with the eigenvalue $\lambda_p(s) = p^{\frac{1}{2}-s} + p^{s-\frac{1}{2}}$. The operator T_p is a scalar operator on the space Eis_s (or in fact on any automorphic representation of $G_{\mathbb{R}}$ coming from an adele automorphic representation). It turns out that, as a result, the functional ℓ_{2s-1} is an eigenfunction of some operators with the same eigenvalue $\lambda_p(s)$ (or rather with $\lambda_p(1-s) = \lambda_p(s)$) with respect to the usual action on the right by $G_{\mathbb{R}}$ on the automorphic representation Eis_{1-s} (the dual of Eis_s)[1].

1.5.5 Hecke and Frobenius

Let $\nu : V_\pi \to C^\infty(X)$ be an automorphic representation. We consider only representations with the trivial central character. Hence a vector $v \in V_\pi$ (in an abstract representation π) has the corresponding automorphic realization $\varphi_v(x) = \nu(\pi(g)v) \in C^\infty(X)$. Let ℓ_ν be the given corresponding automorphic functional. We write ℓ_π suppressing ν and view it as a (generalized) vector in the dual representation π^*. We denote by $\langle \cdot, \cdot \rangle : \pi \otimes \pi^* \to \mathbb{C}$ the natural pairing. We assume that it coincides with the pairing on X for automorphic realizations of π and π^* (at least for cuspidal ν).

[1] There is no "action" of T_p on π_{1-2s}. It acts as a scalar operator on the automorphic realization Eis_s! The formula (1.7) does not come from the group algebra action of $GL_2(\mathbb{R})$. However, on the special vector ℓ_{2s-1} this scalar action *coincides* with the action of an operator (given by the same elements appearing in T_p) coming from the *group action* of $GL_2(\mathbb{R})$ on π_{2s-1}.

It turns out that the functional ℓ_π is an eigenfunction of some similarly looking operators with the same eigenvalue $\lambda_p(\pi)$ with respect to the *usual* action of G on the right on functions on X.

Let $\mathcal{T}_p = \frac{1}{\sqrt{p}} \sum_i \gamma_i^{-1}$ be the element in the group algebra of G, where elements γ_i are as in (1.7). We have

$$\begin{aligned}\langle \pi^*(\mathcal{T}_p)\ell_\pi, v\rangle &= \frac{1}{\sqrt{p}} \sum_i \langle \pi^*(\gamma_i^{-1})\ell_\pi, v\rangle = \frac{1}{\sqrt{p}} \sum_i \langle \pi(\gamma_i)^*\ell_\pi, v\rangle \\ &= \frac{1}{\sqrt{p}} \sum_i \langle \ell_\pi, \pi(\gamma_i)v\rangle = \Big[\frac{1}{\sqrt{p}} \sum_i \nu(v)(x\gamma_i)\Big]\Big|_{x=e} \\ &= \big[T_p(\nu(v))(x)\big]\big|_{x=e} = \lambda_p(\pi)\cdot \nu(v)(e) = \lambda_p(\pi)\cdot \langle \ell_\pi, v\rangle\end{aligned}$$

for any $v \in V_\pi$. Hence we have

$$\pi^*(\mathcal{T}_p)\ell_\pi = \frac{1}{\sqrt{p}} \sum_i \pi^*(\gamma_i)\ell_\pi = \lambda_p(\pi)\cdot \ell_\pi \ . \tag{1.8}$$

We also have (essentially from the definition)

$$\langle \pi^*(\mathcal{T}_p)\ell_\pi, v\rangle = \langle \ell_\pi, \pi(\mathcal{T}'_p)v\rangle \ , \tag{1.9}$$

where $\pi(\mathcal{T}'_p) = \frac{1}{\sqrt{p}} \sum_i \pi(\gamma_i)$ is the action of an element in the group algebra (and not the Hecke operator acting on the automorphic representation (π, ν) by the scalar $\lambda_p(\pi)$).

1.5.6 Approximate Eigenvectors

It will be crucial for us that all elements γ_i appearing in the description of Hecke operators above could be chosen in the same Borel subgroup. The (convenient to us) classical choice for these elements is $\gamma_i = \begin{pmatrix} p & 0 \\ i & 1 \end{pmatrix}$ for $0 \leqslant i \leqslant p-1$ and $\gamma_p = \begin{pmatrix} 1 & 0 \\ 0 & p \end{pmatrix}$.

The main observation is that all elements γ_i preserve the point $\theta = 0$ (under the natural action on S^1), the singularity of the vector u_w. As a result, the vector u_w is essentially an eigenvector of certain operators related to T_p. We have the following elementary

Lemma 1.1. *Let π_τ be a principal series representation of $PGL_2(\mathbb{R})$, p an integer prime and $\mathcal{T}'_p = p^{-\frac{1}{2}} \sum \gamma_i$ the element in the group algebra defined for γ_i as above. For any $\sigma \in \mathbb{C}$ and a smooth function $g \in C^\infty(S^1)$ with a small enough support around $\theta = 0$, the following relation holds*

$$\pi_\tau(\mathcal{T}'_p)(|\theta|^\sigma \cdot g(\theta)) = \beta_p(\tau, \sigma)\cdot |\theta|^\sigma g_{\tau,\sigma}(\theta) \ , \tag{1.10}$$

where the function $g_{\tau,\sigma}$ is a smooth function (in θ) holomorphically depending on τ and σ, and $\beta_p(\tau,\sigma) = p^{\tau/2-\sigma} + p^{\sigma-\tau/2}$. Moreover, we have $g_{\tau,\sigma}(0) = g(0)$ for all τ and σ.

Here we view all functions of the variable θ (possibly depending on complex parameters τ and σ) as (a family of) vectors in π_τ (realized in the same space $C^\infty_{\text{even}}(S^1)$, but with the action of $PGL_2(\mathbb{R})$ depending on τ).

Proof. It is easier to write formulas in the line (or unipotent) model of π_τ. We recall that the principal series representation π_τ of $GL_2(\mathbb{R})$ with the trivial character (i.e., representation of $PGL_2(\mathbb{R})$) has a realization in the space of functions on the real line (or more appropriately on the unipotent subgroup), and the action is given by $\pi_\tau(g)f(t) = f(g^{-1}t)|\det g|^{\frac{\tau-1}{2}}$. Specializing to the (lower) Borel subgroup, we have

$$\pi_\tau\left(\begin{pmatrix} a & 0 \\ b & c \end{pmatrix}\right) f(t) = f(a^{-1}t/(-bt/ac + c^{-1}))| - bt/ac + c^{-1}|^{\tau-1}|ac|^{\frac{\tau-1}{2}}.$$

Consider a vector $v(\theta) = |\theta|^\sigma \cdot g(\theta)$ in the circle model of the representation π_τ. Assume that g is a smooth function and has small support around $\theta = 0$. Clearly in the line model, such a vector is given by $v(t) = |t|^\sigma \tilde{g}_{\tau,\sigma}(t)$ for some smooth function $\tilde{g}_{\tau,\sigma}$ supported near $t = 0$, and depending holomorphically on τ and σ.

Hence for γ_i as above and $f \in C^\infty(\mathbb{R})$ supported in a small enough neighborhood of $0 \in \mathbb{R}$, we have $\pi_\tau(\gamma_i)\left(|t|^\sigma f(t)\right) = p^{\sigma-\frac{\tau-1}{2}}|t|^\sigma f_{\tau,\sigma,i}(t)$ for $0 \leqslant i \leqslant p-1$, and $\pi_\tau(\gamma_p)\left(|t|^\sigma f(t)\right) = p^{-\sigma+\frac{\tau-1}{2}}|t|^\sigma f_{\tau,\sigma,p}(t)$, where functions $f_{\tau,\sigma,i}$ are smooth compactly supported functions on $\mathbb{R}$ (depending on τ, σ and γ_i). Hence we have $\pi_\tau(\mathcal{T}'_p)\left(|t|^\sigma f(t)\right) = (p^{\sigma-\tau/2}+p^{-\sigma+\tau/2})|t|^\sigma F_{p,\tau,\sigma}(t)$ for some smooth function $F_{p,\tau,\sigma}$. Taking the limit $t \to 0$ on both sides, we obtain the last claim in the lemma. □

1.5.7 Meromorphic Continuation

Methods of Analytic Continuation:

1. Integral representations (Riemann, Hecke, Rankin-Selberg).

2. Differential equations (Riemann, Selberg, Langlands, Bernstein).

3. Strip by Strip (Euler, Riemann, M. Riesz-Bernstein, Minakshisandaram-Pleijel, Seeley).

We will follow the last method. The idea is to find an operator T in a group algebra of $PGL_2(\mathbb{R})$ such that $T \cdot f(s) = b(s)f(s+1)$. Such an operator exists and is related to Hecke operators.

We have the following main result.

Theorem 1.2. *Let $E(s,z)$ be the classical (normalized) Eisenstein series for $\Gamma = PGL_2(\mathbb{Z})$ and ℓ_{2s-1} the corresponding automorphic functional on the irreducible representation π_{1-2s} of the principal series of $PGL_2(\mathbb{R})$. Let $v_{w,z}$ be a vector such that in the line model of π_{1-2s} it is given by $v_{w,z}(t) = |t|^{w+1}f_z(t)$, where $w \in \mathbb{C}$, $\mathrm{Re}(w) \gg 1$, and $f_z \in C^\infty(\mathbb{R})$ is a smooth function of compact support* *holomorphically depending on* $z \in \mathbb{C}$. *Then the function defined for* $\mathrm{Re}(w) \gg 1$, *by* $\ell_{2s-1}(v_{w,z})$ *is a function meromorphic in* s, w *and* z.

In other words, the value at the identity for the Eisenstein series $E(s, v_{w,z}, g)$ is meromorphic in s, w and z function.

Proof. We proceed by induction. Let $\{F_z\}_{z \in Z}$ be an analytic family of compactly supported smooth functions on $\mathbb{R}$. It follows from the theory of smooth Eisenstein series ([2, 7, 15]) that the functional ℓ_{2s-1} belongs to some Sobolev space completion of the representation[2] π_{2s-1}. This implies that for fixed s ($s \neq 1$), the value of the Eisenstein series at the identity, which is given by the pairing

$$E(s, |t|^{w+1}F_z(t), e) = \langle \ell_{2s-1}, |t|^{w+1}F_z(t) \rangle \ ,$$

is well-defined for $\mathrm{Re}(w) \geqslant T$ for some $T = T_s > 0$ which depends on s. Moreover, the function $E(s, |t|^{w+1}F_z(t), e)$ is analytic in the parameter $z \in Z$.

Consider the operator $\mathcal{T}'_p$ from Lemma 1.1, and the function $g_{w+1}(t) = \beta_p(1 - 2s, w) \cdot v_{w,z} - \pi_{1-2s}(\mathcal{T}'_p)(v_{w,z})$. Lemma 1.1 implies (via the computation of the germ at $t = 0$) that we can write $g_{w+1}(t) = |t|^{w+2}\tilde{g}_{p,s,w,z}(t)$, where $\tilde{g}_{p,s,w,z}$ is a smooth compactly supported function analytically depending on all parameters. Hence the function g_{w+1} belongs to the Sobolev space on which the functional ℓ_{2s-1} is well-defined for w in a bigger region $\mathrm{Re}(s) \geqslant T - 1$. This implies the meromorphic continuation to a bigger strip. Namely, it follows from (1.9) and (1.8) that

$$\begin{aligned}\langle \ell_{2s-1}, g_{w+1} \rangle &= \langle \ell_{2s-1}, \beta_p(1-2s, w) \cdot v_{w,z} - \pi_{1-2s}(\mathcal{T}'_p)(v_{w,z}) \rangle \\ &= \beta_p(1-2s, w)\langle \ell_{2s-1}, v_{w,z} \rangle - \langle \pi_{2s-1}(\mathcal{T}_p)\ell_{1-s}, v_{w,z} \rangle \\ &= \left[\beta_p(1-2s, w) - \lambda_p(s)\right] \cdot \langle \ell_{2s-1}, v_{w,z} \rangle \ .\end{aligned}$$

The left hand side is defined for $\mathrm{Re}(w) > T - 1$. Hence we obtain the meromorphic continuation of $\langle \ell_{2s-1}, v_{w,z} \rangle$ to the half-plane $\mathrm{Re}(w) > T - 1$ which is to the left of the half-plane $\mathrm{Re}(w) \geqslant T$ where $\langle \ell_{2s-1}, v_{w,z} \rangle$ was originally well-defined. □

2 Bounds on Periods and Representation Theory

In this section, we consider periods of automorphic functions. The aim is to use representation theory in order to obtain meaningful bounds on certain periods.

2.1 A Typical Question

We consider as before $G = PGL_2(\mathbb{R})$, a lattice $\Gamma \subset G$, which we assume for simplicity to be co-compact, if not stated otherwise. Consider the (compact) hyperbolic Riemann surface $Y = X/K = \Gamma \setminus G/K$. Let $\ell \subset Y$ be a curve (e.g., a

[2] Of course, in this particular case the smooth extension of Eisenstein series follows directly from its Fourier expansion in terms of the Riemann zeta function and the well-known bounds from analytic number theory.

closed geodesic). Let φ_λ be a Maass form of *norm* one (i.e., a norm one K-fixed vector in irreducible automorphic representations π_λ) with the eigenvalue

$$\mu = \frac{1-\lambda^2}{4}.$$

A typical problem appearing in various areas of Mathematics (Number Theory, PDE, Mathematical Physics) is to bound the following quantities:

(a) $\int_\ell \varphi_\lambda d\ell$ as $\lambda \to \infty$.

(b) $\int_\ell |\varphi_\lambda|^2 d\ell$ as $\lambda \to \infty$.

(c) $\int_\ell \varphi_\lambda(\theta) e^{-in\theta} d\theta$ as $|n|$ and/or $|\lambda|$ tend to ∞. Here we assume that ℓ is closed, and denote by θ the natural parameter on $\ell \simeq S^1$.

This is a sample of analytic questions and there are other similar interesting problems. These questions are of interest in Number Theory (as they involve L-functions) and in the theory of PDE (related to Quantum Chaos).

We must admit that for questions (a) and (b) we do not have techniques to obtain nontrivial results which go beyond general results in PDE theory (but Analytic Number Theory can answer these in special situations). For the question in (c), we have some methods to advance beyond the "trivial" bound, although only in the direction $|n| \to \infty$ and λ is fixed.

We translate these problems into representation theory language. For this we have to restrict the choice of curves $\ell \subset Y$. The representation theory methods can only deal with those appearing as orbits of some subgroups in G. Namely, consider the projection $p : X \to Y$. All our curves will be images of curves in X projected to Y under p. Hence we are looking for orbits of a subgroup $H \subset G$: $\ell = x_0 \cdot H \subset X$. For the reasons that will be clear later, we can only deal with closed orbits (although non-closed ones are also extremely interesting). In that case there are three types:

2.1.1 Examples

1. Closed horocycles (Y is not compact).

$$H = \left\{ \begin{pmatrix} 1 & t \\ 0 & 1 \end{pmatrix} \right\}, \quad t \in \mathbb{R}$$

such that $H \cap \Gamma \neq e$. The horocycle is then the curve $\Gamma \cap H \setminus H \simeq S^1$ projected to Y.

2. Closed geodesics.

$$H = \left\{ \begin{pmatrix} a & 0 \\ 0 & a^{-1} \end{pmatrix} \right\}, \quad a \in \mathbb{R}^x .$$

Let γ_0 be a hyperbolic element. Let $g \in G$ be such that

$$g\gamma_0 g^{-1} = \begin{pmatrix} \alpha & 0 \\ 0 & \alpha^{-1} \end{pmatrix}, \quad 1 \neq \alpha \in \mathbb{R}^x .$$

Then we have the orbit $gH \simeq S^1 \subset X$ which is closed. Any closed geodesic is realized in such a way (after the projection to Y).

3. Geodesic circles. H is a compact (connected) subgroup. Let $K = PSO(2)$ and $H = gKg^{-1}$ for some element $g \in G$. Orbits, projected to Y (all of them are closed), of H are geodesic circles $B(y, r) = \{z \in Y \mid \text{dist}(z, y) = r\}$ centered at a point in Y.

2.2 Periods and Uniqueness

Let φ_λ be a Maass form (on Y) (or holomorphic form) and $(\pi_\lambda, V_\lambda, \nu_\lambda)$ the corresponding automorphic representation (i.e., $\nu_\lambda : V_\lambda \to C^\infty(X)$) generated by φ_λ.

We consider first the integral $\int_\ell \varphi_\lambda d\ell$. It clearly defines a functional $d_\ell : V_\pi \to \mathbb{C}$ given by the same formula: $d_\ell(v) = \int_\ell \nu(v) d\ell$. The functional d_ℓ is H-invariant, i.e., $d_\ell \in \text{Mor}_H(V_\pi, \mathbb{C})$. More generally, let $\chi : H \to \mathbb{C}$ be a character of H such that it descends to a function on ℓ. This means that $\chi \mid_{H \cap \text{Cent}_H(\ell)} \equiv 1$ (it is easy to see that the subgroup $\text{Cent}_H(\ell) \subset H$ acting *trivially* on ℓ is of the form $\text{Cent}_H(\ell) = g\Gamma g^{-1} \cap H$ for some $g \in G$). Then $d_\ell^\chi(v) = \int_{l \in \ell} \nu(v)(l)\chi^{-1}(l) d\ell \in \text{Mor}_H(V_\pi, \chi)$ gives a χ-equivariant functional.

2.2.1 Uniqueness

One of the very fruitful observations is that in many interesting cases the corresponding vector space $\text{Mor}_H(V_\pi, \chi)$ is 1-*dimensional*, i.e., such a functional is unique (up to a multiplicative constant). This phenomenon usually is called multiplicity one, multiplicity free or a Gelfand pair. The uniqueness allows one to define *numerical* invariants of automorphic functions, e.g., Fourier coefficients of automorphic forms, etc. This is done by choosing the standard (or sometimes called model) functional in $\text{Mor}_H(V_\pi, \chi)$.

Let (π, ν), H, ℓ and χ be as before. The integral $d_\chi^{\text{aut}}(v) = \int_\ell \nu(v)(\ell)\chi^{-1}(\ell) d\ell$ defines the corresponding equivariant functional on V_π. We call it the *automorphic functional* corresponding to ℓ and χ. Note that d_χ^{aut} depends on ν, and on the measure $d\ell$. Suppose we *choose* a functional $d_\chi^{\text{mod}} \in \text{Mor}_H(V_\pi, \chi)$ in some other way. We call such a functional a *model functional.* It is defined in terms of the abstract representation π without any use of the automorphic realization ν. Because of the multiplicity one, we obtain a number $a_{\chi,\nu} = a_\chi$ given by

$$d_\chi^{\text{aut}} = a_\chi \cdot d_\chi^{\text{mod}} .$$

(Of course, the coefficient a_χ depends on the choice of d_χ^{mod}!) We consider the familiar example of Fourier coefficients of Maass forms.

Examlpe. Let $H = N = \left\{\begin{pmatrix} 1 & t \\ 0 & 1 \end{pmatrix}\right\}$ be the unipotent subgroup. Let $\chi : \mathbb{R} \to \mathbb{C}$ be the additive *nontrivial* character which is trivial on $\mathbb{Z}$ (i.e., $\chi(t) = \chi_n(t) = e^{int}$). We view it as the character of N in the obvious way. Let (π, V_π, ν) be an automorphic representation of $PGL_2(\mathbb{R})$ of principal series, and assume that Γ has the reduced cusp at ∞ (e.g., $\Gamma = PGL_2(\mathbb{Z})$). We choose the familiar model $V_\pi \simeq C^\infty_\lambda(\mathbb{R})$.

As N acts by translations $x \mapsto x + t$, the functional d_χ^{mod} could be chosen to coincide with the standard Fourier transform:

$$d_\chi^{\text{mod}}(v) = \int_{\mathbb{R}} v(x) e^{-inx} dx.$$

In particular, let us compute it on a K-fixed vector for the representation $\pi \simeq \pi_\lambda$. It is easy to see that in such a model the K-fixed vector is given by

$$e_\lambda(x) = c_0(1 + x^2)^{\frac{\lambda-1}{2}},$$

with $c_0 = \dfrac{1}{\sqrt{\pi}}$ being the normalization constant.

Hence we have

$$d_\lambda^{\text{mod}}(e_\lambda) = \int (1 + x^2)^{\frac{\lambda-1}{2}} e^{-inx} dx = \frac{|n|}{2}^{-\frac{\lambda}{2}} K_{\frac{\lambda}{2}}(n),$$

where K_s is the classical K-Bessel function. Hence $d_\chi^{\text{aut}}(e_\lambda) = a_{\chi,\lambda} \cdot d_\chi^{\text{mod}}(e_\lambda)$, where $a_{\lambda,\chi} = a_n$ is the classical Fourier coefficient of Maass form φ_λ.

Let us return to questions (a), (b), (c). We can try to bound a_n by a pointwise bound on $\varphi_\lambda = \nu(e_\lambda)$ using the improved Sobolev Restriction Theorem. However, there is a simpler general argument (see [3]), which gives more information. Consider the L^2 norm of the restriction to the orbit ℓ for a general vector $v \in V_\pi$:

$$\int_\ell |\varphi_v|^2 d\ell = \sum_{\chi_k} |d_{\chi_k}^{\text{aut}}(\varphi_v)|^2 = \sum_{\chi_k} |a_{\chi_k}|^2 |d_{\chi_k}^{\text{mod}}(v)|^2 ,$$

where $d_{\chi_k}^{\text{mod}}(v) = \int_{\mathbb{R}} v(x) e^{-ikx} dx$ and $\varphi_v = \nu(v)$. Let us integrate this identity over a small fixed ball $B \subset G$. We obtain

$$\|v\|^2_{L^2} \geqslant \int_{\ell \cdot B \text{ is open}} |\varphi_v|^2 d\ell = \sum_{\chi_k} |a_{\chi_k}|^2 \int_B |d_{\chi_k}^{\text{mod}}(\pi(g)v)|^2 dg.$$

Let $v = v_T \in V_\lambda$ be a test vector of the form $v_T(x) = T \cdot v_0(Tx)$ for a real parameter $T \geqslant 1$, where v_0 is a fixed smooth non-negative function of small support around 0. Note that $\|v_T\|^2_{L^2}$ is of order of T.

It is easy to see that there are no cancelations in the integral $d_{\chi_k}^{\text{mod}}(v_T) = \int_{\mathbb{R}} v_T(x) e^{-ikx} dx$ as long as $\max\{|\lambda|, |k|\} \leqslant AT$ for $A > 0$ some fixed constant independent of k, T and λ. Hence we arrive at

Proposition 2.1 (Mean-value bound). *There exists an effective constant $c > 0$ such that*

$$\sum_{|k|\leqslant T} |a_{\lambda,\chi_k}|^2 \leqslant c \cdot \max\{|\lambda|, T\}.$$

For a single coefficient, this gives "convexity" bounds $|a_{\lambda,\chi_k}| \leqslant c|k|^{1/2}$ and $|a_{\lambda,\chi_k}| \leqslant c|\lambda|^{1/2}$.

Conjecture 2.1 (Ramananjun-Peterson, Lindelöf).

$$|a_{\lambda,\chi_k}| \ll (1 + |\lambda| + |k|)^{\varepsilon}, \quad \forall \varepsilon > 0.$$

The central question is how to break the bound $|a_{\lambda,\chi_k}| \leqslant c|k|^{1/2}$.

Only for the case (c) and λ fixed, there is a general result. However, for all cases (a), (b), (c), P. Michel and A. Venkatesh [9] obtained nontrivial bounds for Hecke-Maass and Hecke holomorphic forms using deep arithmetic.

Using uniqueness of some other automorphic periods, one can prove the following result.

Theorem 2.1 ([12]). *For a fixed π (i.e., for a fixed λ), the following bound holds*

$$|a_{\pi,\chi_k}| \ll_{\Gamma,\pi} |k|^{1/3+\varepsilon}, \quad \forall \varepsilon > 0.$$

We note that this theorem holds uniformly for all three types of orbits in Section 2.1.1. In fact, the proof is the same in all three cases, and could be viewed as an extension of the classical Rankin-Selberg method.

Acknowledgments

These notes mostly cover results which are joint with J. Bernstein whom I would like to thank for numerous fruitful discussions. It is also a pleasure to thank CIMPA for the invitation to lecture at the Wehai campus of Shandong University, and Jianya Liu for making the visit possible.

The research was partially supported by a BSF grant, by the Excellency Center of the Israel Science Foundation (grant 1691/10), and by the Emmy Noether Institute for Mathematics (a Center of the Minerva Foundation of Germany).

References

[1] J. Bernstein, Analytic continuation of generalized functions with respect to a parameter, Funct. Anal. Appl. **6** (1972), 273–285.

[2] J. Bernstein and B. Kroetz, Smooth Frechet globalizations of Harish-Chandra modules, preprint.

[3] J. Bernstein and A. Reznikov, Estimates of automorphic functions, Moscow Math. J. **4** (2004), No. 1, 19–37.

[4] J. Bernstein and A. Reznikov, Sobolev norms of automorphic functionals, Int. Math. Res. Not., (2002), No. 40, 2155–2174.

[5] P. Garrett, Standard compact periods for Eisenstein series, notes on homepage.

[6] I. Gelfand, M. Graev and I. Piatetski-Shapiro, Representation Theory and Automorphic Forms, Saunders, 1969.

[7] E. Lapid, A remark on Eisenstein series, in: Eisenstein Series and Its Applications, Progress in Math., Vol. 258, Birkhauser, 2008, 239–249.

[8] S. Lang, $SL_2(\mathbb{R})$, GTM105, Springer, 1985.

[9] P. Michel and A. Venkatesh, The subconvexity problem for GL_2, Publ. Math. Inst. Hautes Études Sci. **111** (2010), 171–271.

[10] G. Ol'shanski, On the fuality theorem of Frobenius, Funct. Anal. Appl. **3** (1969), No. 4, 295–302.

[11] A. Reznikov, Torus periods coefficients on $PGL(2)$ and Dirichlet series, J. of Number Theory, **133** (2013), No. 1, 328–342.

[12] A. Reznikov, Rankin-Selberg identities without unfolding and bounds for spherical Fourier coefficients of Maass forms, J. Amer. Math. Soc., **21** (2008), No. 2, 439–477.

[13] A. Reznikov, A uniform bound for geodesic periods of eigenfunctions on hyperbolic surfaces, to appear in Forum Mathematicum.

[14] A. Reznikov, Geodesic restrictions for the Casimir operator, J. of Functional Analysis, **261** (2011), No. 9, 2437–2460.

[15] J. -L. Waldspurger, Sur les valeurs de certaines fonctions L automorphes en leur centre de symétrie, Compositio Math. **54** (1985), No. 2, 173–242.

[16] N. Wallach, Real Reductive Groups I, Pure and Applied Mathematics, Vol. 132, Academic Press, Boston, MA, 1988.

Automorphic Forms and L-functions
ALM 30, pp. 149–178

Eisenstein Series, L-functions and Representation Theory

Freydoon Shahidi*

Abstract

There notes provide a brief survey of the Langlands-Shahidi method. This is a method which exploits the analytic properties of Eisenstein series to study certain automorphic L-functions; those that appear in constant terms of these series. The survey also covers several applications of the method, including some new cases of functoriality as well as local consequences of the method in representation theory of reductive groups over local fields.

2010 Mathematics Subject Classification: 11F70, 11R39, 11R42, 11S37, 22E55.

Keywords and Phrases: Eisenstein series, Automorphic L-functions, Functoriality.

These notes present a survey of the Langlands-Shahidi method on the study of automorphic L-functions with applications to functoriality and group representations. They were delivered as a series of four lectures by James Cogdell at the CIMPA-UNESCO-CHINA Research School at Weihai, China, in August of 2010, on behalf of the author who could not attend due to a car accident. I would like to thank him very much for taking up the task on such a short notice, as well as for his brief survey of these notes in his contribution to this volume [15].

The lectures and notes are an updated and refined version of the lectures the author presented at PCMI/IAS Summer School in the Summer of 2002 (cf. [65]). In particular, further results on functoriality, which were not available at the time of [65] are included. We have also included a section on representation theoretic implications of the method. Moreover, based on recent developments in the method, we have removed all the assumptions in [65] (e.g., Assumption 4.1 of [65]) as they are now fully proved (cf. Remark 9.1.1 of [66]). For example, Conjecture 3.10 of [65], originally conjectured in [53], has now been completely

*Department of Mathematics, Purdue University, West Lafayette, IN 47907, USA, shahidi@math.purdue.edu. Partially supported by NSF grant DMS-0700280.

proved as we have stated it here as Theorem 7.1. We refer to our book [66] for a more detailed discussion of the method and references, as well as an overview of recent developments.

I am indebted to Jim Cogdell for presenting these lectures. Thanks are also due to Jianya Liu for organizing the Summer School and preparing this volume.

1 Preliminaries

Let F be a number field. For each place v of F, let F_v be the completion of F at v. For each finite v, let O_v be the ring of integers of F_v and denote by $\mathcal{P}_v$ its maximal ideal. Let ϖ_v be a generator of $\mathcal{P}_v$. Let $q_v = [O_v \colon \mathcal{P}_v]$ and fix an absolute value $|\cdot|_v$ such that $|\varpi_v|_v = q_v^{-1}$.

Let $\mathbf{G}$ be a split connected reductive algebraic group over F. This simply means a Zariski closed subgroup of $GL_N(\overline{F})$ for some N, where $\overline{F}$ is the algebraic closure of F, whose radical (maximal normal connected solvable subgroup) consists of only semisimple elements. The radical is then equal to the connected component of the center of $\mathbf{G}$. Being split simply means that $\mathbf{G}$ has a maximal abelian subgroup consisting entirely of diagonalizable elements, a maximal torus, which is isomorphic over F to some power of $\overline{F}^*$.

Let $\mathbf{B}$ be a Borel subgroup of $\mathbf{G}$ (over F), i.e., a maximal connected solvable subgroup of $\mathbf{G}$. Let $\mathbf{T}$ be a maximally split torus of $\mathbf{G}$ contained in $\mathbf{B}$. Then $\mathbf{B} = \mathbf{TU}$, where $\mathbf{U}$ is the unipotent radical of $\mathbf{B}$. The unipotent subgroup $\mathbf{U}$ determines a set of simple roots Δ and positive roots R^+ for $\mathbf{T}$, upon acting on the Lie algebra $\mathfrak{g}$ of $\mathbf{G}$.

Let $\mathbf{P}$ be a parabolic subgroup of $\mathbf{G}$, i.e., a conjugate of a closed subgroup of $\mathbf{G}$ containing $\mathbf{B}$. We will assume $\mathbf{P}$ is standard by $\mathbf{P} \supset \mathbf{B}$. We let $\mathbf{N}$ be the unipotent radical (maximal connected **normal** unipotent subgroup) of $\mathbf{P}$. Then $\mathbf{P} = \mathbf{MN}$, where $\mathbf{M}$ is a reductive subgroup, called a **Levi** subgroup. We will fix $\mathbf{M}$ by assuming $\mathbf{T} \subset \mathbf{M}$. Let $\mathbf{A}$ be the split component of $\mathbf{M}$, i.e., the connected component of the center of $\mathbf{M}$ (the maximally split subtorus of the center of $\mathbf{M}$, if the group is not necessarily split over F). The parabolic subgroup $\mathbf{P}$ is maximal if the dimension of $\mathbf{A}/\mathbf{A} \cap \mathbf{Z_G}$ is one, where $\mathbf{Z_G}$ is the center of $\mathbf{G}$. Then the adjoint action of $\mathbf{A}$ on the Lie algebra of $\mathbf{N}$ has a unique reduced eigenfunction α, the simple root of $\mathbf{A}$ in $\mathbf{N}$. There exists a unique simple root of $\mathbf{T}$ whose restriction to $\mathbf{A}$ is α. We will denote this root of $\mathbf{T}$ also by α and always identify them with each other. Other roots of $\mathbf{A}$ in $\mathbf{N}$ are simply multiples (considered additively) of α. Throughout these lectures $\mathbf{P}$ is always assumed to be maximal. We refer to [5, 52, 72] as our main references for algebraic groups and their structure theory.

Let $\mathbf{H}$ to be any connected reductive algebraic group defined over F. Considering $\mathbf{H}$ as a group over F_v, for each v, we let $H_v = \mathbf{H}(F_v)$. If $\mathbb{A}_F$ is the ring of adeles of F, we set $H = \mathbf{H}(\mathbb{A}_F)$. It may be considered as a restricted product of groups H_v with respect to $\mathbf{H}(O_v)$ for all v, where $\mathbf{H}$ splits over an unramified extension [4]. There will be no restriction if $\mathbf{H}$ splits over F. Moreover, in this case each $K_v = \mathbf{H}(O_v)$ is a maximal compact subgroup of G_v. We let $K = \prod_v K_v$, where each K_v is a good maximal compact subgroup of H_v and $K_v = \mathbf{H}(O_v)$ for almost all v (cf. [4, 53]). Then $G = PK$.

For every algebraic group $\mathbf{H}$ over F, let $X(\mathbf{H})_F$ denote the group of F rational characters of $\mathbf{H}$. We let $X(\mathbf{H}) = X(\mathbf{H})_{\overline{F}}$. Note that if $\mathbf{T}$ is a split torus over F, then $X(\mathbf{T})_F = X(\mathbf{T})$. We set

$$\mathfrak{a} = \mathrm{Hom}(X(\mathbf{M})_F, \mathbb{R}) = \mathrm{Hom}(X(\mathbf{A})_F, \mathbb{R}).$$

Then $\mathfrak{a}^* = X(\mathbf{M})_F \otimes_{\mathbb{Z}} \mathbb{R} = X(\mathbf{A})_F \otimes_{\mathbb{Z}} \mathbb{R}$ and $\mathfrak{a}^*_{\mathbb{C}} = \mathfrak{a}^* \otimes_{\mathbb{R}} \mathbb{C}$ is the complex dual of $\mathfrak{a}$, via $\langle \lambda, \chi \otimes z \rangle = \lambda(\chi) z$, $\lambda \in \mathfrak{a}$, $\chi \in X(\mathbf{M})_F$ and $z \in \mathbb{C}$. For each v, the embedding $X(\mathbf{M})_F \hookrightarrow X(\mathbf{M})_{F_v}$ induces a map from $\mathfrak{a}_v = \mathrm{Hom}(X(\mathbf{M})_{F_v}, \mathbb{R})$ to $\mathfrak{a}$. There exists a homomorphism $H_M \colon M \to \mathfrak{a}$ defined by

$$\exp\langle \chi, H_M(m) \rangle = \prod_v |\chi(m_v)|_v$$

for every $\chi \in X(\mathbf{M})_F$ and $m = (m_v)$. Extend H_M to H_P on G by making it trivial on N and K.

If $\rho_{\mathbf{P}}$ is half the sum of roots in $\mathbf{N}$, we set $\tilde{\alpha} = \langle \rho_{\mathbf{P}}, \alpha \rangle^{-1} \rho_{\mathbf{P}} \in \mathfrak{a}^*$. It is a fundamental weight for $\mathbf{T}$ (cf. [53, 54]).

Finally having fixed $\mathbf{M}$ with $\mathbf{M} \supset \mathbf{T}$, let $\theta \subset \Delta$ denote the subset of simple roots, generating $\mathbf{M}$. We sometimes write $\mathbf{M}_\theta$ for $\mathbf{M}$. Let W be the Weyl group of $\mathbf{T}$ in $\mathbf{G}$. We use $W_{\mathbf{M}}$ to denote its Weyl group in $\mathbf{M}$. There exists a unique element $\tilde{w}_0 \in W$ such that $\tilde{w}_0(\theta) \subset \Delta$, while $\tilde{w}_0(\alpha) < 0$. We will always choose a representative w_0 for $\tilde{w}_0$ in $\mathbf{G}(F)$ and use w_0 to denote each of its components. We will be more specific about the choice of w_0 later. Finally, let $\mathbf{M}'$ be the Levi subgroup of $\mathbf{G}$ generated by $\tilde{w}_0(\theta)$. There exists a parabolic subgroup $\mathbf{P}' \supset \mathbf{B}$ which has $\mathbf{M}'$ as a Levi factor, in fact the unique one containing $\mathbf{T}$. Let $\mathbf{N}'$ be the unipotent radical of $\mathbf{P}'$ (cf. [43, 55]).

2 L-Groups, L-Functions and Generic Representations

Denote by $X^*(\mathbf{T}) = X(\mathbf{T})$ the character group of $\mathbf{T}$ which is the same as $X(\mathbf{T})_F$. Let $X_*(\mathbf{T})$ be the group of cocharacters of $\mathbf{T}$, i.e., homomorphisms from $\mathbb{G}_m = \overline{F}^*$ into $\mathbf{T} = \mathbf{T}(\overline{F})$. Let $\Delta^\vee = \Delta^\vee(\mathbf{T})$ be the set of simple coroots of $\mathbf{T}$, i.e., $\alpha^\vee \colon \mathbb{G}_m \to \mathbf{T}$ satisfying $\alpha(\alpha^\vee(t)) = t^2$. Let

$$\psi_0(\mathbf{G}) = (X^*(\mathbf{T}), \Delta(\mathbf{T}), X_*(\mathbf{T}), \Delta^\vee(\mathbf{T})) \tag{2.1}$$

denote the **based root datum**. By Chevalley-Grothendieck theorem [72], there exist a complex connected reductive group $G^\vee$ with a maximal torus $T^\vee$ such that

$$\begin{aligned} \psi_0(\mathbf{G})^\vee &= (X_*(\mathbf{T}), \Delta^\vee(\mathbf{T}), X^*(\mathbf{T}), \Delta(\mathbf{T})) \\ &= \psi_0(G^\vee) \\ &= (X^*(T^\vee), \Delta(T^\vee), X_*(T^\vee), \Delta^\vee(T^\vee)). \end{aligned}$$

The L-group ${}^LG$ of $\mathbf{G}$ is ${}^LG = G^\vee \times \Gamma_F$, where $\Gamma_F = \mathrm{Gal}(\overline{F}/F)$. In general, one carries the action of Γ_F on roots and coroots dually to $G^\vee$ and let ${}^LG = G^\vee \rtimes \Gamma_F$. (Observe that $G^\vee = {}^LG^0$.) In fact, we have an exact sequence

$$0 \to \mathrm{Int}(\mathbf{G}) \to \mathrm{Aut}(\mathbf{G}) \to \mathrm{Aut}\psi_0(\mathbf{G}) \to 0, \qquad (2.2)$$

where $\mathrm{Int}(\mathbf{G})$ is the subgroup of inner automorphisms. One can show that if $\{X_\beta\}_{\beta\in\Delta'}$, Δ' being the set of simple roots of $\mathbf{T}$, is a set of simple root vectors invariant under Γ_F, then

$$\mathrm{Aut}\ \psi_0(\mathbf{G}) = \mathrm{Aut}(\mathbf{G}, \mathbf{B}, \mathbf{T}, \{X_\beta\}_{\beta\in\Delta'}).$$

The set $\{X_\beta\}_{\beta\in\Delta'}$ is called a **splitting**, as it splits (2.2). The map

$$\Gamma_F \to \mathrm{Aut}(\mathbf{G}, \mathbf{B}, \mathbf{T}, \{X_\beta\}_\beta)$$

defines a map

$$\Gamma_F \to \mathrm{Aut}\psi_0(\mathbf{G})^\vee = \mathrm{Aut}\psi_0(G^\vee)$$

which is a subset of $\mathrm{Aut}(G^\vee)$. This defines the action of Γ_F on $G^\vee$ and defines ${}^LG$ (cf. [4, 52]).

Given a connected reductive algebraic group $\mathbf{H}$ over F, let ${}^LH$ be its L-group. Considering $\mathbf{H}$ as a group over F_v, we then denote by ${}^LH_v$ its L-group over F_v. If $\mathbf{G}$ is split over F and if we decide to only consider ${}^LG^0 = G^\vee$, then we may assume that the L-groups are all the same, no matter the place v. Finally, the natural map $\Gamma_{F_v} \to \Gamma_F$ leads to a map $\eta_v\colon {}^LM_v \to {}^LM$ for all v.

In our setting ${}^LM$ is a Levi subgroup of ${}^LG$ and one can define a unipotent group ${}^LN$ (cf. [4]) so that ${}^LM{}^LN$ is a parabolic subgroup of ${}^LG$ with unipotent radical ${}^LN$. The L-group ${}^LM$ acts on the (complex) Lie algebra ${}^L\mathfrak{n}$ of ${}^LN$. Let r be this representation. Decompose $r = \bigoplus_{i=1}^{m} r_i$ to its irreducible subrepresentations, indexed according to the values $\langle\tilde{\alpha}, \beta\rangle = i$ as β ranges among the positive roots of $\mathbf{T}$. More precisely, $X_{\beta^\vee} \in {}^L\mathfrak{n}$ lies in the space V_i of r_i if and only if $\langle\tilde{\alpha}, \beta\rangle = i$. Here $X_{\beta^\vee}$ is a root vector attached to the coroots $\beta^\vee$, considered as a root of the L-group. Clearly the integer m is equal to the nilpotence class of ${}^L\mathfrak{n}$. We let $r_{i,v} = r_i \cdot \eta_v$ for each i. Again if $\mathbf{G}$ is split over F, we may assume $r_{i,v} = r_i$ for all v and i (cf. [43, 53, 54]).

Let $\pi = \otimes_v \pi_v$ be a cuspidal representation of $M = \mathbf{M}(\mathbb{A}_F)$. Then for almost all v, π_v is an unramified representation of $M_v = \mathbf{M}(F_v)$. This means that π_v has a vector which remains invariant under $\mathbf{M}(O_v)$. In this case, the class of π_v is determined by a semisimple ${}^LM$ – conjugacy class $A_v \subset {}^LM_v = {}^LM$. Given a complex analytic (finite dimensional) representation ρ of ${}^LM$, we define the local Langlands L-function attached to π_v, ρ and a complex number s by

$$L(s, \pi_v, \rho) = \det(I - \rho(A_v)q_v^{-s})^{-1}. \qquad (2.3)$$

When $\mathbf{M}$ is quasisplit over F_v, to split over an unramified extension L_w/F_v, and τ_v is the unique Frobenius conjugacy class in Gal (L_w/F_v), then A_v must be replaced by $t_v \rtimes \tau_v$, $t_v \in {}^L T^0$. We may moreover assume that t_v is fixed by τ_v (cf. [4, 43, 54]).

In these lectures, we will mainly be concerned with the case where $\rho = r_{i,v}$, $i = 1, \ldots, m$.

We shall now discuss the notion of generic representations. We will first assume that F is a local field. Fix an F-splitting $\{X_\beta\}_{\beta \in \Delta'}$ as before. This then determines a map

$$\phi\colon \mathbf{U} \to \prod \mathbb{G}_a \longrightarrow \mathbb{G}_a, \tag{2.4}$$

where the product runs over all $\beta \in \Delta'$, sending $\exp(x_\beta X_\beta)$ to Σx_β. Let ψ_F be a nontrivial character of F. We shall then define a **generic** character χ of $U = \mathbf{U}(F)$ by

$$\chi(u) = \psi_F(\Sigma_\beta x_\beta), \tag{2.5}$$

where $\phi(u) = (x_\beta)_\beta$ and the β-component of u is $\exp(x_\beta X_\beta)$. Conversely, given a **generic** character χ of U, i.e., one which is nontrivial on every simple root group, there exists an F-splitting $\{X_\beta\}_\beta$ such that χ is defined by (2.4) and (2.5).

If π is an irreducible (admissible) unitary representation of $M = \mathbf{M}(F)$, then π is called **generic**, or more precisely χ-generic for a generic character χ of $U^0 = (\mathbf{U} \cap \mathbf{M})(F)$, if there exists a functional λ on the space of π, called a Whittaker functional, which is continuous with respect to the seminorm topology defined by the Hilbert space norm on the space $\mathcal{H}(\pi)$ of π if F is archimedean (cf. [68, 56, 57]) (continuous with respect to the trivial locally convex topology on $\mathcal{H}(\pi)$ for which every seminorm is continuous if F is non-archimedean) and satisfies

$$\langle \pi(u)x, \lambda \rangle = \chi(u)\langle x, \lambda \rangle, \tag{2.6}$$

$u \in U^0$, $x \in \mathcal{H}(\pi)_\infty$, the subspace of C^∞-vectors. By a theorem of Shalika [68], the space of all the Whittaker functionals on $\mathcal{H}(\pi)$ is at most one dimensional. Changing the splitting, we may assume χ is defined by (2.4) and (2.5).

Now, assume F is global. Let $\psi = \otimes_v \psi_v$ be a nontrivial character of $F \backslash \mathbb{A}_F$. Then each ψ_v is nontrivial. Moreover, for almost all v, ψ_v is unramified, i.e., O_v is the largest ideal on which ψ_v is trivial. The map (2.4) is F-rational and therefore extends to a map from $U = \mathbf{U}(\mathbb{A}_F)$ into $\mathbb{G}_a(\mathbb{A}_F)$, sending $\mathbf{U}(F)$ into $\mathbb{G}_a(F)$. We then define a character χ of $\mathbf{U}(F)\backslash U$ by (2.5). Consequently, if $\chi_v(u_v)$ is defined by (2.5) and ψ_v for each v, then $\chi(u) = \prod_v \chi_v(u_v)$, where $u = (u_v)_v \in U$.

Now, let $\pi = \otimes_v \pi_v$ be a cuspidal representation of $M = \mathbf{M}(\mathbb{A}_F)$. Choose a function φ in the space of π and set

$$W_\varphi(m) = \int_{\mathbf{U}^0(F)\backslash U^0} \varphi(um)\overline{\chi(u)}du. \tag{2.7}$$

We shall say π is (globally) χ-generic if $W_\varphi \neq 0$ for some φ. Then each π_v is χ_v-generic. If $\varphi = \otimes_v \varphi_v$, $\varphi_v \in \mathcal{H}(\pi_v)$, then

$$W_\varphi(m) = \prod_v \langle \pi_v(m_v)\varphi_v,\ \lambda_v \rangle = \prod_v W_{\varphi_v}(m_v) \tag{2.8}$$

for some χ_v-Whittaker functional λ_v on $\mathcal{H}(\pi_v)$.

We finally point out that every generic character χ of $\mathbf{U}(F)(\mathbf{U}(F)\backslash\mathbf{U}(\mathbb{A}_F)$ if F is global) points to a F-splitting and conversely, each F-splitting defines a generic character, both by means of (2.4) and (2.5).

Now assume F is local and π_v is an irreducible admissible representation of M_v. Let $s \in \mathbb{C}$ and denote by $I(s,\pi_v) = I(s\tilde{\alpha},\pi_v)$, the induced representation

$$I(s,\pi_v) = \mathrm{Ind}_{M_vN_v\uparrow G_v}\pi_v \otimes q_v^{\langle s\tilde{\alpha},H_{M_v}(\cdot)\rangle} \otimes \mathbf{1}. \tag{2.9}$$

This is the right regular action of G_v on the space of smooth functions f from G_v to $\mathcal{H}(\pi_v)$, satisfying:

$$f(mng) = \pi_v(m)q_v^{\langle s\tilde{\alpha}+\rho_{\mathbf{P}},H_{M_v}(m)\rangle}f(g). \tag{2.10}$$

If $v=\infty$, smooth would mean that f is infinitely many times differentiable. On the other hand for $p<\infty$, it simply means that f is locally constant, i.e., $f(gk)=f(g)$ for k in some open compact subgroup depending on f (cf. [10, 9]).

One can now define a Whittaker functional $\lambda_{\chi_v}(s,\pi_v)$ for $I(s,\pi_v)$ as follows:

$$\lambda_{\chi_v}(s,\pi_v)(f) = \int_{N'_v}\langle f(w_0^{-1}n'),\lambda\rangle\overline{\chi(n')}dn', \tag{2.11}$$

where λ is a fixed Whittaker functional on $\mathcal{H}(\pi_v)$. The integral converges as a principal value integral and stabilizes as N'_v is approached by an increasing sequence of open compact subgroups (cf. [13, 55]), i.e., we may replace N'_v with a sufficiently large open compact subgroup depending on f.

3 Eisenstein Series and Intertwining Operators; The Constant Term

Let $\pi = \otimes_v\pi_v$ be a cusp form on M. Given a K_M-finite function φ in the space of π, we extend φ to a function $\tilde{\varphi}$ on G as follows. The representation π is a subrepresentation of $L_0^2(Z_M^0\mathbf{M}(F)\backslash M,\rho)$, where Z_M^0 is the $\mathbb{A}_F$-points of the connected component $\mathbf{Z}_M^0$ of the center of $\mathbf{M}$ and ρ is a character of $\mathbf{Z}_{\mathbf{M}}^0(F)\backslash Z_M^0$. The function φ is then in this L^2-space and being K_M-finite, its right translations by elements in $K_M = \prod_v K_{M,v}, K_{M,v} = K_v \cap M_v$, generate a finite dimensional representation τ of K_M (cf. [9]). We may assume φ is so chosen that τ is irreducible and write $\tau = \otimes_v\tau_v$, where for almost all v, τ_v is trivial. Next we will choose irreducible (finite dimensional) representations $\tilde{\tau}_v$ of each K_v, containing τ_v. Moreover, we assume $\tilde{\tau}_v$ is the trivial representation for almost all K_v. Set $\tilde{\tau} = \otimes_v\tilde{\tau}_v$ (cf. [58]).

Let P_τ be the projection on the space of τ and fix measures dk_v on each $K_{M,v}$ whose total mass is 1. Let dk be the product measure on K_M. Set

$$\tilde{\varphi}(m) = \int_{K_M}\varphi(mk)\tilde{\tau}(k)dk\cdot P_\tau. \tag{3.1}$$

Observe that $\tilde{\varphi}(mk^{-1}) = \tilde{\tau}(k)\tilde{\varphi}(m)$. This is a $\tilde{\tau}$-function on M in Harish-Chandra's terminology [26]. We extend $\tilde{\tau}$ to all of G by $\tilde{\varphi}(nmk) = \tilde{\tau}(k^{-1})\tilde{\varphi}(m)$. It is easily checked to be a well-defined operator valued function on G ([58]).

Next, set

$$\tilde{\Phi}_s(g) = \tilde{\varphi}(g)\exp\langle s\tilde{\alpha} + \rho_{\mathbf{P}}, H_P(g)\rangle \tag{3.2}$$

and let Φ_s be a matrix coefficient of this operator valued function. (See (3.7) below.) The corresponding **Eisenstein series** is then defined by

$$E(s, \Phi_s, g, P) = \sum_{\gamma \in \mathbf{P}(F)\backslash \mathbf{G}(F)} \Phi_s(\gamma g). \tag{3.3}$$

We will also define the **operator valued Eisenstein series** by

$$\tilde{E}(s, \tilde{\Phi}_s, g, P) = \sum_{\gamma \in \mathbf{P}(F)\backslash \mathbf{G}(F)} \tilde{\Phi}_s(\gamma g). \tag{3.4}$$

They both converge for $\mathrm{Re}(s) \gg 0$ and have a finite number of simple poles for $\mathrm{Re}(s) \geqslant 0$, none with $\mathrm{Re}(s) = 0$ (cf. [26, 44, 48]).

Let $I(s, \pi) = \otimes_v I(s, \pi_v)$ be the representation of $G = \mathbf{G}(\mathbb{A}_F)$ induced from

$$\pi \otimes \exp\langle s\tilde{\alpha} + \rho_{\mathbf{P}}, H_M(\cdot)\rangle \otimes \mathbf{1}. \tag{3.5}$$

Let $f \in V(s, \pi) = \otimes_v V(s, \pi_v)$ be defined by

$$\begin{aligned} &f(n_0 m_0 k_0) \\ &= \exp\langle s\tilde{\alpha} + \rho_{\mathbf{P}}, H_P(m_0)\rangle \cdot \int_{K_M} \langle \tilde{\tau}(k_0^{-1})\tau(k)x,\ \hat{x}\rangle \pi(m_0 k)\varphi dk, \end{aligned} \tag{3.6}$$

with $m_0 \in M$, $n_0 \in N$ and $k_0 \in K$. Here $x \in \mathcal{H}(\tau)$ and $\hat{x} \in \mathcal{H}(\hat{\tau})$, where $\hat{\tau}$ is the contragredient of τ. Moreover

$$f(n_0 m_0 k_0)(e) = \langle \tilde{\Phi}_s(g)x,\ \hat{x}\rangle = \Phi_s(g), \tag{3.7}$$

where the left hand side is the value of the cusp form $f(n_0 m_0 k_0)$ at identity and $g = n_0 m_0 k_0$ (cf. [58]).

Observe that

$$E(s, \Phi_s, g, P) = \langle \tilde{E}(s, \tilde{\Phi}_s, g, P)x,\ \hat{x}\rangle. \tag{3.8}$$

Given $f \in V(s, \pi)$ and $\mathrm{Re}(s) \gg 0$, define the global intertwining operator $A(s, \pi)$ by

$$A(s, \pi)f(g) = \int_{N'} f(w_0^{-1} n' g) dn'. \tag{3.9}$$

Finally, if at each v we define a local intertwining operator by

$$A(s, \pi_v, w_0) f_v(g) = \int_{N'_v} f_v(w_0^{-1} n' g) dn', \tag{3.10}$$

then

$$A(s, \pi) = \otimes_v A(s, \pi_v, w_0). \tag{3.11}$$

Observe that

$$A(s,\pi)\colon I(s,\pi) \longrightarrow I(-s, w_0(\pi)) \tag{3.12}$$

and

$$A(s,\pi_v,w_0)\colon I(s,\pi_v) \longrightarrow I(-s, w_0(\pi_v)). \tag{3.13}$$

Using (3.7), we now define

$$(M(s,\pi)\Phi_s)(g) = A(s,\pi)f(g)(e), \tag{3.14}$$

where by the left hand side we understand the value of the cusp form $A(s,\pi)f(g)$ at e. This is basically the Langlands' $M(s,\pi)$ introduced in [44], or as denoted by Harish-Chandra in [26], his function $c(s,\pi)$.

Constant Term Theorem. *The constant term*

$$E_{\mathbf{P}'}(s,\Phi_s,g,\mathbf{P}) = \int_{\mathbf{N}'(F)\backslash N'} E(s,\Phi_s,n'g,\mathbf{P})dn' \tag{3.15}$$

is equal to

$$E_{\mathbf{P}'}(s,\Phi_s,g,\mathbf{P}) = \delta_{\mathbf{M},\mathbf{M}'}\Phi_s(g) + (M(s,\pi)\Phi_s)(g). \tag{3.16}$$

Here $\delta_{\mathbf{M},\mathbf{M}'}$ is the Kronecker δ-function.

Its analytic properties and therefore those of $M(s,\pi)$ are exactly the same as $E(s,\Phi_s,-,\mathbf{P})$.

In the split case the proof of the first part is in [43], as in elsewhere. But the rest is among the main properties of Eisenstein series and included in several places [44, 26, 48].

To prove the first part, one just substitutes (3.3) in (3.15) and expands and uses Bruhat decomposition and cuspidality of φ.

We finally express the functional equation of Eisenstein series by (cf. [26, 44])

$$E(-s, M(\pi,s)\Phi_s, g, P') = E(s,\Phi_s,g,P). \tag{3.17}$$

4 Constant Term and Automorphic L-Functions

It follows immediately from the Constant Term Theorem that $\otimes_v A(s,\pi_v,w_0)$ is a meromorphic function of s with a finite number of simple poles for $\mathrm{Re}(s) > 0$, none on $\mathrm{Re}(s) = 0$.

Assume v is an unramified place for π, i.e., that π_v is spherical. Take $f_v^0 \in V(s,\pi_v)$ such that $f_v^0(k)$ is a fixed vector invariant under $\mathbf{M}(O_v)$ for all $k \in \mathbf{G}(O_v)$. With notation as in Section 2, let $r = \bigoplus_{i=1}^{m} r_i$ be the adjoint action of ${}^LM$ on ${}^L\mathfrak{n}$. We have

Lemma 4.1 ([43]). *Assume π_v is unramified. Then*

$$\begin{aligned} &A(s,\pi_v,w_0)f_v^0(e_v) \\ &= \prod_{i=1}^{m} L(is,\pi_v,\tilde{r}_i)/L(1+is,\pi_v,\tilde{r}_i)f_v^0(e_v). \end{aligned}$$

With a clever induction [43, 44, 55] the problem reduces to that of SL_2, which we will now explain.

Here $\mathbf{G} = SL_2$ and $\mathbf{M} = \mathbf{T} = \mathbb{G}_m = \overline{F}_v^*$. We need to calculate

$$\begin{aligned}&\int_{F_v} f_v^0\left(\begin{pmatrix} 0 & 1 \\ -1 & 0 \end{pmatrix}\begin{pmatrix} 1 & x \\ 0 & 1 \end{pmatrix}\right) dx \\ &= \int_{F_v} f_v^0\left(\begin{pmatrix} 1 & 0 \\ x & 1 \end{pmatrix}\right) dx,\end{aligned} \tag{4.1}$$

since $R_w f_v^0 = f_v^0$, where $w = \begin{pmatrix} 0 & -1 \\ 1 & 0 \end{pmatrix}$. We then have

$$\begin{aligned}&\int_{F_v} f_v^0\left(\begin{pmatrix} 1 & 0 \\ x & 1 \end{pmatrix}\right) dx \\ &= \int_{|x|_v \leqslant 1} dx + \int_{|x|_v > 1} f_v^0\left(\begin{pmatrix} 1 & 0 \\ x & 1 \end{pmatrix}\right) dx.\end{aligned} \tag{4.2}$$

We now write

$$\begin{pmatrix} 1 & 0 \\ x & 1 \end{pmatrix} = \begin{pmatrix} x^{-1} & 1 \\ 0 & x \end{pmatrix}\begin{pmatrix} 0 & -1 \\ 1 & x^{-1} \end{pmatrix} \tag{4.3}$$

and therefore for $|x|_v > 1$, which implies $x^{-1} \in O_v$,

$$\begin{aligned}f_v^0\left(\begin{pmatrix} 1 & 0 \\ x & 1 \end{pmatrix}\right) &= f_v^0\left(\begin{pmatrix} x^{-1} & 1 \\ 0 & x \end{pmatrix}\right) \\ &= \eta_v^{-1}(x)|x|_v^{-1-s},\end{aligned}$$

where η_v is the character of F^* (with cusp form $\eta = \otimes_v \eta_v$ on $\mathbb{A}_F^*$) defining the induced representation. Then (4.2) equals

$$\begin{aligned}&1 + \sum_{n=1}^{\infty} \eta_v(\varpi_v^n) q_v^{-n-ns} \int_{(\varpi_v^{-n})-(\varpi_v^{-n+1})} |x|_v d^*x \\ &= 1 + \sum_{n=1}^{\infty} \eta_v(\varpi_v^n) q_v^{-n-ns} |\varpi_v^{-n}|_v \int_{O_v^*} d^*x \\ &= 1 + \sum_{n=1}^{\infty} \eta_v(\varpi_v^n)(1 - q_v^{-1}) q_v^{-ns}.\end{aligned}$$

Write $\eta_v(x) = |x|_v^{\mu_v}$ to get

$$\begin{aligned}&1 + \sum_{n=1}^{\infty} q_v^{-n\mu_v - ns}(1 - q_v^{-1}) \\ &= 1 + (1 - q_v^{-1}) \cdot q_v^{-\mu_v - s} \cdot \frac{1}{1 - \frac{1}{q_v^{\mu_v + s}}} \\ &= \frac{1 - \frac{1}{q_v^{\mu_v + 1 + s}}}{1 - \frac{1}{q_v^{\mu_v + s}}}.\end{aligned} \tag{4.4}$$

Now if $\alpha^\vee$ is the standard coroot of SL_2, then

$$q_v^{\mu_v} = \alpha^\vee(A_v),$$

following our H_T identification. (See the remark below.) Here $A_v \in PGL_2(\mathbb{C})$ represents the semisimple conjugacy class parametrizing π_v and $\alpha^\vee$ is the root of $PGL_2(\mathbb{C})$, or the coroot of SL_2. Clearly $\alpha^\vee(A_v)$ is the eigenvalue for the adjoint action of ${}^LT$ on ${}^L\mathfrak{n}$, evaluated at A_v. Therefore

$$q_v^{-\mu_v} = \alpha^\vee(A_v)^{-1} = \tilde{r}(A_v). \tag{4.5}$$

We thus get that (4.1) equals

$$(1 - \alpha^\vee(A_v)^{-1} q_v^{-s})^{-1}/(1 - \alpha^\vee(A_v)^{-1} q_v^{-s-1})^{-1} \tag{4.6}$$

which equals

$$L(s, \eta_v, \tilde{r}_1)/L(1+s, \eta_v, \tilde{r}_1) \tag{4.7}$$

since $m = 1$, i.e., $r = r_1$.

$\mathbf{SL}_2(\mathbb{R})$. It is instructive to also compute the case of $SL_2(\mathbb{R})(GL_2(\mathbb{R})$, respectively). We again need to calculate

$$\int_{\mathbb{R}} f_v^0\left(\begin{pmatrix} 1 & 0 \\ x & 1 \end{pmatrix}\right) dx. \tag{4.8}$$

Here $K = SO_2(\mathbb{R})(O_2(\mathbb{R})$, respectively) and we can write

$$\begin{pmatrix} 1 & 0 \\ x & 1 \end{pmatrix} = \begin{pmatrix} a & y \\ 0 & b \end{pmatrix} k(\theta),$$

where

$$k(\theta) = \begin{pmatrix} \cos\theta & \sin\theta \\ -\sin\theta & \cos\theta \end{pmatrix}.$$

We can then take $\tan\theta = -x$, $b = a^{-1} = \sqrt{x^2+1}$ and $y = x/\sqrt{x^2+1}$. We need to calculate

$$\int_{-\infty}^{\infty} \left(\frac{1}{\sqrt{x^2+1}}\right)^{s_1} (\sqrt{x^2+1})^{s_2} (x^2+1)^{-1/2} dx,$$

where $\eta\begin{pmatrix} a & 0 \\ 0 & b \end{pmatrix} = |a|^{s_1}|b|^{s_2}$, or

$$2\int_0^\infty (x^2+1)^{-(s_1-s_2+1)/2} dx.$$

Let $s = s_1 - s_2$ and set $x = \tan\theta$, we need to calculate

$$2\int_0^{\pi/2} \cos^{s-1}\theta d\theta.$$

Using the standard formula

$$\int_0^{\pi/2} \cos^p\theta \sin^q\theta d\theta = \Gamma\Big(\frac{p+1}{2}\Big)\Gamma\Big(\frac{q+1}{2}\Big)\Big/ 2\Gamma\Big(\frac{p+q}{2}+1\Big),$$

where

$$\Gamma(s) = \int_0^\infty t^s e^{-t} d^*t,$$

(4.8) then equals

$$\Gamma(1/2)\Gamma(s/2)/\Gamma((s+1)/2). \tag{4.9}$$

Using $\Gamma(1/2) = \sqrt{\pi}$, (4.9) equals

$$\pi^{-s/2}\Gamma(s/2)/\pi^{-(s+1/2)}\Gamma(s+1/2)$$

which is again $L(s)/L(s+1)$, where $L(s)$ is the archimedean Hecke-Tate L-function attached to the character $|x|^s$, the $\mathbb{R}$-component of our cusp form on T.

The main result of Langlands in [43] can be stated as follows. Let S be a finite set of places with the property that if $v \notin S$, π_v is an unramified representation. Every $f \in V(s,\pi)$ is of the form

$$f \in \otimes_{v\in S} V(s,\pi_v) \bigotimes \otimes_{v\notin S}\{f_v^0\},$$

where f_v^0 is K_v-spherical for some S. To be precise, the decomposition $\pi = \otimes_v \pi_v$ depends on a choice of $\mathbf{M}(O_v)$-invariant vectors $\{x_v\}$ for all the unramified places which one fixes once for all (cf. [68]). The functions f_v^0 must then satisfy $f_v^0(k_v) = x_v$ for all $k_v \in K_v$ and all $v \notin S$. Assume further that $f = \otimes_v f_v$, with $f_v = f_v^0$ for all $v \notin S$. For each i, let

$$L^S(s,\pi,r_i) = \prod_{v\notin S} L(s,\pi_v,r_i). \tag{4.10}$$

Then by Lemma 4.1

$$\begin{aligned} A(s,\pi)f(e) = \Bigg(\prod_{i=1}^m L^S(is,\pi,\tilde{r}_i)/L^S(1+is,\pi,\tilde{r}_i)\Bigg) \\ \cdot \otimes_{v\notin S} f_v^0(e_v) \bigotimes \otimes_{v\in S} A(s,\pi_v,w_0)f_v(e_v). \end{aligned} \tag{4.11}$$

It now follows from the properties of the constant term $A(s,\pi)$ that

Theorem 4.1 ([43]). *The product quotient*

$$\prod_{i=1}^m L^S(is,\pi,\tilde{r}_i)/L^S(1+is,\pi,\tilde{r}_i)$$

is meromorphic on all of $\mathbb{C}$.

Clearly one needs an induction to get this to lead to meromorphic continuation of each L-function in the product. We will soon discuss this induction.

5 Examples

We shall now give a number of important examples of L-functions which appear in constant terms for appropriate pairs $(\mathbf{G}, \mathbf{M})$. We refer to [43, 54] for the complete list.

5.1. Let $\mathbf{G} = GL_{n+t}$, $\mathbf{M} = GL_n \times GL_t$, $\pi = \otimes_v \pi_v$ cusp form on $GL_n(\mathbb{A}_F)$ and $\pi' = \otimes_v \pi'_v$ one on $GL_t(\mathbb{A}_F)$. Then $m = 1$ and we get $L(s, \pi \times \tilde{\pi}')$, the Rankin-Selberg product L-function for the pair $(\pi, \tilde{\pi}')$ (cf. [30, 62]). It will be discussed by Cogdell in more length.

5.2. Let $\mathbf{G}$ be a classical group, split over F and let $\mathbf{M} = GL_n \times \mathbf{G}'$, where $\mathbf{G}'$ is a classical group of the same type, but lower rank. Let π and π' be cuspidal representations of $GL_n(\mathbb{A}_F)$ and G'. Then $m = 2$. One gets $L(s, \pi \times \tilde{\pi}')$ as its first L-function. For $i = 2$, we get $L(s, \pi, \rho)$, where $\rho = \Lambda^2$ if ${}^LG$ is orthogonal and $\rho = \mathrm{Sym}^2$ if ${}^LG$ is symplectic (cf. [20, 21].

5.3. Let $\mathbf{G} = GSpin_{n+t}$, $\mathbf{M} = GL_n \times GSpin_t$, π and π' cusp forms on $GL_n(\mathbb{A}_F)$ and $GSpin_t(\mathbb{A}_F)$, respectively. Then $m = 2$. Again we get $L(s, \pi \times \tilde{\pi}')$ as our first L-function. The second L-function is then an appropriate twist of either $L(s, \pi, \Lambda^2)$ or $L(s, \pi, \mathrm{Sym}^2)$.

5.3.a. Let $\mathbf{G} = GSpin_{5+2n}$, $\mathbf{M} = GL_n \times GSpin_5 = GL_n \times GSp_4$, and let (π, π') be a cusp form on $GL_n(\mathbb{A}_F) \times GSp_4(\mathbb{A}_F)$. Again we get $L(s, \pi \times \tilde{\pi}')$ as our first L-function. This is very important.

5.3.b. $\mathbf{G} = GSpin_{6+2n}$, $\mathbf{M} = GL_n \times GSpin_6$. We have $0 \to \{\pm 1\} \to GL_4(\mathbb{C}) \to GSO_6(\mathbb{C}) \to 0$. Suppose π is on $GL_n(\mathbb{A}_F)$ and π' on $GL_4(\mathbb{A}_F)$. We then get $L(s, \pi \otimes \tilde{\pi}', \rho_n \otimes \Lambda^2 \rho_4)$ as our first L-function (cf. [36]).

5.4. Let $\mathbf{G}$ be a simply connected group of either type E_6 or E_7. Choose $\mathbf{M}$ such that $\mathbf{M}_D$, the derived group of $\mathbf{M}$, is either $SL_3 \times SL_2 \times SL_3$ or $SL_3 \times SL_2 \times SL_4$, respectively. There exist F-rational injections from $\mathbf{M}$ into $GL_3 \times GL_2 \times GL_t$, $t = 3$ or 4, which are identity on $SL_3 \times SL_2 \times SL_t$. Let $\pi_1 \otimes \pi_2 \otimes \sigma$ be a cuspidal representation of $GL_3(\mathbb{A}_F) \times GL_2(\mathbb{A}_F) \times GL_t(\mathbb{A}_F)$. Then $m = 3$ or 4 according as if $\mathbf{G} = E_6$ or E_7. The first L-function is then $L(s, \pi_1 \times \pi_2 \times \tilde{\sigma})$ (cf. [39]).

All these L-functions will be revisited later in connection with functoriality.

6 Local Coefficients, Nonconstant Term and the Crude Functional Equation

Changing the splitting in $\mathbf{U}$, we may assume each χ_v is defined by means of ψ_v through (2.5). At each v, let $\lambda_{\chi_v}(s, \pi_v)$ be the Whittaker functional defined by (2.11). Next, let $A(s, \pi_v, w_0)$ be the local intertwining operator defined by (3.10). Finally, let $\lambda_{\chi_v}(-s, w_0(\pi_v))$ be the corresponding functional defined for $I(-s, w_0(\pi_v))$ by means of (2.11). Using our assumption on χ_v, Rodier's theorem points to the existence of a complex function $C_{\chi_v}(s, \pi_v)$ of s, depending on π_v, χ_v and w_0 such that

$$\lambda_{\chi_v}(s, \pi_v) = C_{\chi_v}(s, \pi_v) \lambda_{\chi_v}(-s, w_0(\pi_v)) \cdot A(s, \pi_v, w_0). \tag{6.1}$$

This is what we call the **"Local Coefficient"** attached to s, π_v, χ_v and w_0 (cf. [54, 55]). The choice of w_0 is now specified by our fixed splitting as in [56].

Now, let

$$\begin{aligned} &E_\chi(s, \Phi_s, g, P) \\ &= \int_{\mathbf{U}(F)\backslash U} E(s, \Phi_s, ug, P)\overline{\chi(u)}du, \end{aligned} \tag{6.2}$$

the χ-nonconstant term of $E(s, \Phi_s, g, P)$ (cf. [54, 55, 58]), where $\chi = \otimes_v \chi_v$.

If we substitute for $E(s, \Phi_s, ug, P)$ its definition in (3.3) and do some telescoping, we get, using orthogonality of χ, that

$$E_\chi(s, \Phi_s, e, f) = \prod_v \lambda_{\chi_v}(s, \pi_v)(f_v), \tag{6.3}$$

where f_v is the local component of f defined by (3.6) in which $\varphi = \otimes_v \varphi_v$ is identified with $W_\varphi = \otimes_v W_v$, where W_φ is defined by (2.7). As explained in [68], $W_v(e_v) = 1$ for almost all v. We now appeal to the following formula of Casselman-Shalika [13]:

Theorem 6.1 ([13]). *Assume π_v and ψ_v are both unramified and if $f_v(e_v)$ defines a Whittaker function W_v in the Whittaker model of π_v, assume $W_v(e_v) = 1$. Observe that this is the case for almost all v. Then*

$$\lambda_{\chi_v}(s, \pi_v)(f_v) = \prod_{i=1}^{m} L(1 + is, \pi_v, \tilde{r}_i)^{-1}. \tag{6.4}$$

In fact, if $W_{f_v}(g_v) = \lambda_{\chi_v}(s, \pi_v)(I_v(g_v)f_v)$ is the Whittaker function attached to f_v, then (6.3) can be written as

Theorem 6.2. *One has*

$$E_\chi(s, \Phi_s, e, f) = \prod_{v \in S} W_{f_v}(e_v) \prod_{i=1}^{m} L^S(1 + is, \pi, \tilde{r}_i)^{-1}. \tag{6.5}$$

Remark. As opposed to intertwining operators, Whittaker functions are by no means multiplicative and therefore proof of Theorem 6.1 cannot be reduced to rank one calculations by means multiplicativity (cocycle relations). It should be pointed out that in the case of SL_2, Theorem 6.1 is an easy exercise.

Corollary 6.1 ([55]). *The product*

$$\prod_{i=1}^{m} L^S(1 + is, \pi, r_i) \neq 0$$

for $\mathrm{Re}(s) = 0$. *In particular, if π and π' are cusp forms on $GL_n(\mathbb{A}_F)$ and $GL_t(\mathbb{A}_F)$, then*

$$L(1, \pi \times \pi') \neq 0,$$

where local L-functions at every place are corresponding Artin ones (cf. [27, 28]).

Proof. Modulo non-vanishing of $W_{f_v}(e)$ for $\mathrm{Re}(s) = 1$, which is highly nontrivial if $v = \infty$ (cf. [11, 74]), this follows from unitarity (and therefore holomorphy) of $M(s, \pi)$ for $\mathrm{Re}(s) = 0$, and Theorem 6.2. Observe that the integration in (6.2) is over a compact set. □

Now, computing the nonconstant terms from the two sides of the functional equation (3.17), Lemma 4.1 and Theorems 6.1 and 6.2, together with (6.1) implies:

Theorem 6.3 ([55, 58, Crude Functional Equation]). *We have*

$$\prod_{i=1}^{m} L^S(is, \pi, r_i) = \prod_{v \in S} C_{\overline{\chi}_v}(s, \tilde{\pi}_v) \prod_{i=1}^{m} L^S(1 - is, \pi, \tilde{r}_i). \tag{6.6}$$

We just point out that by Lemma 4.1, Theorem 6.1 and (6.1)

$$C_{\overline{\chi}_v}(s, \tilde{\pi}_v) = \prod_{i=1}^{m} L(1 - is, \pi_v, \tilde{r}_i)/L(is, \pi_v, r_i), \tag{6.7}$$

whenever π_v is unramified.

7 The Main Induction, Functional Equations and Multiplicativity

To prove the individual functional equations, i.e., for each $L(s, \pi, r_i)$ with precise root numbers and L-functions , we appeal to the following induction statement (cf. [53, 54]). It is crucial in all the results that we prove from now on.

Proposition 7.1 ([53]). *Given $1 < i \leqslant m$, there exists a split group $\mathbf{G}_i$ over F, a maximal F-parabolic subgroup $\mathbf{P}_i = \mathbf{M}_i\mathbf{N}_i$ and a cuspidal automorphic representation π' of $M_i = \mathbf{M}_i(\mathbb{A}_F)$, unramified for every $v \notin S$, such that if the adjoint action of ${}^LM_i$ on ${}^L\mathfrak{n}_i$ decomposes as $r' = \bigoplus_{j=1}^{m'} r'_j$, then*

$$L^S(s, \pi, r_i) = L^S(s, \pi', r'_1).$$

Moreover $m' < m$.

It was observed by Arthur [1], that each $\mathbf{M}_i$ can be taken equal to $\mathbf{M}$ and $\pi' = \pi$. More precisely:

Proposition 7.2 ([1]). *Given i, $1 < i \leqslant m$, there exists a split connected reductive F-group $\mathbf{G}_i$ with $\mathbf{M}$ as a Levi subgroup and $m' < m$ for which $r'_1 = r_i$. Each $\mathbf{G}_i$ can be taken to be an endoscopic group for $\mathbf{G}$. (Its L-group is the centralizer of a semisimple element in ${}^LG$.)*

Next, we need the following variant of a result of Henniart and Vigneras. In it, we assume that the defining additive character ψ for χ is local component of a global one. Here we shift the ramification to archimedean places. Consequently, we need to use a result of Dixmier-Malliavin on convolution algebras for real semisimple groups.

Proposition 7.3 ([53]). *Let σ be an irreducible χ-generic supercuspidal representation of $G = \mathbf{G}(F)$, where F is a non-archimedean local field and $\mathbf{G}$ is defined over F. Let $\mathbf{B} = \mathbf{TU}$ be the Borel subgroup of $\mathbf{G}$ defining χ. Then there exists a number field K with a ring of integers O, a split group $\mathbf{H}$ over K, a non-degenerate character $\tilde{\chi} = \otimes_v \tilde{\chi}_v$ of $\mathbf{U_H}(K)\backslash \mathbf{U_H}(\mathbb{A}_K)$, and a globally $\tilde{\chi}$-generic cusp form $\pi = \otimes_v \pi_v$ on $H = \mathbf{H}(\mathbb{A}_K)$ such that:*

1) *$K_{v_0} = F$ for some place v_0 of K.*

2) *$\tilde{\chi}_{v_0} = \chi$.*

3) *As a group over F, $\mathbf{H} = \mathbf{G}$.*

4) *$\pi_{v_0} = \sigma$.*

5) *For every other finite place v of K, $v \neq v_0$, π_v is of class one with respect to a special maximal compact subgroup Q_v of $\mathbf{H}(K_v)$. Here $\mathbf{U_H}$ is the unipotent radical of a Borel subgroup of $\mathbf{H}$ for which $\mathbf{U_H}$ as a group over F equals $\mathbf{U}$.*

Proposition 7.4 ([53, 56]). *Assume either F_v is archimedean or π_v has a vector fixed by an Iwahori subgroup. Let $\varphi_v \colon W'_{F_v} \to {}^L M_v$ be the homorphism of the Deligne-Weil group parametrizing π_v. For each i, let $L(s, r_i\cdot\varphi_v)$ and $\varepsilon(s, r_i\cdot\varphi_v, \psi_v)$ be the Artin L-function and root number attached to $r_i \cdot \varphi_v$. Then*

$$C_{\chi_v}(s,\pi_v) = \prod_{i=1}^{m} \varepsilon(is, \tilde{r}_i \cdot \varphi_v, \overline{\psi}_v) \frac{L(1-is, r_i \cdot \varphi_v)}{L(is, \tilde{r}_i \cdot \varphi_v)}. \tag{7.1}$$

Remark. If $\mathbf{G}$ is quasisplit, but not split, then a product of Langlands λ-functions (Hilbert symbols) will also appear on the right hand side of (7.1).

Applying these propositions inductively and using the **Crude Functional Equations** (6.6) then implies:

Theorem 7.1 ([53]). *Assume $\mathbf{G}$ is a split reductive algebraic group over a local field F of characteristic zero. Let $\mathbf{P} = \mathbf{MN}$, $\mathbf{P} \supset \mathbf{B}$, be a maximal parabolic subgroup as before. Let χ be a generic character defined by the splitting and $\psi_F \in \hat{F}$. Given an irreducible admissible χ-generic representation σ of $M = \mathbf{M}(F)$, there exist m complex functions $\gamma(s, \sigma, r_i, \psi_F)$, $1 \leqslant i \leqslant m$, such that:*

1) *If F and σ satisfy the conditions of Proposition 7.3, then*

$$\gamma(s, \sigma, r_i, \psi_F) = \varepsilon(s, r_i \cdot \varphi, \psi_F) L(1-s, \tilde{r}_i \cdot \varphi)/L(s, r_i \cdot \varphi). \tag{7.2}$$

2) *Equation (7.1) holds (in the form $\prod_{i=1}^{m} \gamma(is, \sigma, \tilde{r}_i, \overline{\psi}_F)$).*

3) *$\gamma(s,\sigma,r_i,\psi_F)$ is multiplicative under induction (to be discussed below).*

4) *Whenever σ becomes a local component of a globally generic cusp from, then γ's become the local factors needed in their functional equations.*

Moreover 1), 3) *and* 4) *determine the γ-functions uniquely.*

What is multiplicativity? We will discuss this only in examples since the general formulation is complicated. It simply says that the γ-functions are multiplicative under parabolic induction and is a consequence of multiplicativity of intertwining operators (3.10) under that (cf. [55]). This is very deep from the point of view of Rankin-Selberg method and usually quite hard to prove. Here are some examples:

Example 7.1 ([59]). Suppose $\mathbf{G} = Sp(2n+2t)$ and $\mathbf{M} = GL_n \times Sp(2t)$, where n and t are positive integers. Write $\sigma = \sigma_1 \otimes \tau$. Suppose $\mathbf{M}' = GL_{n_1} \times \cdots \times GL_{n_k} \times GL_{t_1} \times \cdots \times GL_{t_\ell} \times Sp(2a)$, where $n_1 + \cdots + n_k = n$ and $t_1 + \cdots + t_\ell + a = t$. By case C_n of [54], r_1 is equal to the tensor product of the standard representation of $GL_n(\mathbb{C})$ and $SO_{2t+1}(\mathbb{C})$.

If $\sigma' = \bigotimes_{j=1}^{k} \sigma_j' \otimes \bigotimes_{b=1}^{\ell} \sigma_b'' \otimes \tau'$, then multiplicativity simply means that, if

$$\sigma \subset \ \mathrm{Ind}_{M'N'\uparrow G}\sigma' \otimes \mathbf{1},$$

then

$$\begin{aligned}&\gamma(s,\sigma_1\times\tau,\psi_F)\\ &=\prod_{j=1}^{k}\prod_{b=1}^{\ell}\gamma(s,\sigma_j'\times\sigma_b'',\psi_F)\gamma(s,\sigma_j'\times\tilde{\sigma}_b'',\psi_F)\prod_{j=1}^{k}\gamma(s,\sigma_j'\times\tau',\psi_F).\end{aligned} \tag{7.3}$$

If ρ_n is the standard representation of $GL_n(\mathbb{C})$, then $r_2 = \Lambda^2\rho_n$. With σ_1 and σ_j' as before, multiplicativity for r_2 means

$$\gamma(s,\sigma_1,\Lambda^2\rho_n,\psi_F) = \prod_{j=1}^{k}\gamma(s,\sigma_j',\Lambda^2\rho_{n_j},\psi_F)\prod_{1\leqslant i<j\leqslant k}\gamma(s,\sigma_i'\times\sigma_j',\psi_F).$$

No more r_i beyond r_2 shows up and this is the case for all the classical groups. Equality of the dimension on both sides of (7.3) simply means the following trivial identity:

$$\left(\sum_{i=1}^{k} n_i\right)^2 - \left(\sum_{i=1}^{k} n_i\right) = \sum_{i=1}^{k}(n_i^2 - n_i) + 2\sum_{1\leqslant i<j\leqslant k} n_i n_j.$$

***L*-function and root number.** L-functions are now defined using γ-functions. When σ is tempered, we define $L(s,\sigma,r_i)$ as the inverse of the normalized polynomial $P(q^{-s})$ in q^{-s} satisfying $P(0)=1$ and

$$\gamma(s,\sigma,r_1,\psi_F) = \varepsilon(s,\sigma,r_i,\psi_F)L(1-s,\sigma,\tilde{r}_i)/L(s,\sigma,r_i). \tag{7.4}$$

The L-function $L(s,\sigma,\tilde{r}_i)$ and the root number $\varepsilon(s,\sigma,r_i,\psi_F)$ are also uniquely defined by (7.4). To proceed we need the following theorem.

Theorem 7.2. *Suppose σ is tempered. Then $L(s,\sigma,r_i)$ are all holomorphic for* $\mathrm{Re}(s)>0$.

With this theorem in hand, $L(s,\sigma,r_i)$ are now also multiplicative if σ is tempered. (See below.)

To define $L(s,\sigma,r_i)$ for any irreducible χ-generic representation, we appeal to Langlands classification [45, 70]. We embed $\sigma \subset \mathrm{Ind}_{M'(N'\cap M)\uparrow M}\sigma'_\nu \otimes \mathbf{1}$, where σ'_ν is quasi-tempered with a negative Langlands parameter ν. Then σ'_0 is tempered. By multiplicativity, we then write $\gamma(s,\sigma,r_i,\psi_F)$ as a product of appropriate γ-functions $\gamma(s,\sigma'_{\nu,j},r_{ij},\psi_F)$, where j runs over a finite index set determined by $\mathbf{M}'$ and $\mathbf{M}$, i.e.,

$$\gamma(s,\sigma,r_i,\psi_F)=\prod_j \gamma(s,\sigma'_{\nu,j},r_{ij},\psi_F). \tag{7.5}$$

More precisely, for each j, there exist Levi subgroups (not necessarily maximal) $\mathbf{M}'_j$ and $\tilde{\mathbf{M}}_j$ of $\mathbf{G}$ with $\mathbf{T}\subset\mathbf{M}'_j\subset\tilde{\mathbf{M}}_j$ as a maximal Levi subgroup. The representation $\sigma'_{\nu,j}$ is a quasi-tempered representation of M'_j for which $\sigma'_{0,j}$ is tempered. The representation r_{ij} of ${}^LM'_j$ is an irreducible constitutent of the action of ${}^LM'_j$ on the Lie algebra of the L-group of $\tilde{\mathbf{M}}_j\cap\mathbf{N}'_j$, where $\mathbf{N}'_j\subset\mathbf{U}$ is the unipotent radical of $\mathbf{P}'_j=\mathbf{M}'_j\mathbf{N}'_j$. Thus the γ-function $\gamma(s,\sigma'_{\nu,j},r_{ij},\psi_F)$ is a γ-function attached to the pair $(\tilde{\mathbf{M}}_j,\mathbf{M}'_j)$.

When $\nu=0$, by Theorem 7.2, the product of the numerators of $\gamma(s,\sigma'_{0,j},r_{ij},\psi_F)$ equals to the numerator of the product and if $L(s,\sigma'_{0,j},r_{ij})^{-1}$ denotes the normalized numerator of $\gamma(s,\sigma'_{0,j},r_{ij},\psi_F)$, i.e., the reciprocal of the tempered L-function attached to $\sigma'_{0,j}$ and r_{ij} by means of the pair $(\tilde{\mathbf{M}}_j,\mathbf{M}'_j)$, we then use $L(s,\sigma'_{\nu,j},r_{ij})$ to denote its analytic continuation to ν. We now set

$$L(s,\sigma,r_i)=\prod_j L(s,\sigma'_{\nu,j},r_{ij}). \tag{7.6}$$

This agrees with the way Artin L-functions are defined [45, 32, 53, 73]. Details are given in [55]; also see the discussion in pages 862 and 863 of [39]. The root number is then defined uniquely to satisfy (7.4). We should point out that in (7.6) we do not need to assume the validity of Theorem 7.2. But if valid, it implies that L-functions are also multiplicative, if the representations are tempered.

Having defined our L-functions and root numbers everywhere, we set

$$L(s,\pi,r_i)=\prod_v L(s,\pi_v,r_i)$$

and

$$\varepsilon(s,\pi,r_i)=\prod_v \varepsilon(s,\pi_v,r_i,\psi_v).$$

We then have:

Theorem 7.3 ([53, Functional Equation]). *For each* i, $1 \leqslant i \leqslant m$,

$$L(s,\pi,r_i) = \varepsilon(s,\pi,r_i)L(1-s,\pi,\tilde{r}_i). \tag{7.7}$$

Exercise 7.1. Use the pair $(\mathbf{G},\mathbf{M})$, $\mathbf{G} = GL_3$ and $\mathbf{M} = GL_2 \times GL_1$, to get the standard L-function for GL_2. Determine L-functions using our method and show that they are equal to those of Jacquet-Langlands.

Exercise 7.2. Let $\mathbf{G} = E_6^{sc}$ and $\mathbf{M}$ be such that $\mathbf{M}_D = SL_3 \times SL_2 \times SL_3$.

Fact 7.1. There exists a F-rational map (injection) $f\colon \mathbf{M} \to GL_3 \times GL_2 \times GL_3$ whose restriction to $\mathbf{M}_D$ is identity.

Fact 7.2. $m = 3$ and if $\pi_2 \otimes \pi_1 \otimes \sigma$ is an unramified representation of $GL_3(F) \times GL_2(F) \times GL_3(F)$, where F is a local field, then

$$L(s,\pi_2 \times \pi_1 \times \tilde{\sigma}) = L(s,(\pi_2 \otimes \pi_1 \otimes \sigma)\cdot f,\ r_1).$$

Define $\gamma(s,\pi_2 \times \pi_1 \times \tilde{\sigma})$ to be $\gamma(s,(\pi_2 \otimes \pi_1 \otimes \sigma)\cdot f,\ r_1)$, using our method for arbitrary local representations $\pi_2 \otimes \pi_1 \otimes \sigma$. Assume $\tilde{\sigma} = \mathrm{Ind}_{(F^*)^3 \times U \uparrow GL_3(F)}(\mu_1 \otimes \mu_2 \otimes \mu_3) \otimes \mathbf{1}$. Show that multiplicativity implies:

$$\gamma(s,\pi_2 \times \pi_1 \times \tilde{\sigma}) = \prod_{j=1}^{3} \gamma(s,\pi_2 \times (\pi_1 \otimes \mu_j)),$$

where the γ-function on the right are those of Rankin-Selberg for $GL_3(F) \times GL_2(F)$. (This is crucial to Kim and Shahidi's proof of functoriality [39] of the inclusion $GL_2(\mathbb{C}) \otimes GL_3(\mathbb{C}) \hookrightarrow GL_6(\mathbb{C})$, to be discussed later.)

Exercise 7.3. Let $\mathbf{G} = SO(2m+2n+1)$ and $\mathbf{M} = GL_m \times SO(2n+1)$. Let $\sigma \otimes \pi$ be an irreducible admissible χ-generic representation of $GL_m(F) \times SO_{2n+1}(F)$, where F is a local field. Assume $\sigma \hookrightarrow \mathrm{Ind}_{(F^*)^m \times U \uparrow GL_m(F)} \bigotimes_{j=1}^{m} \mu_j$. Show that multiplicativity implies:

$$\gamma(s,\sigma \times \pi) = \prod_{j=1}^{m} \gamma(s,\pi \otimes \mu_j).$$

(This is crucial to Cogdell, Kim, Piatetski-Shapiro and Shahidi's proof [20] of functorial transfer from generic cusp forms on classical groups to $GL_{2n}(\mathbb{A}_F)$.)

8 Twists by Highly Ramified Characters, Holomorphy and Boundedness

Since our aim is to establish those analytic properties of L-functions from our method which are crucial in proving the striking new cases of functoriality, we will limit our discussion on the issue of holomorphy of L-functions only to twists by highly ramified characters. In fact, as we explained in earlier sections, the functional equations for L-functions within our method are proved quite generally and multiplicativity and the related machinary necessary for applying converse theorems to our L-functions are in perfect shape.

Nothing that general can be said about the holomorphy and possible poles of these L-functions. On the other hand, there has recently been some remarkable new progress on the question of holomorphy of these L-functions, mainly due to Kim [34, 35, 38]. They rely on reducing the existence of the poles to that of existence of certain unitary automorphic forms, which in turn point to the existence of certain local unitary representations. One then disposes of these representations, and therefore the pole, by checking the corresponding unitary dual of the local group. In view of the functional equation, this needs to be checked only for $\mathrm{Re}(s) \geqslant 1/2$, if a certain local assumption on normalized local intertwining operators is valid. To explain, let $A(s, \pi_v, w_0)$ be the local intertwining operator attached by (3.10) to our inducing representation π_v. We recall that we are dealing with a pair $(\mathbf{G}, \mathbf{M})$ and a χ-generic cuspidal representation $\pi = \otimes_v \pi_v$ of $M = \mathbf{M}(\mathbb{A}_F)$. Let, for each i, $1 \leqslant i \leqslant m$, $L(s, \pi_v, r_i)$ and $\varepsilon(s, \pi_v, r_i, \psi_v)$ be the corresponding L-function and root number specified earlier. We define a normalized operator $N(s, \pi_v, w_0)$ by

$$N(s, \pi_v, w_0) = r(s, \pi_v, \psi_v) A(s, \pi_v, w_0), \tag{8.1}$$

where the normalizing factor is defined as [53]

$$r(s, \pi_v, \psi_v) = \prod_{i=1}^{m} \varepsilon(is, \pi_v, \tilde{r}_i) L(1 + is, \pi_v, \tilde{r}_i) / L(is, \pi_v, \tilde{r}_i). \tag{8.2}$$

Theorem 8.1. *The operator $N(s, \pi_v, w_0)$ is holomorphic and nonzero for $\mathrm{Re}(s) \geqslant 1/2$.*

It should be mentioned that it is a result of Zhang [75] that, if Theorem 7.2 is applied to the pair $(\mathbf{G}, \mathbf{M})$, then the non-vanishing of $N(s, \pi_v, w_0)$ for $\mathrm{Re}(s) \geqslant 1/2$ follows from its holomorphy over the same range.

Arguments given in [20, 36, 39], then proceed under the validity of Theorem 8.1 for $(\mathbf{G}, \mathbf{M})$ as well as for all other related lower rank pairs (that come into the multiplicativity), which consequently are verified in each of the cases in [20, 36, 39].

The main issue with this argument is that one cannot always get such a contradiction and rule out the pole. In fact, there are many unitary duals whose complementary series extend all the way to $\mathrm{Re}(s) = 1$, making the results far from general.

On the other hand, if one considers a highly ramified twist π_η (see below) of π, then it can be shown quite generally that every $L(s, \pi_\eta, r_i)$ is entire (cf. [60] for its local analogue). In fact, if η is highly ramified, then by checking central characters, $w_0(\pi_\eta) \not\simeq \pi_\eta$, whose negation is a necessary condition for $M(s, \pi_\eta)$ to have a pole, a basic fact from Langlands spectral theory of Eisenstein series (Lemma 7.5 of [44]). This lemma was used by Kim in [34] and in view of the present powerful converse theorems of Cogdell and Piatetski-Shapiro [16–18], that is all one needs concerning the holomorphy of L-functions to prove our cases of functoriality [20, 36, 39]. We formalize this by quoting the following (Proposition 2.1) from [39].

Theorem 8.2. *There exists a rational character $\xi \in X(\mathbf{M})_F = X(\mathbf{M})$, with the following property. Let S be a non-empty finite set of finite places of F. For every globally generic cuspidal representation π of $M = \mathbf{M}(\mathbb{A}_F)$, there exist non-negative integers f_v, $v \in S$, such that for every grössencharacter $\eta = \otimes_v \eta_v$ of F for which the conductor of η_v, $v \in S$, is larger than or equal to f_v, every L-function $L(s, \pi_\eta, r_i)$, $i = 1, \ldots, m$, is entire, where $\pi_\eta = \pi \otimes (\eta \cdot \xi)$. The rational character ξ can be simply taken to be $\xi(m) = \det(Ad(m)|\mathfrak{n})$, $m \in \mathbf{M}$, where $\mathfrak{n}$ is the Lie algebra of $\mathbf{N}$.*

The last ingredient in applying converse theorems is that of boundedness of each $L(s, \pi, r_i)$ in every vertical strip of finite width, away from its finite number of poles. The finiteness of poles is again a consequence of the finiteness of the poles of $M(s, \pi)$ for $\mathrm{Re}(s) \geqslant 0$ and the functional equation satisfied by each $L(s, \pi, r_i)$, together with Theorem 8.1 (cf. [24]). In this full generality, the boundedness in finite vertical strips, away from their poles, were proved by Gelbart and Shahidi in [24], again using our method. The main difficulty in proving this result is having to deal with reciprocals of each $L(s, \pi, r_i)$, $2 \leqslant i \leqslant m$, near and on the line $\mathrm{Re}(s) = 1$, the edge of critical strip, whenever $m \geqslant 2$, which is unfortunately the case for each of the L-functions appearing in our cases of functoriality. We handle this by appealing to (6.5) (Theorem 6.2) and estimating the nonconstant term (6.2) by means of Langlands [44] and Müller [50], and a nontrivial result from complex function theory (Matsaev's theorem). Here is the statement of the main result of [24] as formulated for π_η to avoid the issue of poles.

Theorem 8.3. *Let ξ and η be as in Theorem 6.2. Assume η is sufficiently ramified so that each $L(s, \pi_\eta, r_i)$ is entire. Then, given a finite real interval I, each $L(s, \pi_\eta, r_i)$ remains bounded for all s with $\mathrm{Re}(s) \in I$.*

Remark. Another proof of Theorem 8.3 is due to Gelbart and Lapid, following some ideas of Sarnak.

9 Examples of Functoriality with Applications

Consider the embedding

$$i \colon GL_2(\mathbb{C}) \otimes GL_3(\mathbb{C}) \hookrightarrow GL_6(\mathbb{C}).$$

This is a homomorphism from ${}^LGL_2 \times {}^LGL_3$ into ${}^LGL_6$. Accordingly functoriality predicts a map

$$\mathrm{Aut}\ (GL_2(\mathbb{A}_F)) \times\ \mathrm{Aut}\ (GL_3(\mathbb{A}_F)) \to\ \mathrm{Aut}\ (GL_6(\mathbb{A}_F)).$$

More precisely, let $\pi_1 = \otimes_v \pi_{1v}$ and $\pi_2 = \otimes_v \pi_{2v}$ be cusp forms on $GL_2(\mathbb{A}_F)$ and $GL_3(\mathbb{A}_F)$, respectively, with π_{1v} given by $t_{1v} \in GL_2(\mathbb{C})$ and π_{2v} by $t_{2v} \in GL_3(\mathbb{C})$

for almost all v. Let Π_v be the irreducible admissible spherical representation of $GL_6(F_v)$ defined by $\{t_{1v} \otimes t_{2v}\} \subset GL_6(\mathbb{C})$. Then we can state the functoriality in this case as:

Functoriality. *There exists an automorphic representation* $\Pi' = \otimes_v \Pi'_v$ *of* $GL_6(\mathbb{A}_F)$ *such that* $\Pi'_v = \Pi_v$ *for almost all* v.

More precisely, let $\rho_{iv}\colon W'_{F_v} \to GL_{i+1}(\mathbb{C})$, $i = 1, 2$, parametrize π_{iv} (Harris and Taylor [27], Henniart [28]) for all v. Let $\rho_{1v} \otimes \rho_{2v}$ be the six dimensional representation of W'_{F_v}, i.e., the homomorphism

$$\rho_{1v} \otimes \rho_{2v}\colon W'_{F_v} \to GL_6(\mathbb{C}).$$

Denote by $\pi_{1v} \boxtimes \pi_{2v}$ the irreducible admissible representation of $GL_6(F_v)$ attached to $\rho_{1v} \otimes \rho_{2v}$. Let $\pi_1 \boxtimes \pi_2 = \otimes_v (\pi_{1v} \boxtimes \pi_{2v})$.

Theorem 9.1 ([39]). *The irreducible admissible representation* $\pi_1 \boxtimes \pi_2$ *of* $GL_6(\mathbb{A}_F)$ *is automorphic. Thus* $GL_2(\mathbb{C}) \otimes GL_3(\mathbb{C}) \hookrightarrow GL_6(\mathbb{C})$ *is functorial.*

For the proof, one applies an appropriate version of the converse theorem [17] to the following cases of our method. In each case $\mathbf{G}$ and $\mathbf{M}_D$, the derived group of $\mathbf{M}$, are given as follows.

(1) $\mathbf{G} = SL(5)(or\mathrm{GL}(5))$, $\quad \mathbf{M}_D = SL_2 \times SL_3$;

(2) $\mathbf{G} = Spin(10)$, $\quad \mathbf{M}_D = SL_3 \times SL_2 \times SL_2$;

(3) $\mathbf{G} = E_6^{sc}$, $\quad \mathbf{M}_D = SL_3 \times SL_2 \times SL_3$;

(4) $\mathbf{G} = E_7^{sc}$, $\quad \mathbf{M}_D = SL_3 \times SL_2 \times SL_4$.

We then get the necessary analytic properties of the highly ramified twisted L-functions $L(s, \pi_1 \times \pi_2 \times (\sigma \otimes \eta))$, $\sigma \otimes \eta = \sigma \otimes \eta \cdot \det$, η a highly ramified grössencharacter, where σ's are appropriate cusp forms on $GL_j(\mathbb{A}_F)$, $j = 1, 2, 3, 4$, respectively. Observe that

$$L(s, \pi_{1v} \times \pi_{2v} \times \sigma_v) = L(s, (\pi_{1v} \boxtimes \pi_{2v}) \times \sigma_v)$$

for almost all v. Similarly for root numbers. We in fact prove these equalities for all v.

This is immediate from the fact that the local components of the weak transfer established through the converse theorem (Theorem 3.8 of [39]) is in fact $\pi_{1v} \boxtimes \pi_{2v}$ for each v (Theorem 5.1 of [39]). Proof of this is quite delicate and beside the techniques already discussed (functional equations, multiplicativity, ...), relies on several other techniques such as base change, both normal [2] and non-normal [31], as well as certain results from the theory of types [8].

Next, let $\pi = \otimes_v \pi_v$ be a cusp form on $GL_2(\mathbb{A}_F)$. Let $\mathrm{Ad}(\pi)$ be its Gelbart-Jacquet transfer [23]. Then

$$\pi \boxtimes Ad(\pi) = \mathrm{Sym}^3(\pi) \otimes \omega_\pi^{-1} \boxplus \pi$$

implies that $\mathrm{Sym}^3(\pi)$ is an automorphic representation of $GL_4(\mathbb{A}_F)$.

More precisely, for every v let $\rho_v\colon W'_{F_v} \to GL_2(\mathbb{C})$ be the two dimensional representation of the Deligne-Weil group W'_{F_v} attached to π_v (cf. [41]).

Consider:

$$\mathrm{Sym}^3 \cdot \rho_v = \mathrm{Sym}^3\rho_v\colon W'_{F_v} \to GL_4(\mathbb{C}).$$

Let $\mathrm{Sym}^3\pi_v$ be the irreducible admissible representation of $GL_4(F_v)$ attached to $\mathrm{Sym}^3\rho_v$. Set

$$\mathrm{Sym}^3\pi = \otimes_v \mathrm{Sym}^3\pi_v.$$

Theorem 9.2 ([39]). *$\mathrm{Sym}^3\pi$ is an automorphic representation of $GL_4(\mathbb{A}_F)$. It is cuspidal unless π is of dihedral or tetrahedral type.*

Next, let $\Pi = \otimes_v \Pi_v$ be a cuspidal representation of $GL_4(\mathbb{A}_F)$. Let $\Lambda^2\colon GL_4(\mathbb{C}) \to GL_6(\mathbb{C})$ be the exterior square map. Let $\varphi_v\colon W'_{F_v} \to GL_4(\mathbb{C})$ parametrize Π_v for all v [27, 28]. Then $\Lambda^2\varphi_v\colon W'_{F_v} \to GL_6(\mathbb{C})$ parametrizes $\Lambda^2\Pi_v$, an irreducible admissible representation of $GL_6(F_v)$.

Theorem 9.3 ([36]). *There exists an automorphic representation $\Pi' = \otimes_v \Pi'_v$ of $GL_6(\mathbb{A}_F)$ such that $\Pi'_v = \Lambda^2\Pi_v$.*

Corollary 9.1 ([36]). *The representation $\mathrm{Sym}^4(\pi)$ is automorphic, where*

$$\mathrm{Sym}^4(\pi) = \otimes_v \mathrm{Sym}^4(\pi_v)$$

and $\pi = \otimes_v \pi_v$ is a cusp form on $GL_2(\mathbb{A}_F)$.

Proof. Apply Λ^2 to $\mathrm{Sym}^3\pi$;

$$\Lambda^2(\mathrm{Sym}^3\pi) = \mathrm{Sym}^4\pi \otimes \omega_\pi \boxplus \omega_\pi^3. \qquad \square$$

Proposition 9.1 ([40]). *$\mathrm{Sym}^4(\pi)$ is cuspidal unless π is of dihedral, tetrahedral or octahedral type.*

Theorem 9.3 is proved by using

$$\mathbf{G} = Spin_{2k+8}, \quad \mathbf{M}_D = SL_{k+1} \times SL_4, \ k = 0, 1, 2, 3.$$

We get $L(s, \pi \otimes \sigma, \ \Lambda^2\rho_4 \otimes \rho_{k+1})$ for each cusp form σ on $GL_{k+1}(\mathbb{A}_F)$, $k = 0, 1, 2, 3$.

If we use the isogeny $SL_4 \to SO_6 \to 0$, we note that functoriality established in Theorem 9.3 is a special case of functoriality for the embedding

$$GSO_{2n}(\mathbb{C}) \hookrightarrow GL_{2n}(\mathbb{C}).$$

The result is the transfer

$$\mathrm{Aut}(GSpin_{2n}(\mathbb{A}_F)) \rightsquigarrow \mathrm{Aut}(GL_{2n}(\mathbb{A}_F)).$$

This is now established by Asgari-Shahidi in Duke J. in the weak form. The full transfer is now being written. It needs generalization of the descent of Ginzburg-Ralis-Soundry to *GSpin* groups. This is half-done by Hundley-Sayag. We have now completed the rest of it and expect to prove results as strong as those of [20] for $GSpin_m$, m = odd or even. We need some results on $L^S(s, \pi, \Lambda^2 \otimes \chi)$ and $L^S(s, \pi, \mathrm{Sym}^2 \otimes \chi)$ when π is a cusp form on $GL_n(\mathbb{A}_F)$. (We need holomorphy for both for $\mathrm{Re}(s) > 1$. This implies non-vanishing for both for $\mathrm{Re}(s) > 1$. Much stronger results now seem to be available through the work of D. Belt for $\Lambda^2 \otimes \chi$ and S. Takeda for $\mathrm{Sym}^2 \otimes \chi$, using Jacquet-Shalika and Bump-Ginzburg, respectively.)

In the cases of Asgari-Shahidi we have $G = GSpin_{2(m+n)}$, $M = GL_m \times GSpin_{2n}$ for $1 \leqslant m \leqslant 2n - 1$. The L-functions are product L-functions. Apply the converse theorem.

Applications:

Theorem 9.4 ([40]). *Let F be an arbitrary number field. Let $\pi = \otimes_v \pi_v$ be a cusp form on $GL_2(\mathbb{A}_F)$. Assume π_v is parametrized by*

$$t_v = \begin{pmatrix} \alpha_v & 0 \\ 0 & \beta_v \end{pmatrix} \in GL_2(\mathbb{C}).$$

Then

$$q_v^{-1/9} < |\alpha_v| \text{ and } |\beta_v| < q_v^{1/9}.$$

Similar inequality holds at archimedean places ($\lambda > 0.23765432...$).

This is proved using the techniques of [54] (which led to $1/5$ using $\mathrm{Sym}^2\pi$ and groups of either type F_4 or E_6, by applying a general theorem from [54] which implies $L(s, \pi_v, \mathrm{Sym}^5(\rho_2))$ is holomorphic for $\mathrm{Re}(s) \geqslant 1$ for all such v). When this general theorem is applied to a simply connected group of type E_8 with a Levi $\mathbf{M}$ for which $\mathbf{M}_D = SL_5 \times SL_4$, together with a representation related to $\mathrm{Sym}^4\pi \otimes \mathrm{Sym}^3\pi$ leading to $L(s, \pi_v, \mathrm{Sym}^9(\rho_2))$, one gets $1/9$. We refer to [19] for an important application of Theorem 9.4. Next, we have

Theorem 9.5 ([37]). *Suppose $F = \mathbb{Q}$, then*

$$p^{-7/64} \leqslant |\alpha_p| \text{ and } |\beta_p| \leqslant p^{7/64}.$$

At the archimedean places we get the estimate $\lambda \geqslant \dfrac{975}{4096} = 0.2380371$ *for the first positive eigenvalue of Δ.*

Proof. This is proved by means of analytic methods of Duke and Iwaniec [22] applied to $L(s, \mathrm{Sym}^4\pi, \mathrm{Sym}^2)$, (cf. [7]) along the lines of Bump, Duke, Hoffstein and Iwaniec [6] which led to $5/28 + \varepsilon$ over $\mathbb{Q}$, when applied to $L(s, \mathrm{Sym}^2\pi, \mathrm{Sym}^2)$. □

10 Applications to Representation Theory

Let F be a p-adic local field of characteristic zero. Let G be a quasisplit connected reductive group over F. Fix a Borel subgroup $B = TU$, where T is a maximal torus and U is the unipotent radical of B, both defined over F. Let P be a maximal parabolic subgroup of G defined over F. Write $P = MN$, a Levi decomposition, with $T \subset M$ and $N \subset U$. Let α be the unique simple root of T in Lie(N), the Lie algebra of N, as in the preliminaries. We then have $\tilde{\alpha}$ (cf. Section 2). Finally, let $w_0 = w_\ell w_{\ell,M}^{-1}$ be as before with w_ℓ and $w_{\ell,M}$ long elements the Weyl group W and W_M of the maximal split component A_0 of T in G and M, respectively.

Let σ be an irreducible supercuspidal representation of $M(F)$. Fix $s \in \mathbb{C}$ and let

$$I(s,\sigma) \colon = I(s\tilde{\alpha},\sigma)$$

as in Section 2.

The adjoint action r of ${}^LM$ on ${}^L\mathfrak{n} = \mathrm{Lie}({}^LN)$ decomposes as $r = \bigoplus_{i=1}^{m} r_i$ as in Section 2. We then have the local L-functions $L(s,\sigma,r_i)$ when σ is generic. One then has the following lemma from [53].

Lemma 10.1. *Assume σ is irreducible supercuspidal and generic. The $L(s,\sigma,r_i) = 1$ for $i \geqslant 3$.*

Next let $A(s\tilde{\alpha},\sigma,w_0)$ be the standard intertwining operator defined in Section 3. It is quite important to determine the poles of this operator. Combining results from [53] and [55] we have

Theorem 10.1. *As a meromorphic operator on $\mathbb{C}$*

$$L(s,\sigma,\tilde{r}_1)^{-1}L(2s,\sigma,\tilde{r}_2)^{-1}A(s\tilde{\alpha},\sigma,w_0)$$

is holomorphic and nonzero for all $s \in \mathbb{C}$.

This theorem can be used both ways:

(a) To determine L-functions.

(b) To determine the poles of $A(s\tilde{\alpha},\sigma,w_0)$, which are on the unitary axis $\mathrm{Re}(s) = 0$ if σ is unitary, i.e., $|\omega_\sigma| = 1$, since σ is supercuspidal [71].

Now, assume σ is unitary supercuspidal, or even more generally in the discrete series. Then

Theorem 10.2 (Harish-Chandra). *Let $I(\sigma) = I(0,\sigma)$. Then $I(\sigma)$ is irreducible unless $w_0(\sigma) \simeq \sigma$. Assume $w_0(\sigma) \simeq \sigma$. Then $I(\sigma)$ is reducible iff $A(s\tilde{\alpha},\sigma,w_0)$ is holomorphic at $s = 0$.*

Thus determining poles of $A(s\tilde{\alpha},\sigma,w_0)$ is quite important. In fact,

(a) Together with the theory of R-groups of Knapp-Stein one can determine points of reducibility for $I(\sigma)$ for an arbitrary parabolic subgroup P. Note that this is an approach that connects very well with global theory and the trace formula of Arthur as these R-groups are supposed to be those defined by Langlands and Arthur (non-tempered). Already the relation with L-functions sets up deep connections between harmonic analysis and number theory which is in complete agreement with the local Langlands correspondences, one of the essence of Langlands Program. Full reducibility theorems can also be proved for all s (cf. [53]).

(b) The poles can be computed by determining the residues of $A(s\tilde{\alpha},\sigma,w_0)$ at $s = 0$. In the cases of classical groups these are done by Shahidi, Goldberg-Shahidi, Spallone, Shahidi-Spallone. One should also mention Xiaoxiang Yu's work on the structural problems involved here. Here is a recent example by Shahidi and Spallone [67].

Let E/F be a quadratic extension of p-adic fields. Let τ be the nontrivial element of the Galois group $\mathrm{Gal}(E/F)$ of E/F. For our example, we consider the group $G = SO_6^*$, the quasisplit special orthogonal group defined by E/F. We will take the parabolic subgroup $P = MN$ of G for which $M = GL_2 \times SO_2^*$. Then $SO_2^*(F) = E^1$, the subgroup of elements of norm 1 in E. Let π be an irreducible unitary supercuspidal representation of $GL_2(F)$. If χ is a character of E^1, then $\sigma = \pi \otimes \chi$ is an irreducible supercuspidal generic representation of $M(F)$. Next, let

$$G_+ = \{g \in GL_2(F) | \det g \in N_{E/F}(E^*)\}.$$

Then

$$\pi | G_+ = \pi_+ \oplus \pi_-.$$

Let $\theta_\pm$ be the representation theoretic character of $\pi_\pm$. Given $\gamma \in E^1$ take $\tilde{\gamma} \in E^*$ such that $\gamma = \tau(\tilde{\gamma})/\tilde{\gamma}$ by Hilbert's Theorem 90. Denote by Δ_E the discriminant function of E^*, considered as F-points of a Cartan subgroup of GL_2. Let

$$R \colon = \mathrm{Res}_{s=0} A(s\tilde{\alpha}, \sigma, w_0)$$

be the residue of $A(s\tilde{\alpha}, \sigma, w_0)$ at $s = 0$, where it may have at most a simple pole (cf. [71]). Then the isomorphism $w_0(\sigma) \simeq \sigma$ which is part of Theorem 10.2 implies that $R = 0$ unless $\tilde{\pi} \simeq \pi$. This implies that $\omega_\pi^2 = 1$, where ω_π is the central character of π. We have

Theorem 10.3 ([67]). *The residue $R = 0$ unless $\omega_\pi^2 = 1$. If $\omega_\pi \equiv 1$, then $R \neq 0$ for all χ. Suppose $\omega_\pi \neq 1$. Then $R = 0$ unless ω_π is the quadratic character attached to E/F by class-field theory. In this case R is proportional to*

$$\int_{E^1} \chi(\gamma) \Delta_E(\tilde{\gamma})(\theta_+(\tilde{\gamma}) - \theta_-(\tilde{\gamma})) d\gamma,$$

in which one can apply Labesse-Langlands character identity (cf. [42])

$$\Delta_E(\tilde{\gamma})(\theta_+(\tilde{\gamma}) - \theta_-(\tilde{\gamma})) \sim \chi(\gamma) + \overline{\chi}(\gamma).$$

Consequently $R \neq 0$ if and only if π is attached to

$$\mathrm{Ind}_{W_E}^{W_F} \tilde{\chi},$$

where $\tilde{\chi}(x) = \chi(x/\tau(x))$, $x \in E^* = W_E$, and E/F is defined by ω_π; or $\omega_\pi \equiv 1$.

This result agrees completely with what is predicted by L-functions [64] and the theory of twisted endoscopy of Kottwitz, Langlands and Shelstad.

References

[1] J. Arthur, Endoscopic L-functions and a combinatorial indentity, Dedicated to H.S.M. Coxeter, Canad. J. Math. **51** (1999), 1135–1148.

[2] J. Arthur and L. Clozel, Simple Algebras, Base Change, and the Advanced Theory of the Trace Formula, Princeton University Press, 1989.

[3] M. Asgari, Local L-functions for split spinor groups, Canad. J. Math. **54** (2002), 673–693.

[4] A. Borel, Automorphic L-functions, in: Automorphic Forms and Automorphic Representations, Amer. Math. Soc., Providence, RI, 1979, 27–61.

[5] A. Borel, Linear Algebraic Groups, Springer-Verlag, New York, 1997.

[6] D. Bump, W. Duke, J. Hoffstein and H. Iwaniec, An estimate for the Hecke eigenvalues of Maass forms, IMRN **4** (1992), 75–81.

[7] D. Bump and D. Ginzburg, Symmetric square L-functions for $GL(r)$, Ann. Math. **136** (1992), 137–205.

[8] C. J. Bushnell and G. Henniart, On certain dyadic representations, Appendix to [39], Ann. Math. **155** (2002), 883–893.

[9] P. Cartier, Representations of p-adic groups: A survey, in: Automorphic Forms and Automorphic Representations, Amer. Math. Soc., Providence, RI, 1979, 111–155.

[10] W. Casselman, Introduction to the theory of admissible representations of p-adic reductive groups, preprint.

[11] W. Casselman, Canonical extensions of Harish-Chandra modules, Canad. J. Math. **41** (1989), 315–438.

[12] W. Casselman and F. Shahidi, On irreducibility of standard modules for generic representations, Ann. Scient. Éc. Norm. Sup. **31** (1998), 561–589.

[13] W. Casselman and J. A. Shalika, The unramified principal series of p-adic groups II; The Whittaker function, Comp. Math. **41** (1980), 207–231.

[14] L. Clozel, Spectral theory of automorphic forms, Automorphic Forms and Applications, IAS/Park City Math. Ser., Vol. 12, Eds P. Sarnak and F. Shahidi, Amer. Math. Soc., Providence, 2007, 43–93.

[15] J. W. Cogdell, L-functions and functoriality, in: Automorphic Forms and L-functions, ALM 30, 2014, 1–47.

[16] J. W. Cogdell and I. I. Piatetski-Shapiro, Converse theorems for GL_n, Publ. Math. IHES **79** (1994), 157–214.

[17] J. W. Cogdell and I. I. Piatetski-Shapiro, Converse theorems for GL_n II, J. Reine Angew. Math. **507** (1999), 165–188.

[18] J. W. Cogdell and I. I. Piatetski-Shapiro, Converse theorems, functoriality and applications to number theory, in: Proceedings of the International Congress of Mathematicians, Vol. II, Higher Education Press, Beijing, 2002, 119–128.

[19] J. W. Cogdell, I. I. Piatetski-Shapiro and P. Sarnak, Estimates for Hilbert modular L-functions and applications, in preparation.

[20] J. W. Cogdell, H. Kim, I. I. Piatetski-Shapiro and F. Shahidi, On lifting from classical groups to GL_N, Publ. Math. IHES **93** (2001), 5–30.

[21] J. W. Cogdell, H. Kim, I. I. Piatetski-Shapiro and F. Shahidi, Functoriality for quasisplit classical groups, in: On certain L-functions, Clay Math. Proc., Vol. 13, Amer. Math. Soc., Providence, RI, 2011, 117–140.

[22] W. Duke and H. Iwaniec, Estimates for coefficients of L-functions I, in: Automorphic Forms and Analytic Number Theory, Ed. R. Murty, CRM, Montreal, 1990, 43–48.

[23] S. Gelbart and H. Jacquet, A relation between automorphic representations of $GL(2)$ and $GL(3)$, Ann. Scient. Éc. Norm. Sup. **11** (1978), 471–552.

[24] S. Gelbart and F. Shahidi, Boundedness of automorphic L-functions in vertical strips, J. AMS **14** (2001), 79–107.

[25] S. Gelbart and F. Shahidi, Analytic properties of automorphic L-functions, in: Perspectives in Mathematics, Vol. 6, Eds. J. Coates and S. Helgason, Academic Press, Inc., Boston-San Diego-New York, 1988.

[26] Harish-Chandra, Automorphic Forms on Semisimple Lie Groups, SLN, Vol. 62, Springer-Verlag, Berlin-Heidelberg-New York, 1968.

[27] M. Harris and R. Taylor, On the Geometry and Cohomology of Some Simple Shimura Varieties, Princeton University Press, 2001.

[28] G. Henniart, Une preuve simple des conjectures de Langlands pour $GL(n)$ sur un corps p-adique, Invent. Math. **139** (2000), 439–455.

[29] G. Henniart, Progrès récents en fonctorialité de Langlands, Seminaire Bourbaki, Juin 2001, Exposé 890, 890-1 to 890-21.

[30] H. Jacquet, I. Piatetski-Shapiro and J. Shalika, Rankin-Selberg convolutions, Amer. J. Math. **105** (1983), 367–464.

[31] H. Jacquet, I. Piatetski-Shapiro and J. Shalika, Relvement cubique non-normal, C.R. Acad. Sci. Paris Sr. I Math. **292** (1981), 567–571.

[32] D. Keys and F. Shahidi, Artin L-functions and normalization of intertwining operators, Ann. Scient. Éc. Norm. Sup. **21** (1988), 67–89.

[33] H. Kim, On local L-functions and normalized intertwining operators, Canad. J. Math., **57** (2005), No. 3, 535–597.

[34] H. Kim, Langlands-Shahidi method and poles of automorphic L-functions: Application to exterior square L-functions, Canad. J. Math. **51** (1999), 835–849.

[35] H. Kim, Langlands-Shahidi method and poles of automorphic L-functions II, Israel J. Math. **117** (2000), 261–284.

[36] H. Kim, Functoriality for the exterior square of GL_4 and symmetric fourth of GL_2, J. Amer. Math. Soc. **16** (2002), 139–183.

[37] H. Kim and P. Sarnak, Refined estimates towards the Ramanujan and Selberg conjectures, Appendix 2 to [36].

[38] H. Kim and F. Shahidi, Symmetric cube L-functions for GL_2 are entire, Ann. Math. **150** (1999), 645–662.

[39] H. Kim and F. Shahidi, Functorial products for $GL_2 \times GL_3$ and the symmetric cube for GL_2, Ann. Math. **155** (2002), 837–893.

[40] H. Kim and F. Shahidi, Cuspidality of symmetric powers with applications, Duke Math. J. **112** (2002), 177–197.

[41] P. Kutzko, The Langlands conjecture for GL_2 of a local field, Ann. Math. **112** (1980), 381–412.

[42] J.-P. Labesse and R. P. Langlands, L-indistinguishability for $SL(2)$, Canad. J. Math. **31** (1979), 726–785.

[43] R. P. Langlands, Euler Products, Yale University Press, 1971.

[44] R. P. Langlands, On the Functional Equations Satisfied by Eisenstein Series, SLN, Vol. 544, Springer-Verlag, 1976.

[45] R. P. Langlands, On the classification of irreducible representations of real algebraic groups, in: Representation Theory and Harmonic Analysis on Semisimple Lie Groups, Eds. P. J. Sally, Jr. and D. A. Vogan, 1989, 101–170.

[46] W. Luo, Z. Rudnick and P. Sarnak, On the generalized Ramanujan conjecture for $GL(n)$, Proc. Sympos. Pure Math. **66** (1999), 301–310.

[47] P. Michel, Analytic number theory and families of automorphic L-functions, IAS/Park City Math. Series **12** (2007), 179–296.

[48] C. Moeglin and J.-L. Waldspurger, Spectral Decomposition and Eisenstein Series, Cambridge University Press, 1995.

[49] C. Moeglin and J.-L. Waldspurger, Le spectre résiduel de $GL(n)$, Ann. Sci. Ećole Norm. Sup. **22** (1989), 605–674.

[50] W. Müller, The trace class conjecture in the theory of automorphic forms, Ann. Math. **130** (1989), 473–529.

[51] D. Ramakrishnan, On the coefficients of cusp forms, Math. Research Letters **4** (1997), 295–307.

[52] I. Satake, Classification Theory of Semi-Simple Algebraic Groups, Marcel Dekker, New York, 1971.

[53] F. Shahidi, A proof of Langlands conjecture on Plancherel measures; Complementary series for p-adic groups, Ann. Math. **132** (1990), 273–330.

[54] F. Shahidi, On the Ramanujan conjecture and finiteness of poles for certain L-functions, Ann. Math. **127** (1988), 547–584.

[55] F. Shahidi, On certain L-functions, Amer. J. Math. **103** (1981), 297–355.

[56] F. Shahidi, Local coefficients as Artin factors for real groups, Duke Math. J. **52** (1985), 973–1007.

[57] F. Shahidi, Whittaker models for real groups, Duke Math. J. **47** (1980), 99–125.

[58] F. Shahidi, Functional equation satisfied by certain L-functions, Comp. Math. **37** (1978), 171–207.

[59] F. Shahidi, On multiplicativity of local factors, in: Festschrift in Honor of I. I. Piatetski-Shapiro, Part II, Israel Math. Conference Proceedings, **3** (1990), 279–289.

[60] F. Shahidi, Twists of a general class of L-functions by highly ramified characters, Canad. Math. Bull. **43** (2000), 380–384.

[61] F. Shahidi, Symmetric power L-functions for $GL(2)$, in: Elliptic Curves and Related Topics, Eds. H. Kisilevsky and M. R. Murty, Montréal, 1994, 159–182.

[62] F. Shahidi, Fourier transforms of intertwining operators and Plancherel measures for $GL(n)$, Amer. J. Math. **106** (1984), 67–111.

[63] F. Shahidi, Automorhic L-functions and functoriality, in: Preceedings of the International Congress of Mathematicians, Vol. II, Higher Education Press, Beijing, 2002, 655–666.

[64] F. Shahidi, L-functions and poles of intertwining operators, Appendix to "Residues of intertwining operators for classical groups" by S. Spallone, International Mathematics Research Notices (IMRN), 2008.

[65] F. Shahidi, Langlands-Shahidi method, in: Automorphic Forms and Applications, Eds. P. Sarnak and F. Shahidi, AMS/IAS, 2007, 299–330.

[66] F. Shahidi, Eisenstein series and automorphic L-functions, Colloquium Publications, Vol. 58, AMS, 2010.

[67] F. Shahidi and S. Spallone, Residues of intertwining operators for SO_6^* as character identities, Composition Math. **146** (2010), 772–794.

[68] J. A. Shalika, The multiplicity one for GL_n, Ann. Math. **100** (1974), 171–193.

[69] J.-P. Serre, Appendix to [61].

[70] A. Silberger, The Langlands quotient theorem for p-adic groups, Math. Ann. **236** (1978), 95–104.

[71] A. Silberger, Introduction to harmonic analysis on reductive p-adic groups, in: Mathematical Notes, Vol. 23, Princeton University Press, Princeton, NJ, 1979.

[72] T. A. Springer, Linear Algebraic Groups, Birkhäuser, Boston, 1981.

[73] J. Tate, Number theoretic background, Proc. Sympos. Pure Math. AMS **33** (1979), 3–26.

[74] N. R. Wallach, Asymptotic expansions of generalized matrix entries of representations of real reductive groups, in: Lie Group Representations I, Springer-Verlag, 1983, 287–369.

[75] Y. Zhang, The holomorphy and non-vanishing of normalized local intertwining operators, Pacific J. Math. **180** (1997), 385–398.

Automorphic Forms and
L-functions
ALM 30, pp. 179–204

Lecture Notes on Some Analytic Properties of Automorphic L-functions for $SL_2(\mathbb{Z})$

Yangbo Ye*

Abstract

These notes focus on some basic analytic properties of L-functions attached to cusp forms for $SL_2(\mathbb{Z})$. The main topics include integral representations of L-functions, functional equations, a converse theorem, the Phragmén-Lindelöf principle, convexity bounds, Rankin-Selberg theory, and a proof of absolute convergence of these L-functions for $\Re(s) > 1$.

2010 Mathematics Subject Classification: 11F70.

Keywords and Phrases: Automorphic L-function, $SL_2(\mathbb{Z})$, Automorphic form, Integral representation, Functional equation, Converse theorem, Phragmén-Lindelöf principle, Convexity bound, Rankin-Selberg theory.

1 Introduction

These notes focus on some basic analytic properties of L-functions attached to cusp forms for $SL_2(\mathbb{Z})$. The main topics include integral representations of L-functions, functional equations, a converse theorem, the Phragmén-Lindelöf principle, convexity bounds, Rankin-Selberg theory, and a proof of absolute convergence of these L-functions for $\Re(s) > 1$.

We will assume that the reader has a basic knowledge of holomorphic and Maass cusp forms. Standard references automorphic forms and their L-functions are Borel [5], Bump [7], Goldfeld [9], Goldfeld and Hundley [10], Iwaniec [11, 12], Iwaniec and Kowalski [13], Iwaniec and Sarnak [14], Li [15], Moreno [19], Motohashi [20], Sarnak [21], Serre [22], Shimura [23], Ye [24], etc. For simplicity,

*Department of Mathematics, The University of Iowa, Iowa city, IA 52242-1419, USA, yey@iowa.uiowa.edu.

all cusp forms in this article are forms with respect to the modular group $SL(2,\mathbb{Z})$. Most proofs here can be generalized to other settings.

These notes are based on a course given at the University of Iowa in 2007 and a short course given at the CIMPA-UNESCO-SDU Research School on Automorphic Forms and L-Functions in 2010 by the author. The author would like to thank Dr. Qingfeng Sun who took lecture notes.

2 An Integral Representation and Functional Equation

Let f be a Maass cusp form with Laplace eigenvalue λ and Hecke eigenvalues $\lambda_f(n)$. Then the Fourier expansion of f is

$$f(z) = \sum_{m\neq 0} \lambda_f(m)\sqrt{y}K_{i\nu}(2\pi|m|y)e^{2\pi imx}$$

for $z = x+iy$, where $\nu = \sqrt{\lambda - 1/4}$. The L-function attached to f is defined as

$$L(s,f) = \sum_{n=1}^{\infty} \frac{\lambda_f(n)}{n^s}.$$

By the trivial bound $\lambda_f(n) \ll_f n^{1/2}$, this L-function converges absolutely for $\Re(s) > 3/2$. Later we will show that it actually converges absolutely for $\Re(s) > 1$.

Theorem 2.1 (Euler Product). *Let f be a Hecke eigenform with $\lambda_f(1) = 1$. For $\Re(s) > 3/2$ we have*

$$L(s,f) = \prod_p \Big(1 - \lambda_f(p)p^{-s} + p^{-2s}\Big)^{-1}.$$

Proof. First we have

$$\begin{aligned}
&\sum_{k=0}^{\infty} \frac{\lambda_f(p^k)}{p^{ks}}\Big(1 - \lambda_f(p)p^{-s} + p^{-2s}\Big)\\
&= \sum_{k=0}^{\infty} \frac{\lambda_f(p^k)}{p^{ks}} - \sum_{k=0}^{\infty} \frac{\lambda_f(p^k)\lambda_f(p)}{p^{(k+1)s}} + \sum_{k=0}^{\infty} \frac{\lambda_f(p^k)}{p^{(k+2)s}}\\
&= \sum_{k=0}^{\infty} \frac{\lambda_f(p^k)}{p^{ks}} - \sum_{k=1}^{\infty} \frac{\lambda_f(p^{k-1})\lambda_f(p)}{p^{ks}} + \sum_{k=2}^{\infty} \frac{\lambda_f(p^{k-2})}{p^{ks}}\\
&= \lambda_f(1) + \lambda_f(p)p^{-s} - \lambda_f(p)\lambda_f(1)p^{-s}\\
&\quad + \sum_{k=2}^{\infty}\Big(\frac{\lambda_f(p^k)}{p^{ks}} - \frac{\lambda_f(p^{k-1})\lambda_f(p)}{p^{ks}} + \frac{\lambda_f(p^{k-2})}{p^{ks}}\Big).
\end{aligned}$$

Because $\lambda_f(1) = 1$ and

$$\lambda_f(p^{k-1})\lambda_f(p) = \sum_{d|p} \lambda_f\Big(\frac{p^k}{d^2}\Big) = \lambda_f(p^k) + \lambda_f(p^{k-2}),$$

we get

$$\sum_{k=0}^{\infty} \frac{\lambda_f(p^k)}{p^{ks}} \Big(1 - \lambda_f(p)p^{-s} + p^{-2s}\Big) = 1,$$

which gives us

$$\sum_{k=0}^{\infty} \frac{\lambda_f(p^k)}{p^{ks}} = \Big(1 - \lambda_f(p)p^{-s} + p^{-2s}\Big)^{-1}.$$

Moreover,

$$\prod_p \Big(\sum_{k=0}^{\infty} \frac{\lambda_f(p^k)}{p^{ks}}\Big) = \prod_p \Big(1 - \lambda_f(p)p^{-s} + p^{-2s}\Big)^{-1}$$

and hence

$$\sum_{n=1}^{\infty} \frac{\lambda_f(n)}{n^s} = \prod_p \Big(1 - \lambda_f(p)p^{-s} + p^{-2s}\Big)^{-1}.$$

This finishes the proof of the theorem. □

In general, if f is not a Hecke eigenform, then $f = \sum_j c_j f_j$ with f_j being Hecke eigenforms normalized by $\lambda_{f_j} = 1$. Thus $L(s,f) = \sum_j c_j L(s, f_j)$. Here each $L(s, f_j)$ has an Euler product, but the sum does not necessarily have an Euler product.

Now we want to prove an integral representation of $L(s, f)$.

Lemma 2.1 (The Mellin transform of $K_{i\nu}(y)$). *If $\Re(s) > |\Re(i\nu)|$, then*

$$\int_0^{\infty} K_{i\nu}(y) y^s \frac{dy}{y} = 2^{s-2}\Gamma\Big(\frac{s+i\nu}{2}\Big)\Gamma\Big(\frac{s-i\nu}{2}\Big). \tag{2.1}$$

Proof. Recall

$$K_{i\nu}(y) = \frac{1}{2}\int_0^{\infty} e^{-y(t+1/t)/2} t^{i\nu} \frac{dt}{t}.$$

Thus

$$\int_0^{\infty} K_{i\nu}(y) y^s \frac{dy}{y} = \frac{1}{2}\int_0^{\infty}\int_0^{\infty} e^{-y(t+1/t)/2} t^{i\nu} y^s \frac{dt}{t}\,\frac{dy}{y}. \tag{2.2}$$

Changing variables to $u = \frac{ty}{2}$ and $v = \frac{y}{2t}$, we get

$$\begin{vmatrix} \frac{\partial u}{\partial t} & \frac{\partial u}{\partial y} \\ \frac{\partial v}{\partial t} & \frac{\partial v}{\partial y} \end{vmatrix} = \begin{vmatrix} \frac{y}{2} & \frac{t}{2} \\ -\frac{y}{2t^2} & \frac{1}{2t} \end{vmatrix} = \frac{y}{2t}$$

and

$$\frac{du}{u}\frac{dv}{v} = \frac{y}{2t}\frac{1}{uv}dt\,dy = 2\frac{dy}{y}\frac{dt}{t}.$$

Therefore (2.2) becomes

$$\begin{aligned}\int_0^\infty K_{i\nu}(y)y^s\frac{dy}{y} &= \frac{2^s}{4}\int_0^\infty\int_0^\infty e^{-u-v}u^{(s+i\nu)/2}v^{(s-i\nu)/2}\frac{du}{u}\frac{dv}{v} \\ &= 2^{s-2}\int_0^\infty e^{-u}u^{(s+i\nu)/2}\frac{du}{u}\int_0^\infty e^{-v}v^{(s-i\nu)/2}\frac{dv}{v} \\ &= 2^{s-2}\Gamma\Big(\frac{s+i\nu}{2}\Big)\Gamma\Big(\frac{s-i\nu}{2}\Big).\end{aligned}$$

This proves the lemma. □

Lemma 2.2. *Let f be a Maass form satisfying $\lambda_f(-n) = \lambda_f(n)$. Then*

$$\int_0^\infty f(iy)y^{s-1/2}\frac{dy}{y} = \frac{\pi^{-s}}{2}\Gamma\Big(\frac{s+i\nu}{2}\Big)\Gamma\Big(\frac{s-i\nu}{2}\Big)L(s,f). \tag{2.3}$$

Proof. First let us look at the convergence of the integral on the left side. Since f is a cusp form, $f(iy)$ decreases rapidly when $y \to \infty$. By

$$f(iy) = f\Big(-\frac{1}{iy}\Big) = f\Big(\frac{i}{y}\Big),$$

$f(iy)$ also decreases rapidly when $y \to 0$. Then the integral on the left side of (2.3) converges absolutely for any s and uniformly for any s in a compact region.

Recall the Fourier expansion

$$f(z) = \sum_{n\neq 0}\lambda_f(n)\sqrt{y}K_{i\nu}(2\pi|n|y)e^{2\pi inx}.$$

Then by (2.1)

$$\begin{aligned}\int_0^\infty f(iy)y^{s-1/2}\frac{dy}{y} &= \int_0^\infty\sum_{n\neq 0}\lambda_f(n)\sqrt{y}K_{i\nu}(2\pi|n|y)y^{s-1/2}\frac{dy}{y} \\ &= \sum_{n\neq 0}\lambda_f(n)\int_0^\infty K_{i\nu}(2\pi|n|y)y^s\frac{dy}{y} \\ &= \sum_{n\neq 0}\lambda_f(n)(2\pi|n|)^{-s}\int_0^\infty K_{i\nu}(t)t^s\frac{dt}{t}.\end{aligned}$$

By Lemma 2.1,

$$\begin{aligned}\int_0^\infty f(iy)y^{s-1/2}\frac{dy}{y} &= \sum_{n\neq 0}\lambda_f(n)(2\pi|n|)^{-s}2^{s-2}\Gamma\Big(\frac{s+i\nu}{2}\Big)\Gamma\Big(\frac{s-i\nu}{2}\Big)\\ &= \frac{\pi^{-s}}{4}\Gamma\Big(\frac{s+i\nu}{2}\Big)\Gamma\Big(\frac{s-i\nu}{2}\Big)\sum_{n\neq 0}\frac{\lambda_f(n)}{|n|^s}.\end{aligned}$$

Since $\lambda_f(-n)=\lambda_f(n)$, we get

$$\begin{aligned}\int_0^\infty f(iy)y^{s-1/2}\frac{dy}{y} &= \frac{\pi^{-s}}{2}\Gamma\Big(\frac{s+i\nu}{2}\Big)\Gamma\Big(\frac{s-i\nu}{2}\Big)\sum_{n\neq 0}\frac{\lambda_f(n)}{n^s}\\ &= \frac{\pi^{-s}}{2}\Gamma\Big(\frac{s+i\nu}{2}\Big)\Gamma\Big(\frac{s-i\nu}{2}\Big)L(s,f).\end{aligned}$$

This proves the Lemma. □

Define an operator ι by $(\iota f)(x+iy)=f(-x+iy)$. Obviously, ι^2 is the identity map.

Lemma 2.3. *The Laplace operator Δ, the operator ι and Hecke operators $T(n)$ with $n\geqslant 1$ commute with each other.*

Therefore we can have an orthogonal basis consisting of eigenfunctions of Δ, ι and $T(n)$, $n\geqslant 1$.

Proof. See a reference book on Hecke operators for a proof of $T(n)\Delta=\Delta T(n)$ for any $n\geqslant 1$.

Then $\iota\Delta=\Delta\iota$ holds because

$$\iota\Delta=\iota\Big(-y^2\Big(\frac{\partial^2}{\partial x^2}+\frac{\partial^2}{\partial y^2}\Big)\Big)=\Delta\iota.$$

Finally, $T(n)\iota=\iota T(n)$ holds because

$$T(m)\iota f(z)=\sum_{n\neq 0}\sqrt{y}K_{i\nu}(2\pi|n|y)e^{-2\pi inx}\sum_{d|(n,m)}\lambda_f\Big(\frac{nm}{d^2}\Big)=\iota T(m)f(z).$$

This proves the Lemma. □

Definition 2.1. A Hecke eigenform f is said to be even if $\iota f=f$. It is said to be odd if $\iota f=-f$.

Case 1 (*f is even*). In this case, $\iota f=f$. Note that

$$\begin{aligned}\iota f(z) &= \sum_{n\neq 0}\lambda_f(n)\sqrt{y}K_{i\nu}(2\pi|n|y)e^{2\pi i(-n)x}\\ &= \sum_{n\neq 0}\lambda_f(-n)\sqrt{y}K_{i\nu}(2\pi|n|y)e^{2\pi inx},\\ f(z) &= \sum_{n\neq 0}\lambda_f(n)\sqrt{y}K_{i\nu}(2\pi|n|y)e^{2\pi ix}.\end{aligned}$$

Comparing the coefficients we have $\lambda_f(n) = \lambda_f(-n)$ for all $n \neq 0$. Recall in this case

$$\int_0^\infty f(iy)y^{s-1/2}\frac{dy}{y} = \frac{\pi^{-s}}{2}\Gamma\Big(\frac{s+i\nu}{2}\Big)\Gamma\Big(\frac{s-i\nu}{2}\Big)L(s,f).$$

Define the complete L-function attached to even f by

$$\Lambda(s,f) = \pi^{-s}\Gamma\Big(\frac{s+i\nu}{2}\Big)\Gamma\Big(\frac{s-i\nu}{2}\Big)L(s,f),$$

while $L(s,f)$ is called the finite-part L-function. In other words, we have the following

$$\Lambda(s,f) = L_\infty(s,f)L(s,f) = \prod_{p\leqslant\infty} L_p(s,f),$$

$$L_\infty(s,f) = \pi^{-s}\Gamma\Big(\frac{s+i\nu}{2}\Big)\Gamma\Big(\frac{s-i\nu}{2}\Big),$$

$$L(s,f) = \sum_{m=1}^{\infty}\lambda_f(m)m^{-s} = \prod_p(1-\lambda_f(p)p^{-s}+p^{-2s})^{-1} = \prod_{p<\infty} L_p(s,f),$$

where $L_p(s,f)$ is called a p-adic local factor.

Therefore by (2.3) we have an integral representation of the complete L-function in the even case

$$\Lambda(s,f) = 2\int_0^\infty f(iy)y^{s-1/2}\frac{dy}{y}. \tag{2.4}$$

Since the integral converges uniformly for any s in a compact region, it provides us an analytic continuation of $\Lambda(s,f)$ to the complex plane. This proves

Theorem 2.2. *For even Maass form f, the complete L-function $\Lambda(s,f)$ is entire.*

Now we can prove its functional equation.

Theorem 2.3. *For even Maass form f, $\Lambda(s,f)$=$\Lambda(1-s,f)$.*

Proof. Because

$$f(iy) = f\Big(-\frac{1}{iy}\Big) = f\Big(\frac{i}{y}\Big),$$

we have

$$\begin{aligned}\Lambda(s,f) &= 2\int_0^\infty f(iy)y^{s-1/2}\frac{dy}{y} = 2\int_0^\infty f\Big(\frac{i}{y}\Big)y^{s-1/2}\frac{dy}{y}\\ &= 2\int_0^\infty f\Big(iu\Big)u^{1/2-s}\frac{du}{u} = \Lambda(1-s,f).\end{aligned}$$

This proves the theorem. □

Case 2 (f *is odd*). In this case, $\iota f = -f$. Note

$$\begin{aligned}\iota f(z) &= \sum_{n\neq 0} \lambda_f(n)\sqrt{y}K_{i\nu}(2\pi|n|y)e^{2\pi i(-n)x}\\ &= \sum_{n\neq 0} \lambda_f(-n)\sqrt{y}K_{i\nu}(2\pi|n|y)e^{2\pi inx},\\ f(z) &= \sum_{n\neq 0} \lambda_f(n)\sqrt{y}K_{i\nu}(2\pi|n|y)e^{2\pi ix}.\end{aligned}$$

Comparing the coefficients we have

$$\lambda_f(n) = -\lambda_f(-n) \quad \text{for all} \quad n \neq 0.$$

Define

$$g(z) = \frac{1}{2\pi i}\frac{\partial}{\partial x}f(z).$$

Using the Fourier expansion of f we know that

$$\begin{aligned}g(z) &= \frac{1}{2\pi i}\frac{\partial}{\partial x}\Big(\sum_{n\neq 0} \lambda_f(n)\sqrt{y}K_{i\nu}(2\pi|n|y)e^{2\pi inx}\Big)\\ &= \sum_{n\neq 0} \lambda_f(n)n\sqrt{y}K_{i\nu}(2\pi|n|y)e^{2\pi inx}\\ &= \sum_{n\geqslant 1} \lambda_f(n)n\sqrt{y}K_{i\nu}(2\pi|n|y)\cos(2\pi nx).\end{aligned} \tag{2.5}$$

On the other hand, because $f(z) = f(-1/z)$, we have

$$\begin{aligned}g(iy) &= \frac{1}{2\pi i}\frac{\partial}{\partial x}f(x+iy)\Big|_{x=0} = \frac{1}{2\pi i}\frac{\partial}{\partial x}f\Big(-\frac{1}{x+iy}\Big)\Big|_{x=0}\\ &= \frac{1}{2\pi i}\frac{d}{dz}f\Big(-\frac{1}{x+iy}\Big)\frac{\partial}{\partial x}\Big(-\frac{1}{x+iy}\Big)\Big|_{x=0}\\ &= -\frac{1}{y^2}\frac{1}{2\pi i}\frac{d}{dz}f(z)\Big|_{z=i/y} = -\frac{1}{y^2}g\Big(\frac{i}{y}\Big).\end{aligned} \tag{2.6}$$

If $y \to \infty$, by (2.5), $g(iy) \to 0$ rapidly; if $y \to 0$, by (2.6), we also have $g(iy) \to 0$ rapidly. Thus

$$\int_0^\infty g(iy)y^{s+1/2}\frac{dy}{y}$$

converges absolutely and uniformly for s in any compact region.

If $\Re(s) > |\Re(1+i\nu)|$ we can compute as follow:

$$\begin{aligned}\int_0^\infty g(iy)y^{s+1/2}\frac{dy}{y} &= \sum_{n\neq 0}\lambda_f(n)n\int_0^\infty \sqrt{y}K_{i\nu}(2\pi|n|y)y^{s+1/2}\frac{dy}{y}\\ &= \sum_{n\neq 0}\lambda_f(n)n(2\pi|n|)^{-s-1}\int_0^\infty K_{i\nu}(y)y^{s+1}\frac{dy}{y}\\ &= \sum_{n\neq 0}\lambda_f(n)n(2\pi|n|)^{-s-1}2^{s-1}\Gamma\Big(\frac{s+1+i\nu}{2}\Big)\Gamma\Big(\frac{s+1-i\nu}{2}\Big),\end{aligned}$$

by Lemma 2.1. Therefore

$$\begin{aligned}\int_0^\infty g(iy)y^{s+1/2}\frac{dy}{y} &= \frac{\pi^{-s-1}}{4}\Gamma\Big(\frac{s+1+i\nu}{2}\Big)\Gamma\Big(\frac{s+1-i\nu}{2}\Big)\sum_{n\neq 0}\lambda_f(n)n|n|^{-s-1}\\ &= \frac{\pi^{-s-1}}{2}\Gamma\Big(\frac{s+1+i\nu}{2}\Big)\Gamma\Big(\frac{s+1-i\nu}{2}\Big)\sum_{n=1}^\infty\lambda_f(n)n^{-s}\\ &= \frac{\pi^{-s-1}}{2}\Gamma\Big(\frac{s+1+i\nu}{2}\Big)\Gamma\Big(\frac{s+1-i\nu}{2}\Big)L(s,f).\end{aligned}$$

Define the complete L-function attached to odd f by

$$\Lambda(s,f) = \pi^{-s-1}\Gamma\Big(\frac{s+1+i\nu}{2}\Big)\Gamma\Big(\frac{s+1-i\nu}{2}\Big)L(s,f).$$

Then we have an integration representation for the complete L-function in the odd case

$$\Lambda(s,f) = 2\int_0^\infty g(iy)y^{s+1/2}\frac{dy}{y}. \tag{2.7}$$

Similar to the even case, we have

Theorem 2.4. *For odd Maass form f, the complete L-function $\Lambda(s,f)$ is an entire function on $\mathbb{C}$.*

Theorem 2.5. *For odd Maass form f, $\Lambda(s,f){=}{-}\Lambda(1-s,f)$.*

Proof. By (2.6) and (2.7) we have

$$\Lambda(s,f) = 2\int_0^\infty \Big(-\frac{1}{y^2}g\Big(\frac{i}{y}\Big)\Big)y^{s+1/2}\frac{dy}{y}.$$

Taking $t = y^{-1}$, we have

$$\begin{aligned}\Lambda(s,f) &= -2\int_0^\infty t^2 g(it)t^{-s-1/2}\frac{dt}{t}\\ &= -2\int_0^\infty g(it)t^{-s+3/2}\frac{dt}{t} = -\Lambda(1-s,f).\end{aligned}$$

This proves the theorem. □

In general, we write

$$\Lambda(s,f) = \Gamma_{\mathbb{R}}(s+\mu_f(1))\Gamma_{\mathbb{R}}(s+\mu_f(2))L(s,f),$$

where $\Gamma_{\mathbb{R}}(s) = \pi^{-s/2}\Gamma(s/2)$. If f is even, then $\mu_f(1) = i\nu$ and $\mu_f(2) = -i\nu$. If f is odd, then $\mu_f(1) = 1+i\nu$ and $\mu_f(2) = 1-i\nu$. We will write

$$\Lambda(s,f) = L_\infty(s,f)L(s,f)$$

with

$$L_\infty(s,f) = \Gamma_{\mathbb{R}}(s+\mu_f(1))\Gamma_{\mathbb{R}}(s+\mu_f(2)).$$

Let $\varepsilon_f = 0$ if f is even and $\varepsilon_f = 1$ if f is odd. Then the functional equation becomes

$$\Lambda(s,f) = (-1)^{\varepsilon_f}\Lambda(1-s,f) \tag{2.8}$$

and

$$\mu_f(1) = \varepsilon_f + i\nu, \quad \mu_f(2) = \varepsilon_f - i\nu. \tag{2.9}$$

Now let us turn to holomorphic cusp forms. Let f be a holomorphic cusp form of weight k. The Fourier expansion of f is

$$f(z) = \sum_{n=1}^{\infty} a_f(n)e^{2\pi inz}.$$

The trivial bound of the Fourier coefficient $a_f(n)$ is $a_f(n) \ll n^{k/2}$. We can define the corresponding L-function as

$$L(s,f) = \sum_{n=1}^{\infty} \frac{a_f(n)}{n^s}$$

which converges absolutely for $\Re(s) > k/2+1$. We will prove that this L-function actually converges absolutely for $\Re(s) > (k+1)/2$.

Since f is holomorphic, it satisfies the Cauchy-Riemann equations. We also have the same Hecke operators as those in the Maass forms case. If f is a Hecke eigenform with $a_f(1) = 1$, then we can have the Euler product

$$L(s,f) = \prod_p (1 - a_f(p)p^{-s} + p^{k-1-2s})^{-1}, \qquad \text{if } \Re(s) > k/2+1.$$

Define the corresponding complete L-function by

$$\Lambda(s,f) = 2(2\pi)^{-s}\Gamma(s)L(s,f). \tag{2.10}$$

It can be proved likewise that $\Lambda(s,f)$ is entire. Recall

$$\Gamma(2z) = \frac{2^{2z-1}}{\sqrt{\pi}}\Gamma(z)\Gamma\Big(z+\frac{1}{2}\Big).$$

Therefore

$$\begin{aligned}\Lambda(s,f) &= 2(2\pi)^{-s}\frac{2^{s-1}}{\sqrt{\pi}}\Gamma\Big(\frac{s}{2}\Big)\Gamma\Big(\frac{s+1}{2}\Big)L(s,f)\\ &= \pi^{-s-1/2}\Gamma\Big(\frac{s}{2}\Big)\Gamma\Big(\frac{s+1}{2}\Big)L(s,f)\\ &= L_\infty(s,f)L(s,f),\end{aligned}$$

where the Archimedean local factor can be written as

$$L_\infty(s,f) = \prod_{j=1}^{2}\Gamma_{\mathbb{R}}(s+\mu_f(j))$$

with $\Gamma_{\mathbb{R}}(s) = \pi^{-s/2}\Gamma(s/2)$ and $\mu_f(1) = 0$ and $\mu_f(2) = 1$. Note that for $SL(2,\mathbb{Z})$, the weight must be even and $\geqslant 4$. At the end of Section 3 we will rescale the Fourier coefficients of f and modify the definition of $\Lambda(s,f)$ to get a functional equation of the form $\Lambda(s,f) = (-1)^{k/2}\Lambda(1-s,f)$.

3 A Converse Theorem

If f is a holomorphic cusp form, we can similarly prove an integral representation of the complete L-function $\Lambda(s,f)$

$$\Lambda(s,f) = 2\int_0^\infty f(iy)y^s\frac{dy}{y}.$$

From this integral representation we can have holomorphic continuation of $\Lambda(s,f)$ to $\mathbb{C}$ and the functional equation

$$\Lambda(s,f) = (-1)^{k/2}\Lambda(k-s,f). \tag{3.1}$$

What we will do in this section is to prove a converse theorem (cf. Bump [6], Cogdell [8]).

Theorem 3.1 (The converse theorem). *Let $A(n)$ be a sequence satisfying $A(n) \ll n^K$ for some $K > 0$. Define*

$$f(z) = \sum_{n=1}^{\infty} A(n)e^{2\pi i n z} \tag{3.2}$$

and

$$L(s,f) = \sum_{n=1}^{\infty}\frac{A(n)}{n^s}, \qquad \Re(s) > K+1. \tag{3.3}$$

Denote $\Lambda(s,f) = 2(2\pi)^{-s}\Gamma(s)L(s,f)$. Suppose that

(i) *$\Lambda(s,f)$ is entire;*

(ii) $\Lambda(s, f) = (-1)^{k/2}\Lambda(k - s, f)$ *for some even positive integer* k;

(iii) $\Lambda(s, f)$ *is bounded in any vertical strip.*

Then f *is a holomorphic cusp form of weight* k. *If* $L(s, f)$ *also has an Eulerean product*

$$L(s, f) = \prod_p (1 - A(p)p^{-s} + p^{k-1-2s})^{-1}, \tag{3.4}$$

then f *is a Hecke eigenform.*

Proof. Absolute convergence of (3.2) follows from the polynomial growth of $A(n) \ll n^K$ and the exponential decay of $e^{2\pi i n z}$ for z in the upper half plane. By the same reasons, $f(z)$ in (3.2) decays rapidly at ∞. (3.2) also implies that $f(z+1) = f(z)$, i.e.,

$$f\Big(\begin{pmatrix} 1 & 1 \\ 0 & 1 \end{pmatrix} z\Big) = f(z).$$

Since

$$SL(2, \mathbb{Z}) = \Big\langle \begin{pmatrix} 1 & 1 \\ 0 & 1 \end{pmatrix}, \begin{pmatrix} 0 & -1 \\ 1 & 0 \end{pmatrix} \Big\rangle,$$

we only need to show that

$$f\Big(\begin{pmatrix} 0 & -1 \\ 1 & 0 \end{pmatrix} z\Big) = z^k f(z).$$

Since $f(z)$ is analytic on $\mathcal{H}$, it suffices to verify this for $z = iy$, $y > 0$, i.e., $f(i/y) = (iy)^k f(iy)$ which is equivalent to

$$f(iy) = \frac{1}{(iy)^k} f\Big(\frac{i}{y}\Big). \tag{3.5}$$

We need an integral representation:

$$\int_0^\infty f(iy) y^s \frac{dy}{y} = \frac{1}{2}\Lambda(s, f). \tag{3.6}$$

Note that if f is a cusp form, we know that the integral in (3.6) converges absolutely as in Section 2. However this time we have not proved that f is a cusp form. We thus have to prove the convergence of (3.6) in a different way.

Since $f(iy)$ decays rapidly at ∞, the integral in (3.6) converges absolutely at ∞. On the other hand, at 0, we have

$$f(iy) = \sum_{n=1}^\infty A(n) e^{-2\pi n y} \ll \sum_{n=1}^\infty n^K e^{-2\pi n y}.$$

Because

$$\frac{1}{1-x} = \sum_{n=0}^\infty x^n, \qquad |x| < 1,$$

we have

$$\sum_{n=1}^{\infty} n^K x^n = \left(x\frac{d}{dx}\right)^K \frac{1}{1-x} \ll \frac{1}{(1-x)^{K+1}}.$$

Taking $x = e^{-2\pi y}$, then

$$f(iy) \ll \sum_{n=1}^{\infty} n^K e^{-2\pi n y} \ll \frac{1}{(1-e^{-2\pi y})^{K+1}} \ll \frac{1}{y^{K+1}}.$$

Hence we know that

$$\int_0^1 f(iy) y^s \frac{dy}{y} \ll \int_0^1 y^s \frac{dy}{y^{K+2}}$$

which converges absolutely for $\Re(s) > K+1$. This proves the absolute convergence of the integral in (3.6) for $\Re(s) > K+1$ and uniform convergence for s in a compact region in $\Re(s) > K+1$.

Consequently we can integrate the following integral term-by-term

$$\int_0^{\infty} f(iy) y^s \frac{dy}{y} = \sum_{n=1}^{\infty} A(n) \int_0^{\infty} e^{-2\pi n y} y^s \frac{dy}{y}.$$

This then gives us

$$\int_0^{\infty} f(iy) y^s \frac{dy}{y} = (2\pi)^{-s}\Gamma(s)L(s,f) = \frac{1}{2}\Lambda(s,f)$$

for $\Re(s) > K+1$.

Since k is even, to prove (3.5) we need to show

$$f(iy) = i^k y^{-k} f\left(\frac{i}{y}\right) \tag{3.7}$$

using (3.6). Let ϕ be a function on $(0,\infty)$. Define the Mellin transform of ϕ by

$$\Phi(s) = \int_0^{\infty} \phi(y) y^s \frac{dy}{y}. \tag{3.8}$$

Assume that there exist $\sigma_1, \sigma_2 \in (-\infty, \infty)$ such that the integral in (3.8) converges absolutely for $\Re(s) = \sigma_1, \sigma_2$.

Lemma 3.1 (Mellin inversion formula). *For $\sigma_1 < \sigma < \sigma_2$,*

$$\phi(y) = \frac{1}{2\pi i}\int_{\sigma-i\infty}^{\sigma+i\infty} \Phi(s) y^{-s} ds.$$

Proof. Write $\phi_\sigma(v) = e^{v\sigma}\phi(e^v)$, $v \in \mathbb{R}$. Recall

$$\Phi(\sigma + it) = \int_0^\infty \phi(y)y^{\sigma+it}\frac{dy}{y} = \int_{-\infty}^\infty \phi(e^v)e^{v(\sigma+it)}dv = \int_{-\infty}^\infty \phi_\sigma(v)e^{ivt}dv.$$

By the Fourier inversion formula, we have

$$\begin{aligned} e^{v\sigma}\phi(e^v) = \phi_\sigma(v) &= \frac{1}{2\pi}\int_{-\infty}^\infty \Phi(\sigma+it)e^{-ivt}dt, \\ \phi(e^v) &= \frac{1}{2\pi}\int_{-\infty}^\infty \Phi(\sigma+it)e^{-v(\sigma+it)}dt, \\ \phi(y) &= \frac{1}{2\pi i}\int_{\sigma-i\infty}^{\sigma+i\infty} \Phi(s)y^{-s}ds. \end{aligned}$$

This proves the Mellin inversion formula. □

In our case, $\sigma_1 = K+1$ and $\sigma_2 = \infty$. Taking $\sigma = \Re(s) > K+1$, then

$$\begin{aligned} f(iy) &= \frac{1}{4\pi i}\int_{\sigma-i\infty}^{\sigma+i\infty} \Lambda(s,f)y^{-s}ds, \\ f\Big(\frac{i}{y}\Big) &= \frac{1}{4\pi i}\int_{\sigma-i\infty}^{\sigma+i\infty} \Lambda(s,f)y^{s}ds. \end{aligned} \tag{3.9}$$

Recall our assumption $\Lambda(s,f) = (-1)^{k/2}\Lambda(k-s,f)$. Thus we can rewrite

$$\begin{aligned} f(iy) &= \frac{(-1)^{k/2}}{4\pi i}\int_{\sigma-i\infty}^{\sigma+i\infty} \Lambda(k-s,f)y^{-s}ds \\ &= \frac{(-1)^{k/2}}{4\pi i}\int_{k-\sigma-i\infty}^{k-\sigma+i\infty} \Lambda(w,f)y^{w-k}dw \\ &= \frac{i^k}{y^k}\frac{1}{4\pi i}\int_{k-\sigma-i\infty}^{k-\sigma+i\infty} \Lambda(w,f)y^{s}ds. \end{aligned} \tag{3.10}$$

Comparing (3.9) and (3.10), (3.7) is reduced to proving

$$\frac{1}{4\pi i}\int_{k-\sigma-i\infty}^{k-\sigma+i\infty} \Lambda(s,f)y^s ds = \frac{1}{4\pi i}\int_{\sigma-i\infty}^{\sigma+i\infty} \Lambda(s,f)y^s ds. \tag{3.11}$$

Since $\Lambda(s,f)$ is entire by our assumption, there is no singularity passed if we shift the integral contour in (3.11) from $x = k-\sigma$ to $x = \sigma$. The shifting, however, has to be done in the following way, where T is a large parameter:

$$\begin{aligned} \int_{k-\sigma-i\infty}^{k-\sigma+i\infty} &= \int_{\sigma-i\infty}^{\sigma+i\infty} + \int_{k-\sigma-i\infty}^{k-\sigma-iT} + \int_{k-\sigma+iT}^{k-\sigma+i\infty} \\ &\quad - \int_{\sigma+iT}^{\sigma+i\infty} - \int_{\sigma-i\infty}^{\sigma-iT} + \int_{k-\sigma-iT}^{\sigma-iT} - \int_{k-\sigma+iT}^{\sigma+iT}. \end{aligned} \tag{3.12}$$

We want to prove the last six integrals tend to zero as $T \to \infty$. For this, we need the Phragmén-Lindelöf principle.

Theorem 3.2 (The Phragmén-Lindelöf Principle). *Let $f(s)$ be a function which is holomorphic in the upper half of a vertical strip $\sigma_1 \leqslant \Re(s) \leqslant \sigma_2$, $\Im(s) > c$. Let $\Re(s) = \sigma$. Suppose that*

$$f(\sigma + it) = O\Big(e^{t^\alpha}\Big) \quad \text{for some } \alpha > 0, \text{ when } \sigma_1 \leqslant \sigma \leqslant \sigma_2; \tag{3.13}$$

$$f(\sigma + it) = O\Big(t^{-M}\Big) \quad \text{for some } M, \text{ when } \sigma = \sigma_1 \text{ and } \sigma = \sigma_2. \tag{3.14}$$

Then

$$f(\sigma + it) = O\Big(t^{-M}\Big) \quad \text{for any } \sigma_1 \leqslant \sigma \leqslant \sigma_2 \text{ uniformly for } \sigma.$$

We will prove Theorem 3.2 in Section 4. Now let us continue our proof of Theorem 3.1. Note that $\Lambda(s,f) = 2(2\pi)^{-s}\Gamma(s)L(s,f)$. On one hand, for $s = \sigma + it$ with $\sigma > K + 1$,

$$|L(s,f)| \leqslant \sum_{n=1}^{\infty} \frac{|A(n)|}{n^\sigma} < \infty, \tag{3.15}$$

and hence $L(s,f) \ll 1$ as $t \to \infty$. On the other hand, by the Stirling formula, we have

$$|\Gamma(\sigma + it)| \sim \sqrt{2\pi}|t|^{\sigma-1/2}e^{-\pi|t|/2}.$$

Therefore,

$$\Lambda(\sigma + it, f) \ll |t|^{\sigma-1/2}e^{-\pi|t|/2}. \tag{3.16}$$

Using the functional equation assumed, we have

$$\Lambda(k - \sigma - it, f) \ll |t|^{\sigma-1/2}e^{-\pi|t|/2}. \tag{3.17}$$

By (3.16) and (3.17) we have

$$\int_{k-\sigma-i\infty}^{k-\sigma+iT} + \int_{k-\sigma+iT}^{k-\sigma+i\infty} + \int_{\sigma-iT}^{\sigma-i\infty} + \int_{\sigma+i\infty}^{\sigma+iT} \to 0 \tag{3.18}$$

as $T \to \infty$.

Now we deal with the last two integrals in (3.12). Note that K and hence $\sigma > K + 1$ may be large. Since $k - \sigma$ may then be negative, we do not have a Dirichlet series expansion of $L(s,f)$ and hence (3.15) for $\Re(s)$ between σ and $k - \sigma$. To overcome this difficulty, we need the Phragmén-Lindelöf principle. We assumed that $\Lambda(s,f)$ is bounded in any vertical strip so that the first condition (3.13) holds. From (3.16) and (3.17) we know that

$$\Lambda(\sigma \pm it, f) \ll |t|^{-M},$$
$$\Lambda(k - \sigma \pm it, f) \ll |t|^{-M}$$

for any $M > 0$, which are (3.14). By Theorem 3.2, we have

$$\Lambda(\gamma \pm iT, f) \ll T^{-M} \tag{3.19}$$

for any γ between $k - \sigma$ and σ. By (3.19) we obtain

$$\int_{k-\sigma-iT}^{\sigma-iT} + \int_{k-\sigma+iT}^{\sigma+iT} \to 0 \quad \text{as } T \to \infty. \tag{3.20}$$

By (3.18) and (3.20) we can finally establish (3.11) and thus prove the converse theorem. □

Now we will rescale the Fourier expansion of a holomorphic cusp form

$$f(z) = \sum_{n=1}^{\infty} a_f(n) e^{2\pi i n z} = \sum_{n=1}^{\infty} \lambda_f(n) n^{(k-1)/2} e^{2\pi i n z},$$

where $a_f(n) = \lambda_f(n) n^{(k-1)/2}$ with $a_f(1) = 1$ and $\lambda_f(1) = 1$. Thus we can define a rescaled L-function

$$L(s, f) = \sum_{n=1}^{\infty} \lambda_f(n) n^{-s} = \sum_{n=1}^{\infty} a_f(n) n^{-s-(k-1)/2} = L_{\text{old}}\Big(s + \frac{k-1}{2}, f\Big).$$

Define a rescaled complete L-function $\Lambda(s, f)$ likewise. Then we can rewrite the functional equation

$$\begin{aligned} \Lambda_{\text{old}}(s, f) &= (-1)^{k/2} \Lambda_{\text{old}}(k - s, f), \\ \Lambda_{\text{old}}\Big(s + \frac{k-1}{2}, f\Big) &= (-1)^{k/2} \Lambda_{\text{old}}\Big(k - s - \frac{k-1}{2}, f\Big) \\ &= (-1)^{k/2} \Lambda_{\text{old}}\Big(\frac{k-1}{2} + 1 - s, f\Big), \\ \Lambda(s, f) &= (-1)^{k/2} \Lambda(1 - s, f). \end{aligned}$$

In the rest of the article, we will use

$$\Lambda(s, f) = \Gamma_{\mathbb{R}}(s + \mu_1) \Gamma_{\mathbb{R}}(s + \mu_2) L(s, f), \tag{3.21}$$

$$\Lambda(s, f) = \varepsilon_f \Lambda(1 - s, f), \quad \varepsilon_f = \pm 1 \tag{3.22}$$

for both holomorphic and Maass cusp form f.

4 The Phragmén-Lindelöf Principle and Convexity

We have used the Phragmén-Lindelöf principle to prove the converse theorem. In this section we will follow Lang [16] to prove it.

Theorem 4.1 (Phragmén-Lindelöf principle). *Let f be holomorphic in $\sigma_1 \leqslant \Re(s) \leqslant \sigma_2$. Suppose that f is bounded by 1 in absolute value on the sides of the strip and bounded by $O(e^{|s|^\alpha})$ in the strip for some $\alpha \geqslant 1$. Then $|f(s)| \leqslant 1$ in this strip.*

Theorem 4.2 (The maximum principle). *Let f be holomorphic in a bounded region and bounded by 1 on the boundary. Then $|f(s)|$ is bounded by 1 in the region.*

Note that the Phragmén-Lindelöf principle is different from the maximal principle, because a vertical strip is not bounded.

Proof of the Phragmén-Lindelöf principle. By our assumptions, for $|t|$ sufficiently large, we have

$$|f(\sigma + it)| \leqslant e^{|t|^\lambda} \quad \text{for any} \quad \lambda \geqslant \alpha.$$

Take $m > \lambda$ with $m \equiv 2 \pmod 4$. Write $s = \sigma + it = re^{i\theta}$, so that $s^m = r^m(\cos m\theta + i \sin m\theta)$. If $|t|$ is sufficiently large, then θ is close to $\pm\pi/2$, $m\theta$ is close to $\pi \pmod{2\pi}$, and $\cos m\theta$ is close to -1.

Define $g_\varepsilon(s) = f(s)e^{\varepsilon s^m}$ with $\varepsilon > 0$. Then for $\sigma_1 \leqslant \Re(s) \leqslant \sigma_2$,

$$|g_\varepsilon(s)| \leqslant e^{|t|^\lambda} e^{\varepsilon r^m \cos m\theta}.$$

Thus for t sufficiently large and $\sigma_1 \leqslant \Re(s) \leqslant \sigma_2$, we have $|t| \sim r$. Since $\cos m\theta \sim -1$, we have $\varepsilon \cos m\theta < 0$. Consequently, $|g_\varepsilon(s)| \leqslant 1$ for $\sigma_1 \leqslant \Re(s) \leqslant \sigma_2$ and $|t|$ sufficiently large.

On the two sides $\Re(s) = \sigma_1$ or σ_2, we have $|f(s)| \leqslant 1$. Therefore

$$|g_\varepsilon(s)| \leqslant e^{\varepsilon r^m \cos m\theta} \leqslant 1$$

for $\Re(s) = \sigma_1$ or σ_2.

By the maximal principle, we have $|g_\varepsilon(s)| \leqslant 1$ for s in $\sigma_1 \leqslant \Re(s) \leqslant \sigma_2$ and $|t| \leqslant T$. Therefore

$$|f(s)| \leqslant e^{-\varepsilon r^m \cos m\theta} |g_\varepsilon(s)| \leqslant e^{-\varepsilon r^m \cos m\theta}$$

for s as above and any $\varepsilon > 0$. Taking $\varepsilon \to 0^+$, we now have $|f(s)| \leqslant 1$. □

Theorem 4.3 (The first convexity theorem). *Let $f(s)$ be holomorphic and bounded in absolute value in $a \leqslant \Re(s) \leqslant b$. For each σ, define $M_f(\sigma) = \sup_t |f(\sigma + it)|$. Then $\log M_f(\sigma)$ is a convex function of σ, i.e.*

$$M_f(\sigma)^{b-a} \leqslant M_f(a)^{b-\sigma} M_f(b)^{\sigma-a}.$$

Proof. See Lang [16, p.263]. □

Theorem 4.4 (The second convexity theorem). *Let f be holomorphic in $a \leqslant \Re(s) \leqslant b$. For each σ, assume that*

$$f(\sigma + it) \ll (1 + |t|)^{N_\sigma}$$

for some N_σ. Let $\psi(\sigma)$ be the least non-negative number for which

$$f(\sigma + it) \ll (1 + |t|)^{\psi(\sigma)+\varepsilon}$$

for every $\varepsilon > 0$. Assume $f(\sigma + it) \ll e^{|t|^\alpha}$ in the strip with $\alpha \geqslant 1$. Then $\psi(\sigma)$ is convex and continuous as a function of σ.

Proof. See Lang [16, p.265]. □

Now we apply the Phragmén-Lindelöf principle to L-functions. First we recall the integral representation of the complete L-function in (2.4) when f is even and in (2.7) when f is odd. By their absolute convergence and uniform convergence in compact regions, we know that $\Lambda(s, f)$ is bounded in any vertical strip.

Recall (3.21)

$$\begin{aligned}\Lambda(s, f) &= \Gamma_{\mathbb{R}}(s + \mu_1)\Gamma_{\mathbb{R}}(s + \mu_2)L(s, f)\\ &= \pi^{-s-(\mu_1+\mu_2)/2}\Gamma\Big(\frac{s + \mu_1}{2}\Big)\Gamma\Big(\frac{s + \mu_2}{2}\Big)L(s, f).\end{aligned}$$

Consequently

$$L(s, f) = \pi^{s+(\mu_1+\mu_2)/2}\Gamma\Big(\frac{s + \mu_1}{2}\Big)^{-1}\Gamma\Big(\frac{s + \mu_2}{2}\Big)^{-1}\Lambda(s, f). \tag{4.1}$$

We need the Stirling formula

$$\Gamma(s) = \Gamma(\sigma + it) = \Big(\frac{2\pi}{s}\Big)^{1/2}\Big(\frac{s}{e}\Big)^s\Big(1 + O\Big(\frac{1}{|s|}\Big)\Big)$$

i.e.

$$\Gamma(\sigma + it) = \sqrt{2\pi}(it)^{\sigma-1/2}e^{-\pi|t|/2}\Big(\frac{|t|}{e}\Big)^{it}\Big(1 + O\Big(\frac{1}{1 + |t|}\Big)\Big), \tag{4.2}$$

for $|t| \geqslant t_0 > 0$. Thus

$$\Gamma\Big(\frac{s + \mu_1}{2}\Big)^{-1}\Gamma\Big(\frac{s + \mu_2}{2}\Big)^{-1} \ll |t|^M e^{\pi t} \tag{4.3}$$

for some M for s in any vertical strip, when $|t| \to \infty$. Since Γ functions are nonzero, the left side of (4.3) is bounded when $|t|$ is bounded. Applying these results to (4.1), we conclude that $L(s, f) \ll |t|^M e^{\pi t}$ in any vertical strip and hence is a function of order one.

Next we will assume that the L-function

$$L(s, f) = \sum_{n=1}^{\infty}\frac{\lambda_f(n)}{n^s}$$

converges absolutely for $\Re(s) > 1$. This will be proved in Section 5. Thus

$$L(s,f) \ll 1 \quad \text{as} \quad |t| \to \infty, \quad \sigma = \Re(s) = 1+\varepsilon.$$

For $L(s,f)$, the functional equation is

$$L(s,f) = (-1)^{\varepsilon_f}\pi^{s-1}\frac{\Gamma\big((1-s+\varepsilon_f+i\nu)/2\big)\Gamma\big((1-s+\varepsilon_f-i\nu)/2\big)}{\Gamma\big((s+\varepsilon_f+i\nu)/2\big)\Gamma\big((s+\varepsilon_f-i\nu)/2\big)}L(1-s,f).$$

We know that $L(1+\varepsilon \pm it, f) \ll 1$ and we want to bound $L(-\varepsilon - it, f)$ in the following way.

Let $s = \sigma + it$ and

$$\gamma(s) = \pi^{s-1}\frac{\Gamma\big((1-s+\varepsilon_f+i\nu)/2\big)\Gamma\big((1-s+\varepsilon_f-i\nu)/2\big)}{\Gamma\big((s+\varepsilon_f+i\nu)/2\big)\Gamma\big((s+\varepsilon_f-i\nu)/2\big)}.$$

By the Stirling formula (4.2), we have

$$\begin{aligned}\gamma(s) &= \pi^{s-1}\frac{\Gamma\big((1-\sigma+\varepsilon_f+i(-t+\nu))/2\big)\Gamma\big((1-\sigma+\varepsilon_f+i(-t-\nu))/2\big)}{\Gamma\big((\sigma+\varepsilon_f+i(t+\nu))/2\big)\Gamma\big((s+\varepsilon_f+i(t-\nu))/2\big)}\\ &= \pi^{s-1}\frac{\big(i(-t+\nu)/2\big)^{(-\sigma+\varepsilon_f)/2}\big(i(-t-\nu)/2\big)^{(-\sigma+\varepsilon_f)/2}}{\big(i(t+\nu)/2\big)^{(\sigma+\varepsilon_f-1)/2}\big(i(t-\nu)/2\big)^{(\sigma+\varepsilon_f-1)/2}}\\ &\quad \cdot e^{iA}\Big(1+O\Big(\frac{1}{1+|t+\nu|}\Big)+O\Big(\frac{1}{1+|t-\nu|}\Big)\Big)\\ &= C_{\sigma,\varepsilon_f}\pi^{s-1}(t^2-\nu^2)^{(1-2\sigma)/2}\Big(1+O\Big(\frac{1}{1+|t+\nu|}\Big)+O\Big(\frac{1}{1+|t-\nu|}\Big)\Big),\end{aligned}$$

when $|t \pm \nu| \geqslant t_0 > 0$. Now take $s = -\varepsilon + it$, $\sigma = -\varepsilon$. When $|t\pm\nu| \leqslant t_0$, we note that $\Gamma\big((1-s+\varepsilon_f \pm i\nu)/2\big)$ has no pole. Therefore

$$\gamma(s) \ll (1+|t-\nu||t+\nu|)^{1/2+\varepsilon}$$

for any $t \in \mathbb{R}$. Thus on $\Re(s) = -\varepsilon$,

$$L(s,f) \ll (1+|t-\nu||t+\nu|)^{1/2+\varepsilon}.$$

Now by the second convex theorem, for $-\varepsilon \leqslant \sigma \leqslant 1+\varepsilon$,

$$L(s,f) \ll_{f,\varepsilon} (1+|t-\nu||t+\nu|)^{(1-\sigma)/2+\varepsilon/2}.$$

The most interesting case is $\sigma = 1/2$,

$$L\Big(\frac{1}{2}+it,f\Big) \ll_{f,\varepsilon} (1+|t-\nu||t+\nu|)^{1/4+\varepsilon/2}. \tag{4.4}$$

This is called the convexity bound for $L(s,f)$ on the critical line.

Now let f be a holomorphic cusp form or a Maass form for the Hecke congruence subgroup $\Gamma_0(N)$ with Laplace eigenvalue $\lambda = \frac{1}{4} + \nu^2$. Assume that f is a Hecke eigenform. Then we can define its L-function likewise

$$\Lambda(s,f) = N^{s/2}\pi^{-s}\Gamma\Big(\frac{s+\varepsilon_f+i\nu}{2}\Big)\Gamma\Big(\frac{s+\varepsilon_f-i\nu}{2}\Big)L(s,f).$$

Its functional equation is

$$\Lambda(s,f) = \varepsilon(f)\Lambda(1-s,f), \quad |\varepsilon(f)| = 1.$$

For the finite-part L-function, the functional equation becomes

$$\begin{aligned}\Lambda(s,f) &= \varepsilon(f)N^{1/2-s}\pi^{2s-1} \\ &\cdot \frac{\Gamma\big((1-s+\varepsilon_f+i\nu)/2\big)\Gamma\big((1-s+\varepsilon_f-i\nu)/2\big)}{\Gamma\big((s+\varepsilon_f+i\nu)/2\big)\Gamma\big((s+\varepsilon_f-i\nu)/2\big)}L(1-s,f).\end{aligned}$$

Using the Stirling formula (4.2), we can get the approximate functional equation as the $SL(2,\mathbb{Z})$ case and prove the convexity bound

$$L\Big(\frac{1}{2}+it,f\Big) \ll \Big(1+N|t+\nu||t-\nu|\Big)^{1/4+\varepsilon}. \tag{4.5}$$

Any power saving over (4.4) or (4.5) is called a subconvexity bound (cf. Iwaniec and Sarnak [14], Michel [17], Michel and Venkatesh [18]). The Lindelöf hypothesis claims that (4.4) and (4.5) hold without 1/4 in the exponent.

5 The Rankin-Selberg Theory

Recall that if f is a Hecke eigenform, then

$$\begin{aligned}L(s,f) &= \sum_{n=1}^{\infty}\lambda_f(n)n^{-s} = \prod_p\Big(1-\lambda_f(p)p^{-s}+p^{-2s}\Big)^{-1} \\ &= \prod_p\Big(1-\alpha_f(p)p^{-s}\Big)^{-1}\Big(1-\beta_f(p)p^{-s}\Big)^{-1},\end{aligned}$$

where

$$\alpha_f(p)\beta_f(p) = 1, \quad \alpha_f(p)+\beta_f(p) = \lambda_f(p).$$

Define the Rankin-Selberg L-function attached to two Maass cusp forms f and g as

$$\begin{aligned}L(s,f\times g) = \prod_p\Big[&(1-\alpha_f(p)\alpha_g(p)p^{-s})^{-1}(1-\alpha_f(p)\beta_g(p)p^{-s})^{-1} \\ &\cdot(1-\beta_f(p)\alpha_g(p)p^{-s})^{-1}(1-\beta_f(p)\beta_g(p)p^{-s})^{-1}\Big].\end{aligned}$$

By the trivial bound we have $|\alpha_f(p)\beta_f(p)| \ll p$. Consequently $L(s,f\times g)$ converges for $\Re(s) > 2$.

Lemma 5.1. *Let*

$$\sum_{r=0}^{\infty} A(r)x^r = (1-\alpha_1 x)^{-1}(1-\alpha_2 x)^{-1},$$

$$\sum_{r=0}^{\infty} B(r)x^r = (1-\beta_1 x)^{-1}(1-\beta_2 x)^{-1}.$$

Then

$$\sum_{r=0}^{\infty} A(r)B(r)x^r = (1-\alpha_1\alpha_2\beta_1\beta_2 x^2)\prod_{i=1}^{2}\prod_{j=1}^{2}(1-\alpha_i\beta_j x)^{-1}.$$

Proof. See Bump [7, p.72]. □

Set $\alpha_1 = \alpha_1(p) = \alpha_f(p)$, $\alpha_2 = \alpha_2(p) = \beta_f(p)$, $\beta_1 = \beta_1(p) = \alpha_g(p)$, $\beta_2 = \beta_2(p) = \beta_g(p)$. Then

$$(1-\alpha_1(p))^{-1}(1-\alpha_2(p))^{-1} = \sum_{r=0}^{\infty}\lambda_f(p^r)p^{-rs},$$

$$(1-\beta_1(p))^{-1}(1-\beta_2(p))^{-1} = \sum_{r=0}^{\infty}\lambda_g(p^r)p^{-rs},$$

and hence

$$(1-p^{-2s})\prod_{i=1}^{2}\prod_{j=1}^{2}(1-\alpha_i(p)\beta_j(p)p^{-s})^{-1} = \sum_{r=0}^{\infty}\lambda_f(p^r)\lambda_g(p^r)p^{-rs},$$

$$\prod_p\prod_{i=1}^{2}\prod_{j=1}^{2}(1-\alpha_i(p)\beta_j(p)p^{-s})^{-1} = \prod_p(1-p^{-2s})^{-1}\prod_p\sum_{r=0}^{\infty}\lambda_f(p^r)\lambda_g(p^r)p^{-rs}$$

$$= \zeta(2s)\sum_{n=0}^{\infty}\lambda_f(n)\lambda_g(n)n^{-s}.$$

That is, for $\Re(s) > 2$,

$$L(s, f\times g) = \zeta(2s)\sum_{n=0}^{\infty}\lambda_f(n)\lambda_g(n)n^{-s}.$$

As in the case $L(s,f)$, we have an integral representation for $L(s, f\times g)$. Recall that

$$E(z,s) = \pi^{-s}\Gamma(s)\zeta(2s)\sum_{\gamma\in\Gamma_\infty\backslash\Gamma}\Im(\gamma z)^s = 2\pi^{-s}\Gamma(s)\zeta(2s)\sum_{\gamma\in\Gamma_\infty\backslash\overline{\Gamma}}\Im(\gamma z)^s$$

which converges absolutely in $\Re(s) > 1$ and has meromorphic continuation to $\mathbb{C}$, with only simple poles at $s = 0, 1$. We have $\mathrm{Res}_{s=1}E(z,s) = 1/2$ and the functional equation $E(z,s) = E(z, 1-s)$. We also have

$$E(x+iy, s) \ll y^\sigma, \quad y\to+\infty, \quad \text{where} \quad \sigma = \max\{\Re(s), 1-\Re(s)\}.$$

Now take $\phi(z) = f(z)g(z)$, where f, g are two Maass Hecke eigenforms. For simplicity we will only consider the case of f and g being both even. Note that $\phi(z)$ decays rapidly as $y \to \infty$. Consider the integral

$$\int_{\Gamma\backslash\mathcal{H}} E(z,s)\phi(z)\frac{dxdy}{y^2} = \int_{\overline{\Gamma}\backslash\mathcal{H}} E(z,s)\phi(z)\frac{dxdy}{y^2}$$

which converges absolutely and uniformly for s in any compact region and hence is analytic in $\mathbb{C}$ with only simple poles at $s = 0$, 1.

Since

$$f(z) = \sum_{n\neq 0} \sqrt{y}\lambda_f(n)K_{i\nu_f}(2\pi|n|y)e^{2\pi inx},$$
$$g(z) = \sum_{m\neq 0} \sqrt{y}\lambda_g(n)K_{i\nu_g}(2\pi|m|y)e^{2\pi imx},$$

we have

$$\begin{aligned}\int_{\Gamma\backslash\mathcal{H}} E(z,s)\phi(z)\frac{dxdy}{y^2} &= 2\pi^{-s}\Gamma(s)\zeta(2s)\int_{\overline{\Gamma}\backslash\mathcal{H}} \sum_{\gamma\in\Gamma_\infty\backslash\overline{\Gamma}} \Im(\gamma z)^s\phi(z)\frac{dxdy}{y^2}\\ &= 2\pi^{-s}\Gamma(s)\zeta(2s)\int_{\overline{\Gamma}\backslash\mathcal{H}} \sum_{\gamma\in\overline{\Gamma}_\infty\backslash\overline{\Gamma}} \Im(\gamma z)^s\phi(\gamma z)\frac{dxdy}{y^2},\end{aligned}$$

because $\phi(z)$ is invariant under the action of Γ. Combining the integral and sum, we get

$$\begin{aligned}\int_{\Gamma\backslash\mathcal{H}} E(z,s)\phi(z)\frac{dxdy}{y^2} &= 2\pi^{-s}\Gamma(s)\zeta(2s)\int_{\Gamma_\infty\backslash\mathcal{H}} \Im(z)^s\phi(z)\frac{dxdy}{y^2}\\ &= 2\pi^{-s}\Gamma(s)\zeta(2s)\int_0^\infty\int_0^1 y^s\phi(x+iy)\frac{dxdy}{y^2}.\end{aligned}$$

Expanding f and g into Fourier series, we have

$$\begin{aligned}&\int_{\Gamma\backslash\mathcal{H}} E(z,s)\phi(z)\frac{dxdy}{y^2}\\ &= 2\pi^{-s}\Gamma(s)\zeta(2s)\int_0^\infty y^s \sum_{m,n\neq 0} y\lambda_f(n)\lambda_g(n)K_{i\nu_f}(2\pi|n|y)K_{i\nu_g}(2\pi|m|y)\frac{dy}{y^2}\\ &\quad\cdot\int_0^1 e^{2\pi i(n+m)x}dx\\ &= 4\pi^{-s}\Gamma(s)\zeta(2s)\sum_{n>0}\lambda_f(n)\lambda_g(n)\int_0^\infty y^sK_{i\nu_f}(2\pi|n|y)K_{i\nu_g}(2\pi|n|y)\frac{dy}{y},\end{aligned}$$

because we assumed that both f and g are even. Using Bateman [4, (6.8.47)] which can be proved similar to Lemma 2.1, we get

$$\int_{\Gamma\backslash\mathcal{H}} E(z,s)f(z)g(z)\frac{dxdy}{y^2} = \frac{\pi^{-2s}}{2}\zeta(2s)\sum_{n>0}\frac{\lambda_f(n)\lambda_g(n)}{n^s}\prod_{\pm,\pm}\Gamma\Big(\frac{s\pm i\nu_f\pm i\nu_g}{2}\Big),$$

when $\Re(s) > 2$. Consequently we get an integral representation of the complete Rankin-Selberg L-function

$$\Lambda(s, f \times g) = \pi^{-2s}\Big(\prod_{\pm,\pm}\Gamma\Big(\frac{s \pm i\nu_f \pm i\nu_g}{2}\Big)\Big)L(s, f \times g)$$

as

$$\Lambda(s, f \times g) = 2\int_{\Gamma\backslash\mathcal{H}} E(z, s)f(z)g(z)\frac{dxdy}{y^2}. \tag{5.1}$$

Recall that $E(z, s)$ is analytic on $\mathbb{C}$ with simple poles at $s = 0, 1$ and residues $\text{Res}_{s=1} = 1/2$ and $\text{Res}_{s=0} = -1/2$. By (5.1) we get meromorphic continuation of $\Lambda(s, f \times g)$ to $\mathbb{C}$. $\Lambda(s, f \times g)$ may have poles at $s = 0, 1$, with residues

$$\text{Res}_{s=1}\Lambda(s, f \times g) = \int_{\Gamma\backslash\mathcal{H}} f(z)g(z)\frac{dxdy}{y^2} = \langle f, \overline{g}\rangle.$$

Since $E(z, s) = E(z, 1 - s)$, we have the functional equation

$$\Lambda(s, f \times g) = \Lambda(1 - s, f \times g). \tag{5.2}$$

Thus

$$\text{Res}_{s=0}\Lambda(s, f \times g) = -\langle f, \overline{g}\rangle.$$

In general, let $\varepsilon_{fg} = 0$ if f and g are both even, and $\varepsilon_{fg} = 1$ if one is even and one is odd. Then the complete Rankin-Selberg L-function of f and g is (cf. Bump [6], Motohashi [20])

$$\Lambda(s, f \times g) = \pi^{-2s}\Big(\prod_{\pm,\pm}\Gamma\Big(\frac{s + \varepsilon_{fg} \pm i\nu_f \pm i\nu_g}{2}\Big)\Big)L(s, f \times g) \tag{5.3}$$

and its functional equation is still (5.2). Written in terms of $L(s, f \times g)$ the functional equation becomes

$$L(s, f \times g) = \pi^{4s-2}\Big(\prod_{\pm,\pm}\frac{\Gamma\big((1 - s + \varepsilon_{fg} \pm i\nu_f \pm i\nu_g)/2\big)}{\Gamma\big((s + \varepsilon_{fg} \pm i\nu_f \pm i\nu_g)/2\big)}\Big)L(1 - s, f \times g). \tag{5.4}$$

Applying Stirling's expansion to (5.4) and using the Phragmén-Lindelöf principle as before, we can prove the convexity bound

$$L\Big(\frac{1}{2} + it, f \times g\Big) \ll \prod_{\pm,\pm}\Big(1 + |t \pm \nu_f \pm \nu_g|\Big)^{1/4+\varepsilon}.$$

Now we prove a key theorem we used before.

Theorem 5.1. *The Rankin-Selberg L-function*

$$L(s, f \times g) = \zeta(2s)\sum_{n=1}^{\infty}\frac{\lambda_f(n)\lambda_g(n)}{n^s}$$

converges absolutely for $\Re(s) > 1$.

Proof. By Cauchy's inequality,

$$\Big|\sum_{n=1}^{\infty}\frac{\lambda_f(n)\lambda_g(n)}{n^s}\Big| \leqslant \Big(\sum_{n=1}^{\infty}\frac{|\lambda_f(n)|^2}{n^\sigma}\Big)^{1/2}\Big(\sum_{n=1}^{\infty}\frac{|\lambda_g(n)|^2}{n^\sigma}\Big)^{1/2}.$$

The theorem is thus reduced to showing that

$$\frac{L(s, f\times \overline{f})}{\zeta(2s)} = \sum_{n=1}^{\infty}\frac{|\lambda_f(n)|^2}{n^s} =: F(s) \tag{5.5}$$

converges absolutely for $\Re(s) > 1$. Note that we now only know that (5.4) converges absolutely for $\Re(s) > 2$.

By the integral representation (5.1), $\Lambda(s, f\times g)$ is analytic for $\Re(s) > 1$. By (5.3) we know that $L(s, f\times g)$ is also analytic for $\Re(s) > 1$. Recall that $\zeta(2s) \neq 0$ for $\Re(s) > 1/2$. Thus $F(s)$ is analytic for $\Re(s) > 1$. Since $F(s)$ is analytic at $s = 2+\delta$ for some $\delta > 0$, we have Taylor expansion

$$F(s) = \frac{L(s, f\times \overline{f})}{\zeta(2s)} = \sum_{k=0}^{\infty}\frac{F^{(k)}(2+\delta)}{k!}(s-2-\delta)^k$$

for $|s-2-\delta| < 1+\delta$, where

$$F^{(k)}(2+\delta) = \frac{d^k}{ds^k}\sum_{n=1}^{\infty}\frac{|\lambda_f(n)|^2}{n^s}\Big|_{s=2+\delta}.$$

Since the series in (5.5) and series of its term-by-term derivatives uniformly converge on any compact region in $\Re(s) > 2$, we can differentiate it term-by-term and get

$$\begin{aligned} F^{(k)}(2+\delta) &= \sum_{n=1}^{\infty}\frac{(-1)^k|\lambda_f(n)|^2(\log n)^k}{n^s}\Big|_{s=2+\delta} \\ &= \sum_{n=1}^{\infty}\frac{(-1)^k|\lambda_f(n)|^2(\log n)^k}{n^{2+\delta}}. \end{aligned}$$

Thus

$$\begin{aligned} F(s) &= \sum_{k=0}^{\infty}\sum_{n=1}^{\infty}\frac{(-1)^k}{k!}\frac{|\lambda_f(n)|^2}{n^{2+\delta}}(\log n)^k(s-2-\delta)^k \\ &= \sum_{k=0}^{\infty}\sum_{n=1}^{\infty}\frac{(\log n)^k}{k!}\frac{|\lambda_f(n)|^2}{n^{2+\delta}}(2+\delta-s)^k. \end{aligned}$$

Take $s = 1+\varepsilon$ for any small $\varepsilon > 0$ and hence s is within $|s-2-\delta| < 1+\delta$. Then

$$F(1+\varepsilon) = \sum_{k=0}^{\infty}\sum_{n=1}^{\infty}\frac{(\log n)^k}{k!}\frac{|\lambda_f(n)|^2}{n^{2+\delta}}(1+\delta-\varepsilon)^k$$

converges with positive terms. Thus we can switch summation order so that

$$F(1+\varepsilon) = \sum_{n=1}^{\infty} \frac{|\lambda_f(n)|^2}{n^{2+\delta}} \sum_{k=1}^{\infty} \frac{[(\log n)(1+\delta-\varepsilon)]^k}{k!}$$
$$= \sum_{n=1}^{\infty} \frac{|\lambda_f(n)|^2}{n^{2+\delta}} e^{(\log n)(1+\delta-\varepsilon)} = \sum_{n=1}^{\infty} \frac{|\lambda_f(n)|^2}{n^{1+\varepsilon}}$$

converges. Therefore

$$\frac{L(s, f \times \overline{f})}{\zeta(2s)} = \sum_{n=1}^{\infty} \frac{|\lambda_f(n)|^2}{n^s}$$

converges absolutely for $\Re(s) > 1$. □

Theorem 5.2. *$L(s, f)$ converges absolutely for $\Re(s) > 1$.*

Proof. By Cauchy's inequality, we have

$$\sum_{n=1}^{N} \left| \frac{\lambda_f(n)}{n^s} \right| \leqslant \Big(\sum_{n=1}^{N} \frac{|\lambda_f(n)|^2}{n^\sigma} \Big)^{1/2} \Big(\sum_{n=1}^{N} \frac{1}{n^\sigma} \Big)^{1/2}. \tag{5.6}$$

Since for $\Re(s) > 1$

$$\sum_{n=1}^{\infty} \frac{1}{n^s} \quad \text{and} \quad \sum_{n=1}^{\infty} \frac{|\lambda_f(n)|^2}{n^s}$$

converge absolutely for $\Re(s) > 1$ by Theorem 5.1, the right side of (5.6) is bounded as $N \to \infty$. Therefore the left side of (5.6) converges as $N \to \infty$. □

By Theorems 5.1 and 5.2, the Euler products of $L(s, f \times g)$ and $L(s, f)$ also converge absolutely for $\Re(s) > 1$, when f and g are Hecke eigenforms. Therefore, $L(s, f \times g) \neq 0$ and $L(s, f) \neq 0$ for $\Re(s) > 1$.

References

[1] C. A. Brebbia, J. C. F. Telles and L. C. Wrobe, Boundary Element Techniques: Theory and Applications in Engineering, Springer-Verlag, Berlin, 1984.

[2] M. Haskins, Special Lagrangian cone, Amer. J. Math. **126** (2004), 845–871.

[3] C. LeBrun and M. Einstein, Four-manifolds and differential topology, in: Surveys in Differential Geometry, Vol. III, Ed. S. T. Yau, International Press, Boston, 2003, 235–255.

[4] H. Bateman, Table of Integral Transforms, McGraw-Hill, New York, 1954.

[5] A. Borel, Automorphic Forms on $SL_2(\mathbb{R})$, Cambridge University Press, Cambridge, 1997.

[6] D. Bump, The Rankin-Selberg method: A survey, in: Number Theory, Trace Formulas and Discrete Groups, Academic Press, New York, 1989, 49–109.

[7] D. Bump, Automorphic Forms and Representations, Cambridge University Press, Cambridge, 1998.

[8] J. W. Cogdell, Lectures on L-functions, converse theorems, and functoriality for GL_n, in: Lectures on Automorphic L-functions, Amer. Math. Soc., Providence, 2004, 1–95.

[9] D. Goldfeld, Automorphic Forms and L-Functions for the Group $GL(n, \mathbb{R})$, Cambridge University Press, Cambridge, 2006.

[10] D. Goldfeld and J. Hundley, Automorphic Representations and L-Functions for the General Linear Group, Cambridge University Press, Cambridge, 2011.

[11] H. Iwaniec, Topics in Classical Automorphic Forms, Amer. Math. Soc., Providence, 1997.

[12] H. Iwaniec, Spectral Methods of Automorphic Forms, 2nd ed., Graduate Studies Math., Vol. 53, Amer. Math. Soc. & Revista Matemática Iberoamericana, Providence, 2002.

[13] H. Iwaniec and E. Kowalski , Analytic Number Theory, Amer. Math. Soc., Providence, 2004.

[14] H. Iwaniec and P. Sarnak, Perspectives on the analytic theory of L-functions, in: GAFA 2000 (TelAviv, 1999), Geom. Funct. Anal., Special Volume, Part II, 2000, 705–741.

[15] W. -C. Li, Number Theory with Applications, World Scientific Publ., River Edge, 1996.

[16] S. Lang, Algebraic Number Theory, 2nd ed., Springer, New York, 1994.

[17] P. Michel, Analytic number theory and families of automorphic L-functions, in: Automorphic Forms and Applications, Amer. Math. Soc., Providence, 2007, 181–295.

[18] P. Michel and A. Venkatesh, The subconvexity problem for GL_2, Publ. Math. l'I.H.É.S. **111** (2010), 171-271.

[19] C. J. Moreno, Advanced Analytic Number Theory: L-Functions, Amer. Math. Soc., Providence, 2005.

[20] Y. Motohashi, Spectral Theory of the Riemann Zeta-Function, Cambridge University Press, Cambridge, 1997.

[21] P. Sarnak, Some Applications of Modular Forms, Cambridge University Press, Cambridge, 1990.

[22] J. P. Serre, A Course in Arithmetic, Springer, Berlin, 1973.

[23] G. Shimura, Introduction to the Arithmetic Theory of Automorphic Functions, Princeton University Press, Princeton, 1971.

[24] Y. B. Ye, Modular Forms and Trace Formulas (in Chinese), Peking University Press, Beijing, 2001.

Made in the USA
Monee, IL
01 March 2026

45146624R00122